AF577210

SEDIMENT TRANSPORT IN AQUATIC ENVIRONMENTS

Edited by **Andrew J. Manning**

INTECHOPEN.COM

Sediment Transport in Aquatic Environments
Edited by Andrew J. Manning

Published by InTech

Janeza Trdine 9, 51000 Rijeka, Croatia

Edition 2014

Publishing Process Manager Masa Vidovic

Technical Editor Teodora Smiljanic

Cover Designer InTech Design Team

Additional hard copies can be obtained from orders@intechopen.com

Sediment Transport in Aquatic Environments, Edited by Andrew J. Manning
p. cm.
ISBN 978-953-307-586-0

INTECH
open science | open minds

Contents

Preface

Sediment Transport in Aquatic Environments is a book which covers a wide range of topics. The effective management of many aquatic environments requires a detailed understanding of sediment dynamics. This has both environmental and economic implications, especially where there is any anthropogenic involvement. Numerical models are often the tool used for predicting the transport and fate of sediment movement in these situations, as they can estimate the various spatial and temporal fluxes. However, the physical sedimentary processes can vary quite considerably depending upon whether the local sediments are fully cohesive, non-cohesive, or a mixture of both types. For this reason for more than half a century, scientists, engineers, hydrologists and mathematicians have all been continuing to conduct research into the many aspects which influence sediment transport. These issues range from processes such as erosion and deposition, to how sediment process observations can be applied in sediment transport modelling frameworks. This book reports the findings from recent research in applied sediment transport which has been conducted in a wide range of aquatic environments. The research was carried out by researchers who specialize in the transport of sediments and related issues.

It is a great pleasure to write the preface to this book published by InTech. It comprises 15 chapters written by a truly international group of research scientists, who specialize in areas such as sediment dynamics, hydrology, morphology and numerical sediment transport modelling. The majority of the chapters are concerned with sediment transport related issues in either estuarial, coastal or freshwater environments. For example: suspended sediment transport in the Amazon delta, how hydrodynamic effects influence fluid mud distributions, sediment exchange & dynamics on the continental shelf and sediment transport modelling in a tidal inlet. Other contributions in this book examine: coastal morphological modelling; sedimentation in lakes; the influence of hydrodynamics; fine sediment deposition; water quality associated with sediment transport; and how river bed dynamics can influence potential flooding. Authors are responsible for their views and subsequent concluding statements.

In summary, this book provides an excellent source of information on recent research on sediment transport, particularly from an interdisciplinary perspective. I would like to thank all of the authors for their contributions and I highly recommend this textbook to both scientists and engineers who deal with the related issues.

Andrew J. Manning
University of Plymouth, School of Marine Science & Engineering, Plymouth,
HR Wallingford Ltd, Coasts & Estuaries Group, Wallingford,
United Kingdom

Part 1

Sediment Transport Issues

1

Sediment Transport Patterns in Todos Santos Bay, Baja California, Mexico, Inferred from Grain-Size Trends

Alberto Sánchez[1] and José D. Carriquiry[2]
[1]*Centro Interdisciplinario de Ciencias Marinas, Instituto Politécnico Nacional, La Paz, Baja California Sur*
[2]*Instituto de Investigaciones Oceanológicas, Universidad Autónoma de Baja California, Ensenada, Baja California*
México

1. Introduction

Contaminated sediments do not always remain in the same place in the environment. The mobility of sediments and the contaminants associated to them is a factor that complicates the evaluation of ecological risk, remediation potential and ultimately, the litigation process (e.g., Carriquiry & Sanchez, 1999; Sanchez et al., 2008). A rational decision in this regard must take into consideration the potential stability of contaminated sediments, its sources, transport and final destination (e.g., McLaren & Beveridge, 2006). A technique capable of providing that information in coastal sedimentary environments is the analysis of textural trends of sediments; an empirical method that is based on the relaive changes in the grain size distributrion of the sediments to determine the net direction of sediment transport (e.g., Sunamura & Horikawa, 1971; Mc Laren & Bowles, 1985; Gao & Collins, 1992, 1994; LeRoux, 2002; Poizot & Mear, 2008). Because many contaminants become adsorbed to the sedimentary particles, this information can help in evaluating the relationship between contaminant load and its sources, as well as providing an understanding about the behavior and destination of the contaminants in the sediments (e.g., Carriquiry & Sanchez, 1999; Sanchez et al., 2008).

The original theory used to predict the direction of sediment transport based on the relative change in the particle size distribution was given by Sunamura & Horikawa (1971). Later on, McLaren & Bowles (1985) proposed a one-dimensional approach in which changes in the grain size distribution along a sequence of individual samples that were statistically analyzed by a Z-test could be used to determine the preferred transport direction. Gao & Collins (1991, 1992) and Gao (1996) proposed a two-dimensional model to determine the trend on the basis of vector analysis. A different approach based on analytic geometry and vector analysis to determine the direction of sediment transport was produced by LeRoux (1994) and LeRoux et al. (2002). A summary of all these techniques is provided by Poizot et al. (2008).

In all the sediment transport models based on trend analysis of clastic material, there are considerations that limit the inferences made on the net direction of sediment transport. The most important limitation is the inability to validate the obtained transport trends with hydrodynamic observations in the field. In this chapter we will use the model proposed by LeRoux et al. (2002) to identify the dominant paths of sediment transport based on sediment characteristics and dynamics. The sediment transport model obtained with the LeRoux model is compared with sediment transport inferences made by Sanchez et al. (2009). Subsequently, the model is validated with hydrodynamic observations in the bay. Later, we assess the likely extent of dispersion of specific sources of pollution in the Todos Santos Bay, off Ensenada, Baja California, Mexico. With this goal we will use additional studies of the distribution of organic and inorganic pollutants carried out in this bay.

1.1 Study area and method

Todos Santos Bay (TSB) is located on the west coast of the peninsula of Baja California, Mexico, between 31°41′ and 31°56′N and 116°34′ and 116°51′W (Fig. 1). The natural boundaries are Punta San Miguel to the north, Punta Banda to the South and the Todos Santos Islands (TSI) in the center, which define two entrances to the bay and permit a constant circulation of ocean water. The coastline of the bay consists of a rocky shore that encompasses Punta San Miguel, Punta El Sauzal, Punta Morro and Punta Ensenada, with pocket beaches between them. Punta El Sauzal and Punta Ensenada form natural breakwaters that protect the ports of El Sauzal and Ensenada, respectively. A sandy beach, 14 km long, starts at the South of the port of Ensenada and ends at the base of Punta Banda. This beach is interrupted by an inlet that defines the entrance to a coastal lagoon known as the Punta Banda Estuary. The rocky shore of Punta Banda is very irregular, with almost vertical cliffs and small beaches with little sand between them. The bathymetric configuration of the bay is irregular. The most notable features are: (1) the shoal of San Miguel, with a minimum depth of 5.5 m, located between Punta San Miguel and the Todos Santos Islands; (2) a submarine depression between Punta Banda and the islands, known as the Todos Santos submarine canyon, with depths of 550 m.
In the Todos Santos Bay several studies have been conducted on littoral transport. In the north of the bay, sediment transport is towards the SE and to the central part of the bay, the sediment transport is predominantly in the direction NE-N with a W-component near the mouth of the Punta Banda coastal lagoon (Pérez-Higuera & Chee-Barragán, 1984). The sediment dispersion pattern presents several transport directions within the Todos Santos Bay (Sanchez et al., 2009). In the North of the bay the transport direction is SE (following the isobath contour of 20 m) and towards the NE (near the Todos Santos Island). In the South of the bay, transport direction is NE. In the central region of the bay there is a westerly transport direction. In the Todos Santos canyon a NE trajectory was determined. In the external region of the bay, in front of the Todos Santos Islands and the Peninsula of Punta Banda, the inferred transport is to the W. The central zone of the bay seems to be a convergence zone which in practical terms becomes a depocenter, a site of sediment accumulation.
Sediment samples were collected using a van-Veen grab during cruises OGEO-0893 of the Mexican Navy that consisted of a sampling grid of 51 stations (Fig. 1). The first 5 cm of surface sediments of each grab sample were colleted. Later in the lab the samples were treated to remove organic matter and salts (Mook & Hoskin, 1982). The granulometric analysis was performed by the technique of Ingram (1971) for the sand fraction and Galehouse (1971) for the fine fraction (silt and clay). Sediment textural parameters (mean grain size, sorting and asymmetry) were calculated by the method proposed by McManus (1988).

1.2 Sediment transport model

Sediment transport models have allowed coastal oceanographers to infer the residual sediment transport based on spatial trends of sediment (e.g., McLaren & Bowles, 1985; Gao & Collins, 1992; LeRoux, 1994; Poizot & Méar, 2008). In this study, the model proposed by LeRoux (1994), based on principles of analytic geometry and vector analysis of textural data, was used. With this method, the vector's magnitude and direction (of transport) were obtained by comparison of the textural characteristics of five neighboring sampling stations (one central and four satellites). General considerations of the model are: (1) textural trends result from the hydrodynamic conditions of the environment, (2) it is applicable in coastal zone and shelf where sediment transport is unidirectional, (3) the gradient between textural parameters is constant in the area where we compared the five sampling stations, (4) textural parameters in the model have the same weight and importance, and (5) the distance between the five stations (inter- seasonal) is not critical, especially if there is a clear textural gradient between stations.

2. Results and discussion

2.1 Grain-size trends

Sediment characteristics in the bay have been extensively studied since the 1950s (e.g., Walton 1955; Emery et al. 1957) who emphasized the importance of topographic details and hydrography in controlling grain size. Grain-size decreases from the coastal regions (0.0 ϕ) into the canyon (7.0 ϕ). Also, there was a slight decrease in this parameter in the deepest region of the study area (>8.0 ϕ, Fig. 2). The spatial distribution of grain-size denotes the monotonous tendency to decrease (larger phi values) with the depth of seabed. The silt fraction is dominant in two areas: (1) medium silt in the deepest region in the inner-bay (head of the canyon) and (2) fine silt to clay outside the bay (continental slope). The shallow inner-shelf is characterized by fine to very fine sand with some small areas consisting of medium to coarse sand (Fig. 2). The grain-size distribution is very similar to that reported by Smith et al. (2008), where the grain-size is dominated by silt. The spatial trend of grain size indicates a zone of deposition of fine material in the central area of the bay (a depocenter), consistent with the convergent surface current system, longshore transport and sediment transport in the bay (Sanchez et al., 2009). Although grain size was used as a good indicator of sedimentary dynamics, in most cases is not fully robust due to other sediment sources. Hence, it is necessary consider additional parameters such as sorting and asymmetry of the same sample (e.g., McLaren and Bowles, 1985). The sorting parameter was considered by Sunamura and Horikawa (1971) to be a good estimate of the dispersion of sediments in marine environments, where the contribution of other sources of sedimentary material was insignificant (e.g., from cliffs, streams, etc). Sunamura and Horikawa (1971) indicated that sediments will move in the direction where the average grain size decreases and the sorting of the sedimentary material increases. The surface sediments were well-sorted in shallow stations, where the average grain-size is larger with respect to the deeper stations. In the deepest part of the bay, the sediments were poorly-sorted and grain-size decreases toward the region of the canyon (Fig. 3). Negative skewness values (sediments with an "extra load" of coarse grains) are found in shallow areas, increasing in the central part of the bay; the trend reverses to negative values towards the canyon (Fig. 4).

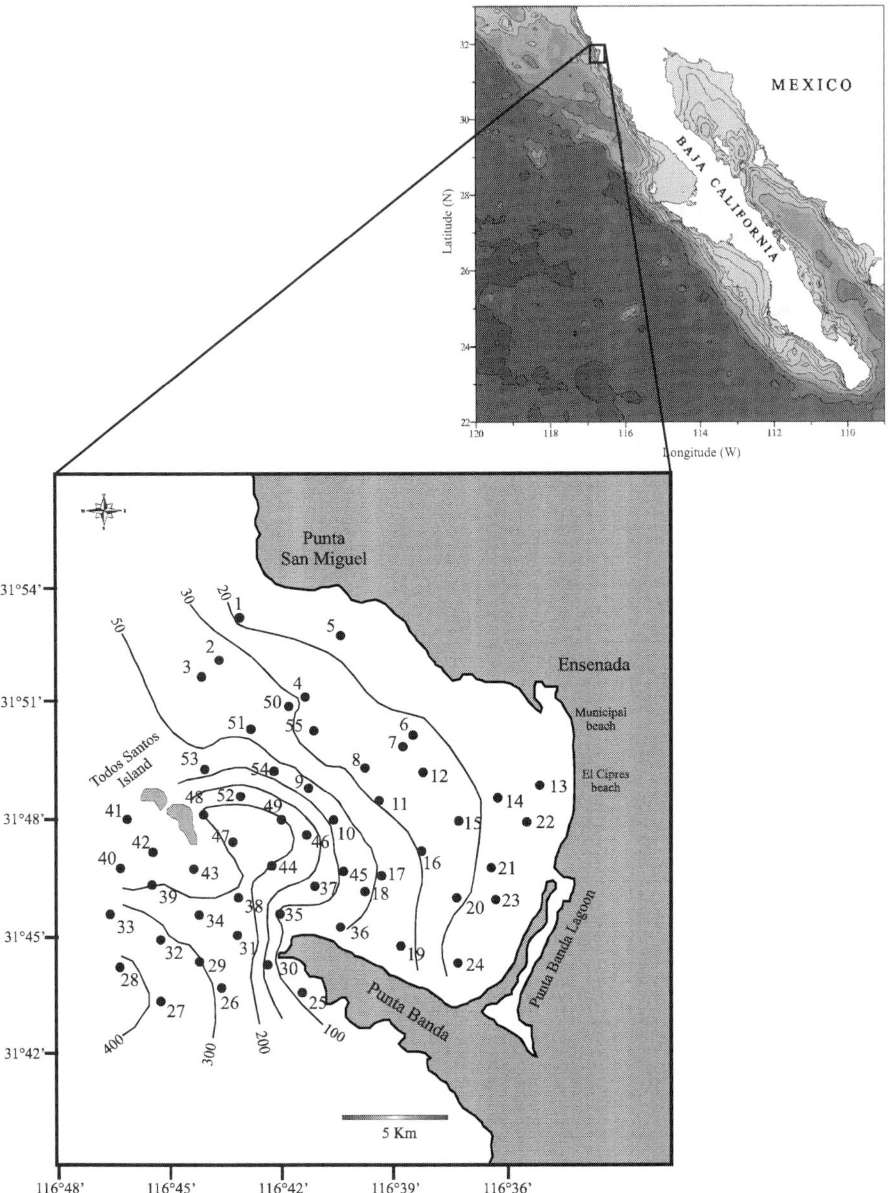

Fig. 1. Stations of collect of surface sediments in the Todos Santos Bay, Baja California, Mexico. The continuous lines correspond to isolines of bathymetry (20, 30, 50, 100, 150, 200, 300 and 400 m depth).

In their model, McLaren and Bowles (1985) also included the parameter of asymmetry in order to yield a more robust estimate of sediment transport paths. Thus, while the combination of sediment textural parameters define the existence of several sediment transport paths, the sorting improves or strengthens the interpretation of sediment transport direction (e.g., McLaren and Bowles, 1985; Gao and Collins, 1992; LeRoux, 1994; Carriquiry and Sánchez, 1999; Carriquiry et al., 2001; Poizot and Méar, 2008; Sánchez et al., 2008, 2009, 2010).

Sanchez et al. (2008, 2010) applied a principal components analysis (PCA) to the sediment textural parameters in both studies, the spatial trends of grain size and sorting explained at least 75% of the variance and 20% of the variance was explained for asymmetry. The remaining 5% of the variance can be related to other factors such as sampling depth, the distance between stations, among others, affecting the distribution of sediment textural parameters (e.g., Poizot et al., 2008). In the case of the surface sediments of the bay, the grain size and sorting explain 50% of the variance and asymmetry, 34% of the variance. These contrast with the results obtained for grain size and asymmetry that describe 95% of the variability of spatial trends of textural parameters in the sediments in the Yellow Sea, China (Cheng et al., 2004). The difficulty of establishing a spatial trend in very fine grained estuarine sediments occurs because flocculation can result in the preferential deposition of fine particles in the environment. Thus, the difference of each factor in the PCA, for each comparison site can be result from processes of flocculation derived from sedimentary material, which allowed the settlement of poorly sorted material in the Yellow Sea, China and Upper Gulf of California, Mexico and Todos Santos Bay, Baja California, Mexico. While, the low discharge of streams in the Bahia Magdalena, Baja California and Hondo River in the Bahia of Chetumal, Quintana Roo, Mexico, caused the well-sorted sediment deposition.

2.2 Sediment transport models

The spatial trend analysis method used to infer the net transport of sediments has been widely applied in various settings of marine and continental environments. In these studies, the net transport and dispersion of sedimentary material have been effectively validated by comparing the resulting transport vectors (defined from the transport models) with the observed ocean currents (*in situ*)(e.g., McLaren and Bowles, 1985; Gao and Collins, 1992; LeRoux, 1994; Carriquiry and Sánchez, 1999; Carriquiry et al., 2001; Poizot et al., 2008; Sánchez et al., 2008, 2009, 2010). Consequently, these results have provided confidence in these models, becoming an excellent tool for inferring the transport direction of sediment particles in places where ocean current studies are limited.

The model of Sunamura and Horikawa (1971) is the most basic and unidirectional, where was identifying sediment source and the final destination of the particles, based on comparison of grain-size and sorting. The asymmetry is not used as an additional comparison criterion; making is less sensitive to identify an exchange of material and sediment transport. In fact, the principal component analysis indicated that the asymmetry explain 34% of the variance of the spatial trend. Therefore, the asymmetry is a criterion that must be considered to define transport paths more robust (e.g., Cheng et al., 2004; Sanchez et al., 2008, 2010).

The one-dimensional method McLaren-Bowles was developed for studies of longitudinal environmental systems, such as rivers, beaches and sand bars (Mc Laren and Bowles, 1985).

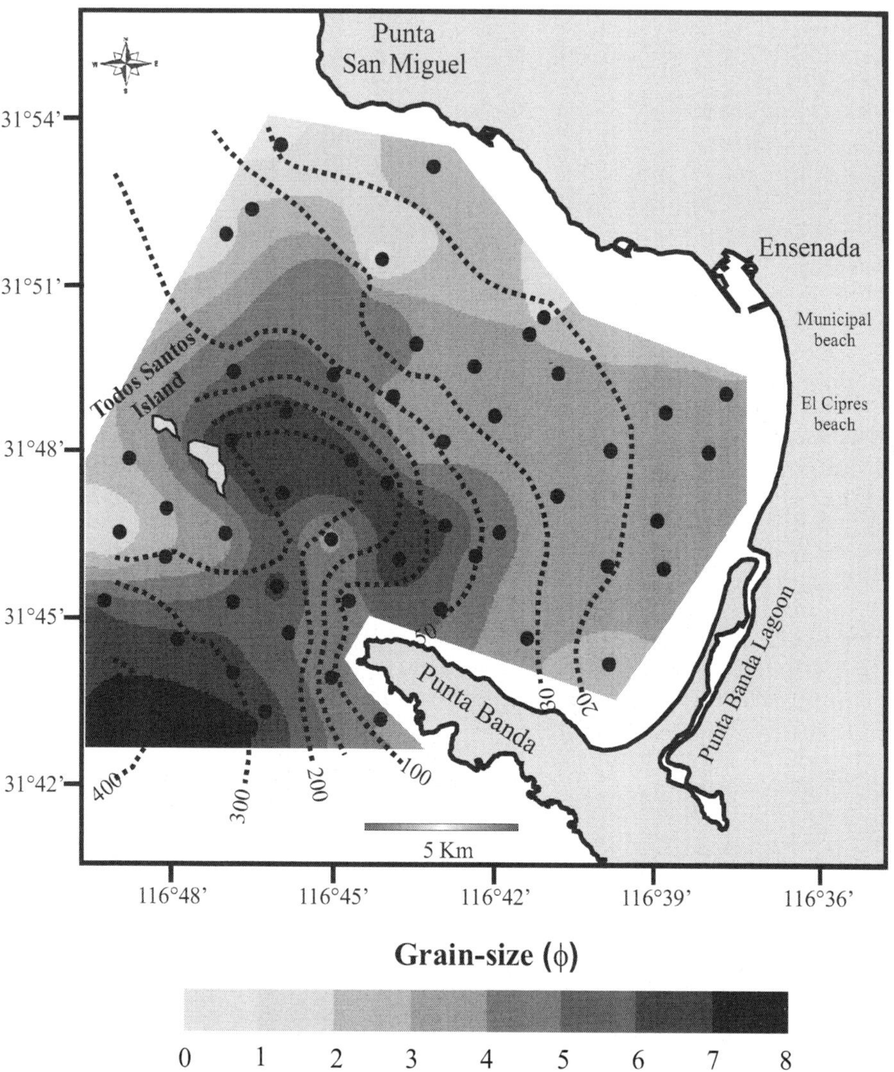

Fig. 2. Spatial distribution of the grain-size (phi units) of sediments in the Todos Santos Bay, Baja California, Mexico. The segmented lines denote the bathymetry (m) of the bay.

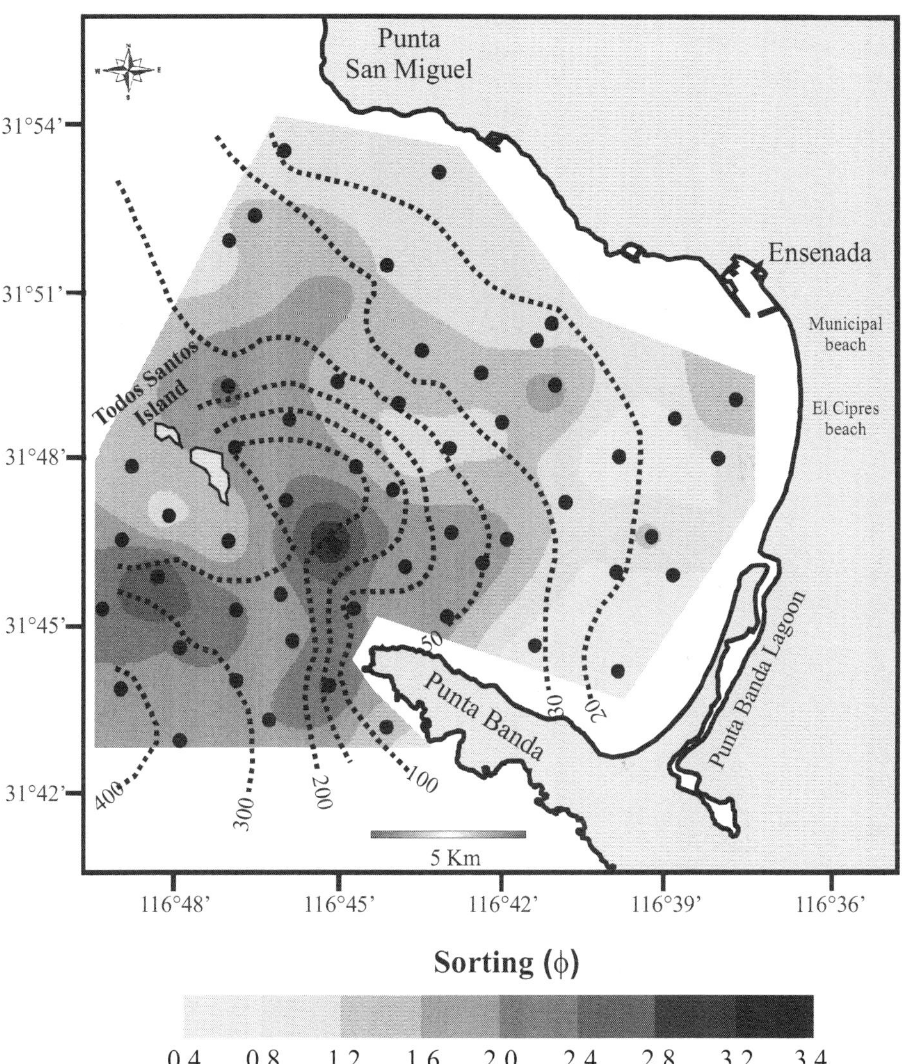

Fig. 3. Spatial distribution of the sorting (phi units) of sediments in the Todos Santos Bay, Baja California, Mexico. The segmented lines denote the bathymetry (m) of the bay.

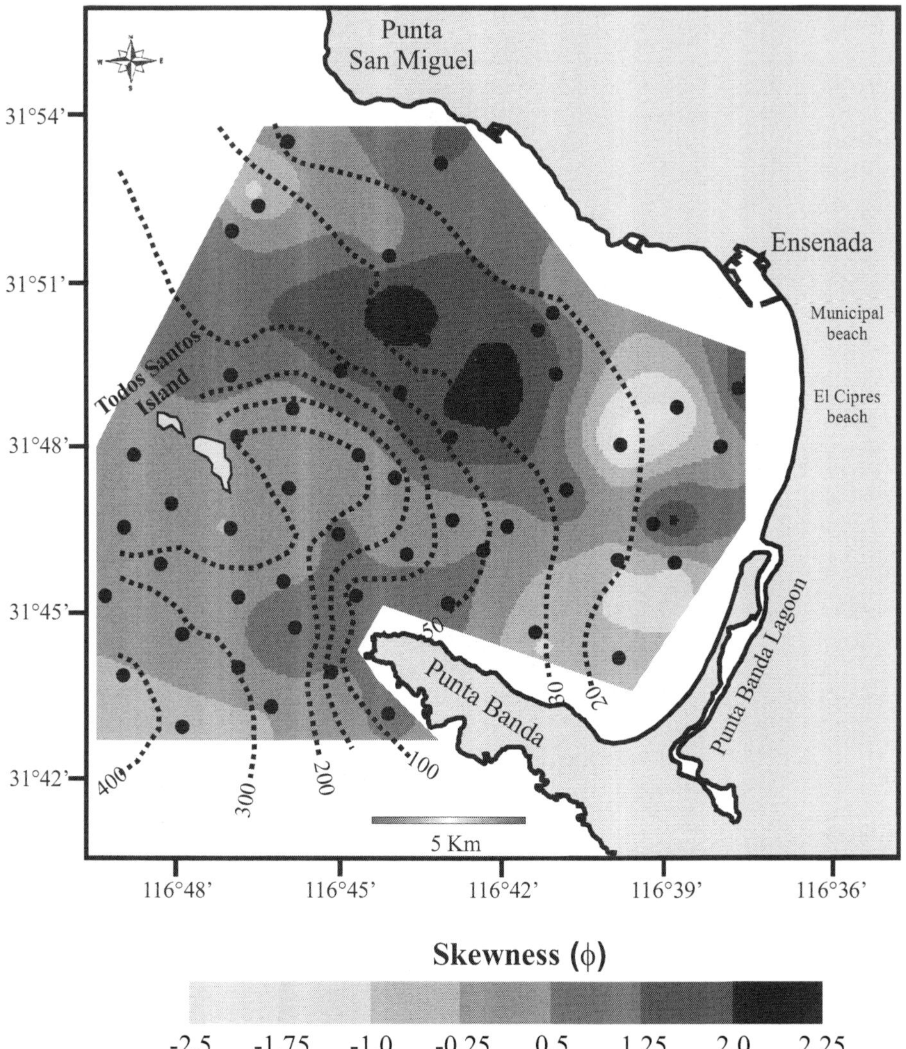

Fig. 4. Spatial distribution of the skewness (phi units) of sediments in the Todos Santos Bay, Baja California, Mexico. The segmented lines denote the bathymetry (m) of the bay.

This model analyzes the grain-size evolution along sampling paths (samples in a linear sequence). To study a more complex area and obtain transport paths in two dimensions requires a 9x9 grid containing 81 samples (Mc Laren and Little, 1987). Masselink (1992) concluded that the unidimensional model is limited and Lanckneus et al. (1992) showed that when sediments become finer and better sorted, the corresponding skewness becomes more positive, which is in contradiction with the sediment transfer function of McLaren and Bowles (1985). The McLaren and Bowles model was applied to 17 locations and were valid in 7 of them; the other cases were partially correct or unacceptable (Poizot et al., 2008). In fact, the results of this study seem to confirm that the one-dimensional model has limitations in complex environments such as the TSB (Sanchez et al., 2009).

Gao and Collins (1992) developed a bidimensional model where vectors are determined by comparing the sedimentary textural parameters in a sampling grid. The distance used for "neighboring" stations is determined on the basis of a characteristic distance that represents the spatial scale of sampling. However, the resultant vector can be affected by the magnitude of such characteristic distance; in other words, if the distance is too great, it may yield a change in the direction of the vectors with one direction that is not representative of sediment transport. For the Todos Santos Bay we applied various distances and sampling densities (± 50%) and the observed effect on the model ouput did not show any significant differences in the model outputs.

In the Le Roux model, sampling sites are rarely located on a regular grid pattern. When the sampling scheme is not regular and there are many discontinuities, edge effects are very noticeable and can become dominant in some parts of the studied areas (Poizot et al., 2006). If an irregular sampling scheme is used, a regular grid can be created by interpolation of the dataset. This procedure was used by Ríos et al. (2003), Friend et al. (2006), Lucio et al. (2006) and Poizot et al. (2006). However, Le Roux and Rojas (2007) argue that interpolation alters the original data distribution and diminishes the reliability of the method. Despite the use of a regular grid, edge effects can persist within the intended mapping area. In the present case, edge effects can affect both the averaging procedure and the definition of the trend vectors. For a site at the centre of a grid, trend vectors are calculated with more sites, lying within a defined characteristic distance, than sites on the edges of the grid (Gao and Collins, 1994a,b). This effect is increased at the corners of the grid. Such effects are likely to introduce some distortions, so sediment trend vectors along the edges of the sampling grid should be applied with caution in the final interpretation.

In this study, the edge effect was estimated by increasing the number of stations around the sampling grid located within a spatial distance-unit (minimum distance between stations). In the model of Sunamura and Horikawa (1971) the edge effect was observed in the residuals of the marginal stations that indicated a transport direction that differed by 45° from expected. Subsequent calibration of the model by using a one-unit distance, we observed an improvement in the residuals showing only a deviation of 15 degrees from expected. In the model of McLaren and Bowles (1985) no edge-effect problems arise due to its unidirectional character of the model (stations along a linear sequence of stations). The models of Gao and Collins (1992) and LeRoux (1994a, b) despite having an edge effect, there appears to be insignificant in our study since the delineated net transport paths match the expected transport direction, and the differences observed with the model before and after applying the calibration series is very small (<5 °) with respect to the calibration standards. The observed edge effect declined when using a space-distance no greater than the spatial unit of the calibration series, and also yields a net transport direction closer to the expected path.

2.3 Dispersion pattern

The dispersion pattern for the TSB is consistent with the general pattern of water circulation (Argote-Espinoza et al., 1975, Alvarez-Sánchez et al., 1988) and the observed longshore currents (Perez-Higuera and Chee -Barragán, 1984) proposed for the site. In shallow areas, the sedimentary material can be transferred and reworked by the longshore current to later be driven out of the surf zone following a transport path perpendicular to the coast (eg, Cruz-Colin, 1994). These observations allow us to identify the existence of materials exchange between those stations near the beach. Carriquiry (1985) found a decrease in the amount of heavy minerals from the north entrance of the bay towards the central bay, coincident with transport direction proposed in this work, assuming that the heavy mineral concentrations are higher in the source area of the material and decreasing in the direction of transport (e.g., Carriquiry and Sánchez, 1999). The transport observed in the central region of the bay to SE direction (Fig. 5) may indicate the transfer of materials to the bay's interior (Cruz-Colin, 1994) resulting from the convergence of longshore currents in the area (Pérez-Higuera and Chee-Barragán, 1984).

In the northern region of the bay near the Punta San Miguel there is a transport to the NE. The component one (daytime) and two (semidiurnal) of the tidal-ellipse indicate the predominant orientation of the tidal wave motion, which coincides with the sediment transport direction obtained in this study (Figure 5).

Hydrodynamic effects of tidal processes, such as residual currents, may be the cause of materials exchange and transport in this area. Towards the center of the bay, the sediment transport direction converges predominantly in a direction towards the southeast (Fig. 5). This same pattern of sediment transport was observed using the model Sunamura and Horikawa (1971), and Gao and Collins (1992); reported by Sanchez et al. (2009). The dispersion of sediments to the central region of the bay coincides with the predominant orientation of the tidal components (components one and two) which converge towards the central part of the bay. In fact, by using the Regional Ocean Modeling System Model (ROMSM), Mateos et al. (2009) observed the persistence of a convergent circulation with a southeast direction, in front of the port of Ensenada, with a clear path towards the Todos Santos submarine canyon (Fig. 5). In the deeper region of the study area, the direction of the sediment transport occurs in two patterns. In the area of the Punta Banda peninsula, the sediment is transported out of the bay with a westward component whereas in the northern entrance of the bay, nearby the Todos Santos islands, the sediment is transported into the bay with a direction east-northeast (Fig. 5). Cruz-Colin (1997) calculated the velocity of the approaching tidal waves, at different depths within the bay, finding an average velocity of 0.20 m s-1. This author also observed that the speed directly increased as a function of depth, and calculated a velocity of 0.32 m s-1 at a depth of 476 m. These observations indicate that hydrodynamics at this depth is capable of resuspending (Shepard and Keller, 1978) and transporting (Komar, 1977) sediments of grain size between 3 to 5 ϕ. These calculations have been corroborated by a series of current measurements which reach 0.30 m s-1 at a depth of 308 m in the canyon (Garcia et al., 1994). In addition to identifying a net direction of flow to the northwest, which agrees with that obtained in this work, Mateos et al. (2009) suggests two modes of circulation for a deep section between Punta Banda and Todos Santos Islands by implementing the ROMSM model (their Figure 2). This deep section can be considered as an exterior and interior threshold system. The external system is characterized by an intense circulation flow out of the bay in the upper 30 m and diminishes down to 1 cm s-1 at 250 m depth, covering 2 / 3 of the cross depth-section. The flow into the bay adjacent to the islands occurs below 50 m depth with a speed of 1 cm s-1 and increases up to 3 cm s-1 to 150 m deep. In fact, Gavidia-Medina (1988) calculated the tidal residual currents for TSB, identifying an anticyclonic eddy in the canyon area, finding greater magnitudes inward than outwards. Therefore, residual tidal currents can become important enough in the net transport of sedimentary material along the canyon towards the interior of the bay.

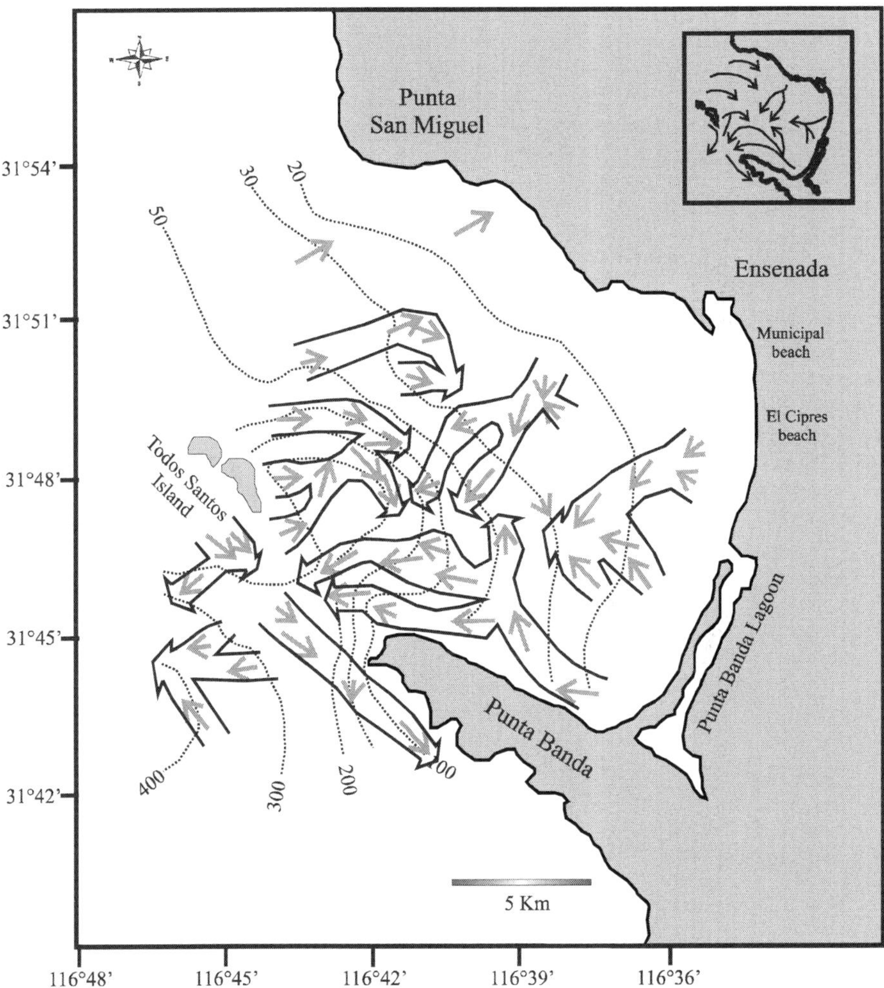

Fig. 5. Dispersion pattern of sediments in the Todos Santos Bay, Baja California, Mexico inferred from transport vectors. The segmented lines denote the bathymetry (m) of the bay.

Submarine canyons can actively function as a structural traps of sedimentary material, especially when there is a net flow downward through the canyon; those with coarse materials (coarse sands) in the deep regions are considered as dead or inactive (Shepard and Marshall, 1973). In the latter case, current speeds of 0.44 m s^{-1} have been observed, capable of transporting sediments upslope (Shepard and Marshall, 1973, Shepard and Keller, 1978). The possibility of sediment transport upslope the submarine canyons may seem intuitively difficult, however, depending on the hydrodynamics prevailing at the canyons' mouth it is possible to promote a net upcanyon transport of sedimentary material. The existence of flows with high enough speed (0.30 m s^{-1}), capable of resuspending and transporting sediment (of 3-4ϕ grain size) suggests that somehow, different processes such as the California Countercurrent, internal waves propagating towards the coast or residual tidal currents, independently or in combination, can promote the movement of particles in the direction proposed by this study (i.e., upslope the canyon).
The dispersal pattern of sediments proposed for TSB seems to promote the development of a site of convergence in the central region (e.g., a depocenter), rendering the particles to environmental stagnation (eg, pollutants) in this convergence zone or depocenter. This zone not only collects the sediments that naturally converge here, but also retains the exogenous materials generated from the dredging of the port that are dumped in the vicinity of the stations 9, 46, 47, 49, 51, 52 and 54. In a study on metals distribution in sediments within Bahía de Todos Santos, Romero-Vargas-Marquez (1995) reported the highest concentrations of several metals, including Cd, Cu, Zn, and Ni in the dredge discharge area and inside the port. Furthermore, in a study in which sediment samples were collected along the coast of Baja California from the México-USA border to Todos Santos Bay, Sandoval-Salazar (1999) reported that the sediments most enriched in metals (e.g., Cd, Ag, Cu, Pb, and Zn) were found around the Todos Santos Islands. Analysis of the 16 polyaromatic hydrocarbons (PAHs) listed as priority pollutants by EPA, was carried out on surface sediments at Bahía de Todos Santos, Baja California, Mexico. The PAHs composition of car exhaust, grass and shrubs combustion, asphalt and tire dust, were all compared to the relative abundance of PAHs signature found on marine sediments of the bay. These studies found that surface distribution of PAHs in TSB coincided with the reported water circulation in the bay. The lower concentrations were found in stations near the coast. Since this zone corresponds to the sedimentary environment with the highest energy relative to the rest of sampling sites, it is unlikely that particles accumulated in this part of the bay, and hence unlikely to find hydrocarbons associated with this material. The higher concentrations found around the Todos Santos island clearly suggest that this is a natural deposit area for particulate material such as this (rich in associated contaminants). This area is probably a good place for preservation of PAH's and other synthetic organic molecules, specially those that are difficult to decompose by bacterial attack (Villegas-Jimenez et al., 1997).

3. Conclusion

The principal component analysis applied to textural parameters indicate that the sediment grain size and sorting significantly contribute to the spatial trends that define the dispersion pattern of sediments in coastal environments. By considering all these parameters in the analysis has the advantage of generating a more robust definition of net sediment transport. The model of sediment transport applied to textural trends of sediments in the Todos Santos Bay indicates that the two-dimensional model of LeRoux defines sediment transport paths that agree with the hydrodynamic characteristics of the bay (surface, subsurface/bottom currents). This model is also consistent with the one-dimensional model of Sunamura-Horikawa and with the bi-dimensional model of Gao-Collins. All these models of sediment transport in BTS broadly define a transport from the coastal areas and from the Todos Santos Islands converging into the center of the bay, indicating the existence of a depocenter that naturally accumulates sediments. Ths depocenter zone is now being used to dump contaminants dredged from the port of Ensenada, creating an environmental health hazard that could be avoided if the dumping site were located outside the bay at greater depths.

4. Acknowledgment

We thank the M.C. Rene Navarro, Director of the Oceanographic Station of the Navy Department, for providing samples for this study. Al M.C. Oscar Delgado's IIO-UABC for contributing data corrientómetros. A reviewer for their valuable suggestions to this work.
IPN-COFAA by financial support in the article processing charge.

5. References

Alvarez-Sánchez, L.G.; Hemández-Walls, R. & Durazo-Arvizu, R. (1988). Patrones de deriva de trazadores lagrangeanos en la Bahía de Todos Santos. *Ciencias Marinas*, Vol. 14, No. 4, (Diciembre, 1988), pp. 135-162, ISSN 0185-3880.

Argote-Espinoza, M. L.; Amador-Buenrostro, A. & Morales-Zuñiga, C. (1975). Distribución de los Parámetros de Salinidad y Temperatura y Tendencias de la Circulación en la Bahía de Todos Santos, B. C. *Memorias de la Primera Reunión de CIBCASIO*, pp. 3-30.

Carriquiry-Beltrán, J. D. (1985). Análisis de la distribución de materiales pesados presentes en los sedimentos clásicos de la Bahía de Todos Santos, B.C.. *Tesis Licenciatura.* FCM. UABC.

Carriquiry, J.D. & Sánchez, A. (1999). Sedimentation in the Colorado River delta and Upper Gulf of California after nearly a century of discharge loss. *Marine Geology*, Vol. 158, No. 1-4, (June, 1999), pp. 125-145, ISSN 0025-3227.

Carriquiry, J.D.; Sánchez, A. & Camacho-Ibar, V.F. (2001). Sedimentation in the Northern Gulf of California after the elimination of Colorado River Discharge. *Sedimentary Geology*, Vol. 144, No. 1, (October, 2001), pp. 37-62, ISSN 0037-0738.

Cheng, P.; Gao, S. & Bokuniewicz, H. (2004). Net sediment transport patterns over the Bohai Strait based on grain size trend analysis. *Estuarine, Coastal and Shelf Science*, Vol. 60, No. 2, (June, 2004), pp. 203-212, ISSN 0272-7714.

Cruz-Colín, M.E. (1994). Balance sedimentario de la Bahía de Todos Santos, B.C., México. Tesis Licenciatura. FMC. UABC.

Cruz-Colín, M.E. (1997). Variabilidad de temperatura del mar en la Bahía de Todos Santos, B.C., México. Tesis Maestría en Ciencias. FCM. UABC.

Emery, K.O.; Gorsline, D.S.; Uchupi, E. & Terry, R.D. (1957). Sediments of three bays of Baja California: Sebastian Viscaino, San Cristobal, and Todos Santos. *Journal of Sedimentary Petrology*, Vol. 27, No. 2, (June, 1957), pp. 95 - 115, ISNN 1527-1404.

Friend, P.L.; Velegrakis, A.F.; Weatherston, P.D. & Collins, M.B. (2006). Sediment transport pathways in a dredged ria system, southwest England. *Estuarine Coastal Shelf Science*, Vol. 67, No. 3, pp. 491-502, ISSN 0272-7714.

Galehouse, L., (1971). Sedimentation Analisys. In: *Procedures in Sedimentary Petrology*, R.E. Carver, (Ed), 69-64, Wiley Intersciense, ISSN 978-0471138556, New York, USA.

Gao, S. (1996). A Fortran program for grain-size trend analysis to define sand sediment transport pathways. *Computers and Geosciences*, Vol. 22, No. 4, (May, 1996), pp. 449–452, ISSN 0098-3004.

Gao, S. & Collins, M.B. (1992). Net sediments transport patterns from grain size trends, based upon definition of 'transport vectors'. *Sedimentary Geology*, Vol. 81, No. 1-2, (November, 1992), pp. 47–60, ISSN 0037-0738.

Gao, S. & Collins, M.B. (1994). Analysis of grain size trends, for defining sediment transport pathways in marine environments. *Journal of Coastal Research*, Vol. 10, No. 1, (Winter, 1994), pp. 70–78, ISSN 0749-0208.

García, C.; Robles, P. M.; Figueroa, C.C. & Delgado, G. O. (1994). Observaciones de corrientes y temperatura en la Bahía de Todos Santos, RC.. Durante Noviembre de 1993-Enero 1994. *Informe Técnico*. Comunicaciones Académicas, Serie Oceanográfica Física, CICESE. 63 p.

Gavidia-Medina, J. (1988). Simulación Numérica de la Circulación Barotrópica en la Bahía de Todos Santos, Francisco. Depto. de Oceanografía Física, CICESE, Tesis de maestría, 95 pp.

Ingram, D.L., 1971. Sieve analysis, In: *Procedures in Sedimentary Petrology*, R.E. Carver, (Ed), 49-68. Wiley Intersciense, ISSN 978-0471138556, New York, USA.

Komar, D.P. (1977). Selective longshore transport rates of different grain-size fractions within a beach. *Journal of Sedimentary Petrology*, Vol. 47, No. 4, (December, 1977), pp. 1444-1453, ISNN 1527-1404.

Lanckneus, J.; De Moor, G.; De Schaepmeester, G.; Meyus, I. & Spiers, V. (1992). Residual sediment transport directions on a tidal sand bank. Comparison of the «Mc Laren model» with bedform analysis. *Bulletin Society Belge d'Etudes Géographique*, Vol. 2, No. 2, pp. 425–446.

Le Roux, J.P. (1994). An alternative approach to the identification of sand sediment transport paths based on a grain-size trends. *Sedimentary Geology*, Vol. 94, No. 1-2, (August, 1994), pp. 97–107, ISSN 0037-0738.

Le Roux, J.P. & Rojas, E.M. (2007). Sediment transport patterns determined from grain-size parameters: overview and state of the art. *Sedimentary Geology*, Vol. 202, No. 3, (December, 2007), pp. 473-488, ISSN 0037-0738.

Le Roux, J.P.; O'Brian, R.D.; Rios, F. & Cisternas, M. (2002). Analysis of sediment transport paths using grain-size parameters. *Computers and Geosciences*, Vol. 28, No. 6, (July, 2002), pp 717–721, ISSN 0098-3004.

Lucio, P.S.; Bodevan, E.C.; Dupont, H.S. & Ribeiro, L.V. (2006). Directional kriging: a proposal to determine sediment transport. *Journal of Coastal Research*, Vol. 22, No. 6, (November, 2006), pp. 1340–1348, ISSN 0749-0208.

Masselink, G. (1992). Longshore variation of grain size distributions along the coast of the Rhone Delta, Southern France: a test of the "Mc Laren Model". *Journal of Coastal Research*, Vol. 8, No. 2, (Spring, 1992), pp. 286–291, ISSN 0749-0208.

Mateos, E.; Marinone, S.G. & Parés-Sierra, A. (2008). Towards the Numerical Simulation of the Summer Circulation in Todos Santos Bay, Ensenada, B.C. Mexico. Ocean Modeling, Vol. 27, No. 1-2, pp. 107-112, ISSN 1463-5003.

Mc Laren, P. & Bowles, D. (1985). The effects of sediment transport on grain size distributions. *Journal of Sedimentary Petrology*, Vol. 55, No. 5, (September, 1985), pp. 457–470, ISNN 1527-1404.

Mc Laren, P. & Little, D.I. (1987). The effects of sediment transport on contaminant dispersal: an example from Milford Haven. *Marine Pollution Bulletin*, Vol. 18, No. 11, (November, 1987), pp. 586–594, ISSN 0025-326X.

Mc Laren, P. & Beveridge, R.P. (2006). Sediment Trend Analysis of the Hylebos Waterway: Implications for Liability Allocations. *Integrated Environmental Assessment and Management*, Vol. 2, No. 3, (July, 2006), pp. 262–272, ISSN 1551-3793.

McManus, L. (1988). Grain size determinations and interpretation. In: *Techniques in Sedimentology*, M. Tucker (Ed.), 408, Blackwell, ISSN 978-0632013722, Oxford, England.

Pérez, H. R. & Chee, B. A. (1984). Transporte de Sedimentos en la Bahía de Todos santos, B. C. México. *Revista de Ciencias Marinas*, Vol. 10, No. 3, pp. 31-52, ISSN 0185-3880.

Poizot, E.; Méar, Y.; Thomas, M. & Garnaud, S. (2006). The application of geostatistics in defining the characteristic distance for grain size trend analysis. *Computers and Geosciences*, Vol. 32, No 3, (April, 2006), pp. 360–370, ISSN 0098-3004.

Poizot, E. & Méar, Y. (2008). eCSedtrend: a new software to improve Sediment Trend Analysis. *Computers and Geosciences*, Vol. 34, No. pp. 827-837, ISSN 0098-3004.

Poizot, E.; Méar, Y. & Biscara, L. (2008). Sediment Trend Analysis through the variation of granulometric parameters: A review of theories and applications. *Earth-Science Reviews*, Vol. 86, No. 1-4, (January, 2008), pp. 15-41, ISSN 0012-8252.

Ríos, F.; Cisternas, M.; Le Roux, J. & Correa, I. (2002). Seasonal sediment transport pathways in Lirquen Harbor, Chile, as inferred from grain-size trends. *Investigaciones Marinas*, Vol. 30, No., pp. 3–23, ISSN 0716 - 1069.

Romero-Vargas, I.P. (1995). Metales pesados y su fraccionación química en sedimentos de la Bahía de Todos Santos, Baja California, México. Tesis de Licenciatura. Facultad de Ciencias Marinas, Universidad Autónoma de Baja California.

Sandoval-Salazar, G. (1999) Metales pesados en sedimentos superficiales de la cuenca de las Californias: Frontera México-E.U.A. a Bahía de Todos Santos, Ensenada, Baja California, México. MSc thesis, UABC, México.

Sánchez, A.; Alvarez-Legorreta, T.; Sáenz-Morales, R.; Ortiz-Hernández, C.; López-Ortiz, E. & Aguiñiga, S. (2008). Transporte y dispersión de los sedimentos superficiales en la Bahía de Chetumal inferido del análisis de tendencias texturales. *Revista Mexicana de Ciencias Geológicas*, Vol. 25, No. 3, (Octubre, 2008), pp. 523-532, ISSN 1026-8774.

Sánchez, A.; Carriquiry., J.; Barrera, J. & López-Ortiz, B. E. (2009). Comparación de modelos de transporte de sedimento en la Bahía Todos Santos, Baja California, México. *Boletín de la Sociedad Geológica Mexicana*, Vol. 61, No. 3, (Septiembre, 2008), pp. 13-24. ISSN 1405-3322.

Sánchez, A.; Shumilin, E.; López-Ortiz, E.B.; Aguíñiga, S.; Sánchez-Vargas, L.; Romero-Guadarrama, A. & Rodriguez-Meza, D. (2010). Sediment transport in Bahía Magdalena, inferred of grain-size trend analysis. *Journal Latinoamerican of Aquatic Science*, Vol. 38, No. 2, (July, 2010), pp. 167-177, ISSN 0718-560X.

Shepard, F.P. & Marshall, N.F. (1973). Currents along of submarine canyons. *American Association of Petroleum Geologists Bulletin*, Vol. 57, No. 2, pp. 244-264, ISSN 0149-1423.

Shepard, F.P. & Keller, G.H. (1978). Currents and sedimentary processes in submarine canyons off the northeast United States. *Geological Research Division Report.* Scripps Instiution of Oceanography.

Smith S.V.; Ibarra Obando, S.E.; Diaz Castañeda, V.M.; Aranda-Manteca, F.J.; Carriquiry, J.D.; Popp, B.N. & Gonzalez-Yajimovich, O. (2008). Sediment organic carbon in Todos Santos Bay, Baja California, Mexico. *Estuaries and Coastal,* Vol. 31, No. 4, pp. 719-727, ISSN 1559-2723.

Sunamura, T. & Horikawa, K. (1971). Predominant direction of littoral transport along Kujyukuri Beach, Japan. *Coastal Engineering in Japan,* Vol. 14, pp. 107-117, ISSN 0578-5634.

Villegas-Jiménez, A., Macías-Zamora, J.V. & Villaescusa-Celaya, J.A. (1996). Aliphatic and polycyclic aromatic hydrocarbons in surficial sediments of Bahia de Todos Santos, B. C. Hidrobiológica, Vol. 6, No. 1-2, pp. 25-32, ISSN 0188-8897.

Walton, W.R. (1955). Ecology of living benthonic foraminifera, Todos Santos Bay, Baja California. *Journal of Paleontology,* Vol. 29, No. 6, (November, 1955), pp. 952 - 1018, ISSN 0022-3360.

2

Dynamics of Sediments Exchange and Transport in the Bay of Cadiz and the Adjacent Continental Shelf (SW - Spain)

Mohammed Achab
Université Mohammed V-Agdal, Institut Scientifique,
Département des Sciences de la Terre,
Maroc

1. Introduction

In the framework of sediment exchange between the continents and ocean basins, the study of continental margins is of particular interest. Indeed, the deposits sequences represent the sedimentary record of the processes that have taken place in the continent and the ocean (Seibold and Berger, 1982; Agi, 1987). In this context, the continental shelves represent the most sensitive marine areas and of high economic interest, where the deposits are forming a mosaic of relict and modern sediments (Shepard, 1932, Emery, 1968). The littoral zones play an important role in the morphodynamic evolution of the adjacent marine environments, through eustatic changes and the transport of sediments from different input sources (Zazo et al, 1994, 1996, Hernandez-Molina et al, 1996, Gutiérrez-Mas et al, 1998; Achab & Gutierrez-Mas, 1999a). To describe sediment transport dynamics and to understand many land-shelf-ocean interaction processes, the quantification of suspended particulate matter (SPM) and the investigation of its dynamics are of major importance. During the past decades, studies on sediment dynamics have focused on the actual processes that control the sediment transport on continental shelves and the final fate of most particulate matter derived from the continents. (Wegner, 2003).
The study area was the object of numerous multidisciplinary works realized with the objective of determining the recent sedimentary facies distribution and their sources areas, as well as the recognition of the geological formation located in river basin (Achab et al, 1999b; Achab, 2000; Gutierrez Mas et al, 2004; Achab et al, 2005b). Several studies focused on sedimentary dynamics associated with water mass movements have been realized by several authors (Madelian, 1970; Kenyon and Belderson, 1973; Melières, 1974; Palanques et al, 1987; Grousset et al., 1988; Maldonado & Nelson, 1999a; Nelson et al, 1999; Lopez-Galindo et al, 1999; Lobo et al, 2000). Others studies have been focused on the dynamics of fine sediments and clay minerals in the Cadiz bay and the adjacent marine deeper zones (Gutierrez Mas et 1996b, 1997; Achab et al, 1998, 2000b; Achab et al, 2008). Also, they have been approached studies about the dispersal of the suspended matter in the bay of Cadiz and its effects on the inner continental shelf as well as their influence on the recent marine

sedimentation en general (Palanques et al, 1987, Gutierrez Mas et al, 1999; Achab et al, 2000a; Gutierrez Mas et al, 2006). In the bay of Cadiz, the study of the dynamics of sedimentary exchange between the coastal zones and the continental shelf is complex, and presents difficulties due to the diversity of sedimentary processes and environments, as well as to hydrodynamic factors and the physiography of the coast and sea bottoms. The main objective was to elaborate a global hydrodynamic model of exchange and transport of sediments in the Cadiz bay and the adjacent continental shelf. This model take in consideration the sedimentological and mineralogical data complemented by satellite images and suspended matter analyses as well as data obtained from side scan sonar records and other available information.

2. The study area

The study area is located on the margin of the Gulf of Cadiz between the mouth of the Guadalquivir River and the western entrance to the Strait of Gibraltar (Fig.1).

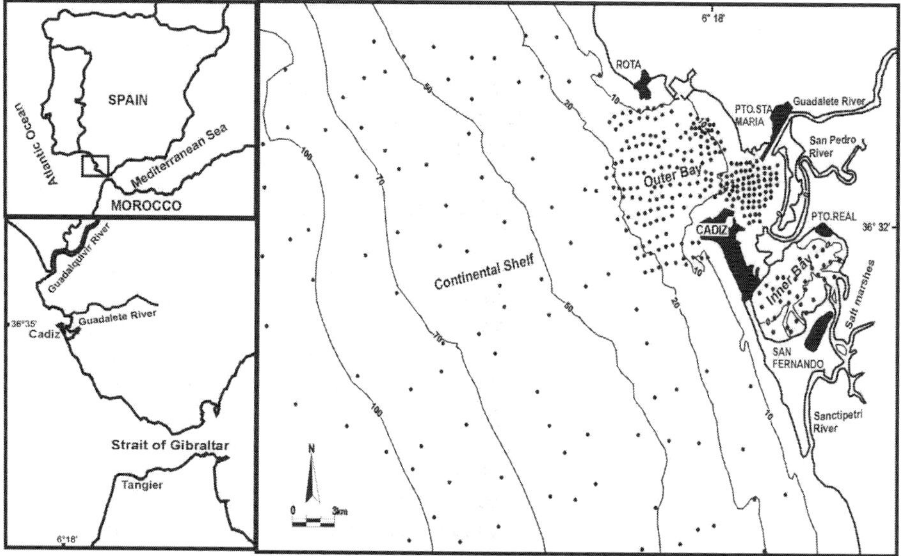

Fig. 1. Geographic situation of the study zone with dept (in m), and location of the samples

2.1 Geological setting

Geologically, the study zone is included in the Tertiary Guadalquivir basin, which represents the foreland basin of the Betic Mountain Ranges. The Bay of Cadiz was generated as a tectonic depression during a distensive tectonic phase in the Late Miocene-Pliocene (Benkhelil, 1976). The depression was occupied by a deltaic system that gives rise to a characteristic stratigraphic sequence (Viguier, 1974; Zazo, 1980). Different geological units and formations constitute the sedimentary strata outcropping in the surrounding area of Cadiz bay (Fig.2).

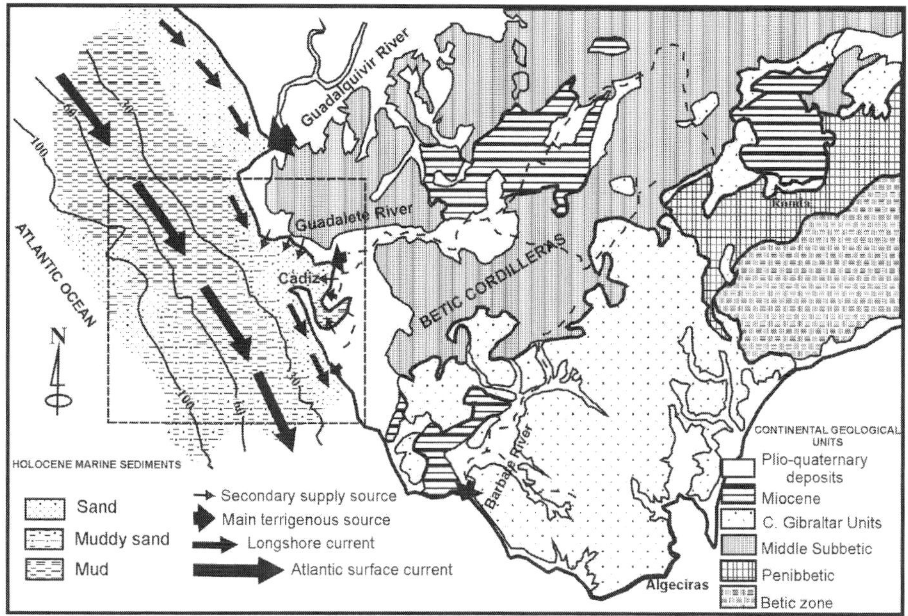

Fig. 2. Schematic geological setting showing supply sources and marine sediments distribution on the study zone

The Post-orogenic materials (autochtonous units) are mainly from the Neogene Guadalquivir depression, and correspond to Plioquaternary and Quaternary sands, clays, marls, sandstones and some limestone and conglomerate levels (Fig.3B). The pre-orogenic materials (Allochtonous units) outcropping in the study area correspond to different units of the Betic Mountain Range, being constituted by calcarenites and marls of Upper Miocene, red marls and gypsum of Subbetic Trias, and the Aljibe sandstones of Oligocene-Lower Miocene (Fig.2). Upon all those materials, appear Quaternary deposits constituted by muddy marshes, beach sands and continental deposits (Mabesoone, 1966; Viguier, 1974; Zazo et al., 1983; Gutiérrez Mas et al., 1990; Moral Cardona et al., 1996; Dabrio et al., 2000; Achab et al., 2005a).

2.2 Physiographic aspects

In this sector of the continental margin of the Gulf, the shelf and the coastline physiography are oriented NNW to SSE, with shorted East-West sections, having a stepped aspect resulting from both old and recent tectonic fractures (Baldy et al, 1977; Sanz de Galdeano, 1990). These, are manifested by several systems of recent fault and diaclase, affecting so much to continental domain as marine bottoms (Fig.3A) (Gracia et al, 1999; Gutierrez-Mas et al, 2004). The tectonic activity had a considerable influence on the distribution, development and nature of the recent marine deposits in the Gulf of Cadiz. Many indicators in areas near the Bay of Cadiz, like faults affecting Quaternary sediments, morphostructural lineaments and other geomorphological features, confirm this neotectonic activity (Benkhelil, 1976; Zazo et al, 1999). An Early Quaternary compressive tectonic episode has been deduced from reverse faults observed in marine sediments of the continental margin (Maldonado & Nelson, 1999; Maldonado et al, 1999b).

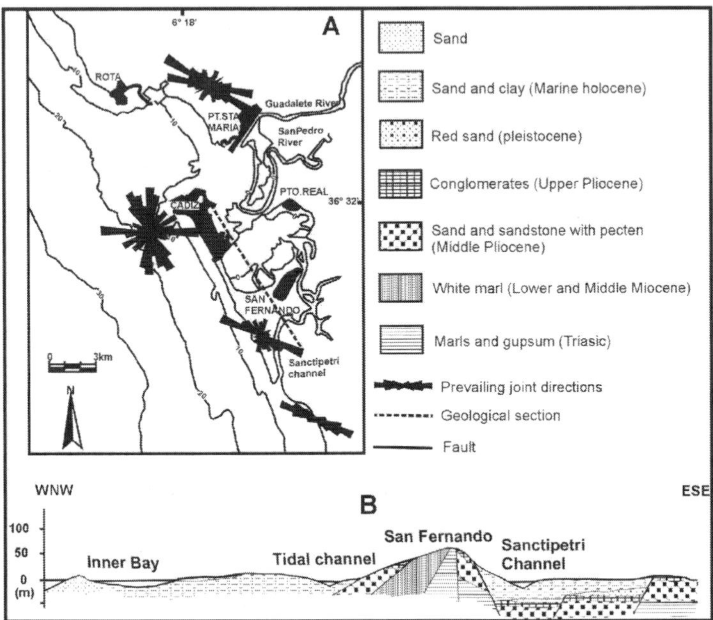

Fig. 3. A: Prevailing joint directions affecting Plio-Quaternary and Pleistocene units outcropping in the coastal zone; B: Cross section showing the main geological formations of the Bay of Cadiz

2.3 Hydrodynamics setting

Different water mass movements and currents control the shelf hydrodynamic regime (Fig.4). The most important being North Atlantic Surface Water (NASW) and littoral currents moving towards the southeast, and are responsible for the dispersal of fine sediments from the Guadalquivir and Guadiana Rivers (Gutierrez-Mas et al, 2006; Achab et al, 2008); and the Mediterranean Outflow Water (MOW) moving west to deeper water (Maldonado & Stanley, 1981;Baringer & Price, 1999). The tidal regimes in the Bay of Cadiz is mesotidal and of a semidiurnal character, with a mean tidal range of 2.39 m, mean spring tidal range of 3.71 m and neep tide of 0.65m (Benavente et al., 2000). The tidal currents are considered to be responsible of fine sediment transport (Achab et al., 1998, 2008). Wind and wave action are also essential factor in the sedimentary dynamics. Western winds are the most frequent blowing with 13.6% of average frequency. Eastern winds are also important with a frequency of 12.3% (Ramos, 1991). Waves present seasonal character and the storm average frequency is of 20 days/year. The strongest storms occur in the fall-winter period and there exists an accused calm during the summer. The significant wave height is 0.6-1m; during storm weather, the maximum wave height can reach 4m (Ramos, 1991). The Sea wave (6.96%) presents a relative predominance of east component, while the Swell wave

(10.26%) dominates the west component (MOPT, 1992). The mean littoral-drift currents are controlled essentially by northwestern waves, generating currents toward the southeast, as a consequence of the coastal orientation, facing westerly and SW winter storms. Easterly winds are also important, generating littoral-drift towards the North and NW.

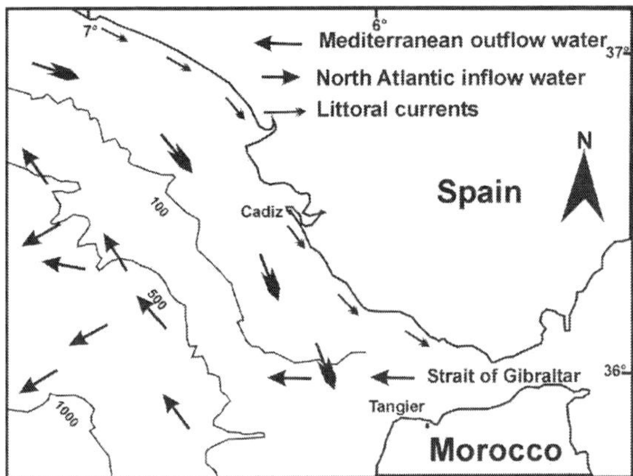

Fig. 4. General circulation patterns of water masses in the Gulf of Cadiz. Modified from Nelson et al (1999).

2.4 Depositional environments

Two main sedimentary environments with different hydrodynamic characteristics can be differentiated: the Cadiz Bay and the adjacent continental shelf. The Bay of Cadiz is about 28.5 km long and 13.5 km wide, it can be defined as a mixed morphodynamic system constituted by a wide bay known as the outer bay, with surface of about 118km^2. This zone is well connected to the continental shelf, and is very affected by the waves, currents and storms, especially of a westerly nature who dominate the sedimentary dynamics. The inner bay (Surface ≈ 40 km2) or Lagoon system located to the South; protected from waves and storms of the West and Southwest. The salt marshes and tidal flat whose surface can reach 227km^2, occupy the most internal and sheltered areas and isolated from the open sea by sandy beach ridges and littoral spits. Its development is a consequence of the sedimentary infilling when the sea level was a few meters above the present, and also by growth of inside tidal deltas. After of the last eustatic fall (5.000-6.000 years B.P), the partial emerging of the bottoms has given cause for a wide marshes zone, occupied by halophyte vegetation and drained by a complex system of tidal creeks and channels; that constituted the hydrodynamics and sedimentological transmission from the inner bay zones and the marine environments (Achab et al., 1998; Dabrio et al, 2000; Achab et al, 2008). The continental shelf has a gentle slope and a slight inclination toward the west, with an average width of 40 km, although there are important variations in the area close to the Straits of Gibraltar, where it narrows to 20-30km. The physiography of the sea-bottom shows a close concordance with the shoreline, the isobaths generally running parallel to the coast. The slope break occurs at 150-200 m water depth and shows significant variation north to south in cross section (Gutierrez-Mas et al, 1996a; Lopez-Galindo et al, 1999).

3. Materials and methods

The study has been carried out on 300 samples of surface sediments distributed across different sectors of the bay and the adjacent continental shelf (Fig.1). Bottom sampling was carried out using gravity cores and *Van Veen* dredging. The sampling stations were positioned with a differential GPS system. The textural and mineralogical analyses were carried out to establish facies distribution and mineralogical composition. The grain size analysis has been made in several phases: i) Humid separation of coarse and fine fractions using a sieve of 0.063 mm, ii) The coarser material was dry-sieved during 15 minutes, iii) The fine fraction sizes analysis was made by use of a laser diffraction analyser (AMD). The characteristic statistic indexes and parameters were calculated using standard method (Folk & ward, 1957; Folk, 1980). The grain size distribution of sediments was used to describe the sedimentary facies and to correlate their physical proprieties with the marine dynamics. The composition of the sand fraction was established by optical microscopy counting 500 grains in each sample. The mineralogical analysis of samples was performed with a Philips PW-1710 X-ray diffractometer, equipped with Cu-Kα radiation, automatic slit and graphite monochromator, using the crystalline powder technique for the bulk mineralogy and oriented aggregates of the < 2 μm for the clay mineralogy. Quantification of different mineralogical phases was calculated by the classic method of area measurement of peaks, considering the different reflection capacities of the minerals (Pevear & Mumpton, 1989; Ortega -Huertas et al, 1991). The factor analysis (Principal Components Analysis) was used to establish the mineral assemblages, as well as the relation between different minerals, grains size fraction and their associations (Reyment & jöreskog, 1993). It was also used to establish possible sediment transport paths from the bay toward the continental shelf. The method by Imbrie (1964) was used for this analysis, which is based on the samples similarity matrix. The associations obtained by this analysis are based on those variables showing the highest scores in each factor; the factor scores represent the weight or influence of each variable and components within the corresponding factor (Mezzadri and Saccani., 1988).
The study of concentration of suspended particulate matter (SPM) was based on the analysis of 30 water samples taken in zones where the concentration of suspended matter was highest, such as the tidal channel, river mouths and the inner zones of the bay. The climatic and hydrodynamic conditions prevailing the sampling time were wind from the north and northeast, average speed of 55km/hr and mean wave height of 0.6 and maximum of 1.5m. The extraction of water samples was executed to specific depths with oceanographic bottles, simultaneously with the passage of the Landsat satellite over the study zone, in order to obtain a synoptic picture of the turbid plumes, by comparing sample data with the satellite images. The separation of SPM has been achieved following the method of Green-Berrg et al (1992), which consists to the filtration of a volume of 5 liters of water through pre-weighed filters by MILLIPORE (0.45 microns). Filters were washed with distilled water, dried at about 60Â° and weighed. The use of satellite images in study of sediment dynamics is based on the utilization of inorganic suspended particulate matter as a natural tracer. Satellite images of the Bay of Cadiz have been recorded by the satellite Landsat TM, using bands 2 and 5, and a spatial resolution of 30x30 m. These images has been analysed to obtain data about extent and direction of turbidity plumes and fine sediments transport in several hydrodynamic situation in Cadiz bay and inner shelf waters. The process of the images has been carried out according to the methodology described by Ojeda et al (1995). In order to

establish the distribution pattern of bottom currents and the flow regime, different bedforms fields have been identified based on the analysis of side scan sonar records, obtained in different oceanographic campaigns, using sonar model Klein 500Khz. The sweeping width is of 100m, resolution > 7.5 cm and overlapping of 30% (Parrado Roman et al, 2000).

4. Results and discussion

4.1 Grain size analysis

4.1.1 Grain-sizes classes

Taking into account the variety of grain size distributions in the modern sediments of Cadiz bay and the continental shelf bottoms, we can differentiate various types of sediment, corresponding to different grain size classes, which reflect the energy level of each depositional environment and the processes of sediment transport. Grain size analysis show that samples are mainly composed of sand and mud, and subordinate amounts of gravel. The grain size distribution histograms and cumulative frecuency diagrams reveal the existance of different sediment types, highlighting fundamentally six classes, as most representative of all Cadiz bay sediments (Fig.5). The class A, is the most frequent with 33% of the total, and related to sandy facies with a very fine sand mode (3.49 Phi; 94 μm) which represent more than 69% of the total sample. The grains size distributions are mesokurtic and coarse skewed. Sands are moderately well sorted with 93% of the grains between 3 and 4 Phi (125 and 63 μm). These sands are mostly located on the outer bay and inner continental shelf, especially in littoral zones, at depths lower than 25m. The class B (28% of the total), characterised by poorly sorted sediments and have a bimodal grain-size

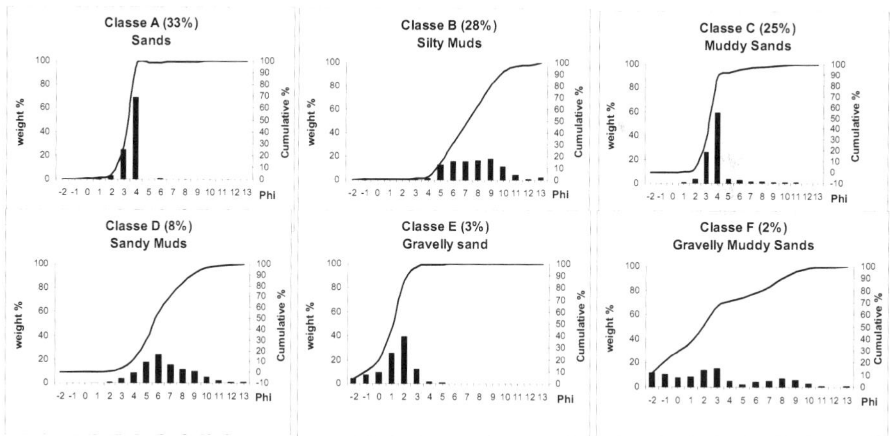

Fig. 5. Histogram distribution of grain size and cumulative frequency curves characteristic of the study zone

distribution, with more than 95% of the samples are finer than 4 phi (63 μm). These types of samples cover a large part of the middle continental shelf in the northern area and inner bay of Cadiz. The class C is quite frequent (25%), beinig associated with muddy sands facies, characterised by a very fine sand mode (59%), symmetrical and very leptokurtic distributions and moderately sorted sediments. Muddy sands distributions contain about 88% of material larger than 4phi and have mud content less than (12%). They are largely located to the south of the continental shelf and in the central and eastern parts of the outer bay. The class D (8%) presents modes of coarser silt (25%), composed of 78.4 % fraction between 5 and 9 phi (31 and 2 μm) and have sand content less than 13%. The samples are very poorly sorted and show mesokurtic and fine skewed distributions. These type of samples correspond to sandy muds facies, located in the occidental part of the outer bay and to the north of the continental shelf at depths between 25 and 40m. The class E (4%), characterised by pseudo-bimodal distributions, the main mode corresponds to medium sand (39.8%) and the secondary to very fine gravel (11%). The grains size distributions are coarse skewed and leptokurtic with poorly sorted sediments. This class was found to be related with energetic coastal environments of the bay of Cadiz, especially near rocky shoals. The class F is very scarce (1%) presents variable modes and very poorly sorted distribution. The samples have gravel content more than 22% and contain about 28% of material finer than 4phi. They correspond to sediments located in isolated pockets off the bay of Cadiz.

4.1.2 Grain-size parameters

The analysis of different grain size parameters shows the prevalence of unimodale distribution, however bimodal and polymodal distributions are also present. The average value of the main mode is 3 phi, corresponds to the boundary between fine and very fine sand, and represents the most common and more stable size fraction for the existing energitic conditions and for the dominant transport processe. The mode values increase in coastal areas and near rocky shoals and decreases towards the central zone of the outer bay and in the inner bay. The mean has an average value of 3.5 phi (very fine sand). The low values of means are located in the center of the outer bay, while the highest appear mainly in coastal areas. The average value of the sorting (1.76 phi) indicate poorly sorted sediments. The distribution of this parameter allows us to distinguish two sectors, an eastern one, characterized by well-sorted sediments, and a western sector where the sorting is generally poorer. The high degre of heterogeneity of grain size distributions is due to the variety of energy of the depositional environments and to the mode of transport affecting the sediments. The average value of skewness is about 0.17, corresponding to grain size distributions of positive skewness (fine skewed), which are predominant in the sediments with symmetrical distributions. The kurtosis (1.56) present heterogeneous and leptokurtic distributions, these prevail together with the mesokurtic distribution. The grain size of marine sediment varies by physiographic zones, it decrease progressively with depth, from coarser-medium sand of shallower area, characterised by well-sorted, negatively skewed and very leptokurtic distribution, to finer sand and mud of deeper zones, where the sediments trend to be poorly sorted and positively skewed. The general trend and the variability of different grain-size parameters of surface sediments of the bay of Cadiz and the continental shelf, reflect the control that physiographic and hydrodynamic factors exert on different types of sediment.

4.1.3 Grain size fractions

The grain size of sediments depends of numerous factors, especially the mineralogical composition of grains, erosive and depositional history of these. The combination of these factors gives place to different grain size distributions; whose interpretation was found to be related with sedimentological analysis. Their distribution in the marine environment is primarily a function of the interaction between the strength of waves and currents and the size of individual sediment grains (Nombella et al., 1987, Gao and Collins, 1994 and Olabarria et al, 1996). The distribution of different grain size fractions shows the predominance of the coarser fractions in the outer bay, more exposed to wave action and currents. The finer fractions such as silt and clay appear in more sheltered zones of the bay and in the continental shelf. (Fig.6A). The gravels are underrepresented in the sediments (less than 5%). These coarser fractions are mainly composed of bioclasts and rock fragments, derived from erosion of rocky shoals and coastal cliffs. The sand is the dominant fraction of the outer bay sediments, with an average of 75%. In some areas, this fraction changes laterally to muddy sand; due to the recent action of transport processes taking place in this area of the Cadiz bay. Its origin and distribution is related to several factors: i) the proximity to coastal areas affected by swell and with the presence of numerous sandy beaches that transport sediments to deeper areas of the bay, especially in storm period. ii) The presence of some significance river mouths (Guadalete River among others), as well as abundant valleys and ravines, which act in times of flood and important runoff. iii) The Existence of sandy sediments of relict or palimpsest character, largely reworked by actual processes, which were deposited when sea level was several meters below the present, or because to the action of special dynamic event that eroded beaches and transported sand to the deeper areas. The silt appears in low proportions (10%), giving the highest values in some sectors of the outer bay and the continental shelf. It distribution is of great importance to understand the modern sediment dynamics in the Bay of Cadiz, especially the sediment exchange between the inner and outer bay. These fine materials are deposited on sandy materials of the outer bay, indicating the path followed by water flows that control the sediment

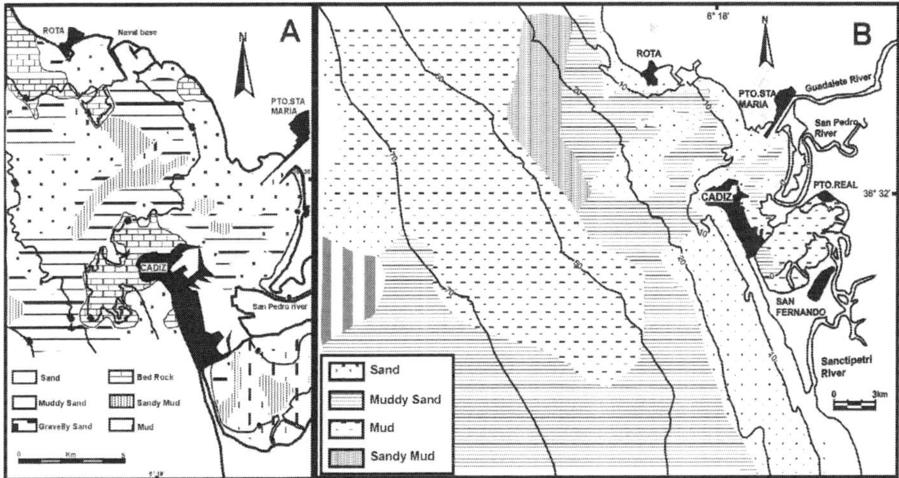

Fig. 6. Recent Holocene marine sediments distribution in the sytudy zone. A: Bay of Cadiz; B: Continental shelf

dynamics between the outer bay and the inner shelf. The clays are an important sedimentary fraction, especially in low-energy sedimentary environments where concentrations reach 90% such as inner bay and the tidal channels that drain the salt marshes, as well as to the northwest sectors of the study area. In the outer bay, the contents are very low (<5%). The abundance of clay fraction in the inner bay and the salt marshes is related to the existence of sedimentary environment of very low hydrodynamic energy. In these areas, predominate the flow and ebb tidal flow, whose action gives rise to a wide intertidal zone dominated by clayey mud deposits. The complex network of tidal channels, the inputs of fine material from the salt marshes, and the abundance of algae (seagrass type) determine the progressive deposition of fine materials in the inner bay bottoms.

4.2 Facies distribution pattern

Recent Holocene marine sediments in Cadiz bay and the adjacent continental shelf have a siliciclastic character (Gutierrez-Mas et al, 1996a; Achab et al, 1999b). Considering the relationships between grain size sediment and different sedimentary environment, the study area can be divided into several specific sectors (Fig.6B): The sandy facies predominate in the outer bay of Cadiz, especially in the littoral zone. Quartzose sands with less amount of gravel and with high content of bioclastic remains, associated to energetic environments, also cover the southern zone of the continental shelf. These transgressive sands of relict character were developed during the Holocene transgression (Rodero et al., 1999). Muddy-sand sediments occupy a central part of the continental shelf; this palimpsest facies is characterized by the mixture of fine materials coming from the Guadalquivir prodelta with relict sands facies. In the outer bay, those sediment facies extend into the 20-30 m deep inner shelf, being configured in two bands, one by the north margin of the bay and another one more to the South, bordering the city of Cadiz. They derive from resuspension of fine-grained materials in the marshlands of the bay and from fine materials supplied by the Guadalete River during periods of rainfall and floods. The presence of mud and sandy mud facies covering sandy bottoms indicates actual processes of deposit and transport of fine sediments from the inner bay toward the external zone reaching the inner continental shelf. Mud and sandy mud facies are the dominant fractions in the inner bay, agrees with sedimentary processes of low-energy characterizing the sheltered zone (Achab et al., 2005b). Their hydrodynamic regime is almost exclusively dominated by tidal currents and wind drifts, especially of the East sector. Muddy facies are also present in the continental shelf as a prodeltaic muddy zone situated to the north and deposited in low energy environment. These fine grained sediments are related to supplies coming from the Guadalquivir River. The salt marsh, tidal creeks and emerged alluvial plain are characterized by the presence of argillaceous-sandy nature sediments in their borders, whereas sandy sediments are present in beaches. The sedimentation is basically controlled by the action of flows and ebb tides. These are responsible for the transport of sediments, and the erosion of tidal creeks and salt marsh border, due to the action of small surge that beats riversides, as well as to the effect of collapses and superficial erosion of argillaceous grounds (Achab et al., 2000b). The pattern of particles size distribution is controlled by the action of hydrodynamic agents, the recent eustatic (sea level) changes during the terminal Holocene as well as the coastal and bottoms physiography (Alvarez et al, 1999; Achab & Gutierrez-Mas 1999a)

The mineralogical analysis (Fig.7) shows that quartz is the most abundant terrigenous mineral; its content is variable, based on the distribution of different grain sized facies, representing an average content of 55%, with maxima of up to 80% in sandy area, while in muddy sands the content is less than 25%. Feldspars are sparse; the content ranges from 5 to 10%, indicating a high degree of compositional maturity of the sediment. Locally they can reach values of 20% in sandy deposits. The calcite, displays average contents of 20 %, being also the second mineral more abundant in sediments after the quartz. Others mineral are the dolomite with very low contents (6%) and the aragonite (<5%), their origin is fundamentally biogenic. Phyllosilicates prevail in the muddy zone and its distribution is very conditioned by grain size. They are mainly constituted by illite (60%), smectite (13%), interstratified illite-smectite (10%), kaolinite (8%) and chlorite (7%).

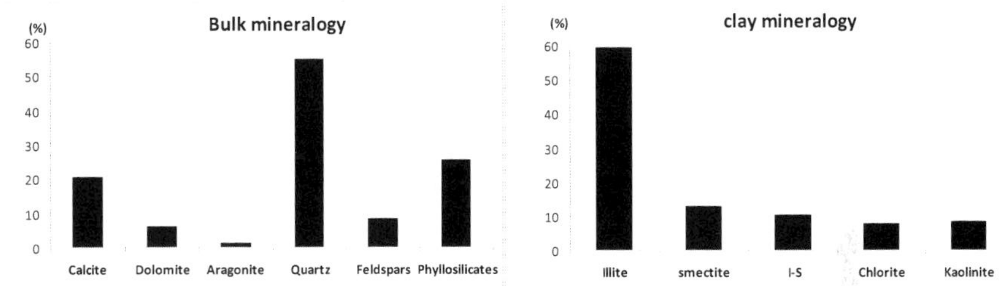

Fig. 7. Bulk and clay mineralogy of surface sediments in the study zone

The microscopic analysis show that the quartz trends to accumulate in very fine and medium grain sizes of the sand fraction (Fig.8), heavy minerals have been also observed specially in the finest fraction (very fine and fine sand). Bioclastic components and rock fragments mainly compose the coarsest sub-fractions. Especially calcareous shell fragments and continental plant remains (pieces of small trunks, small branches, etc.). The rock fragments corresponds to small cobble rocks grains of very diverse nature and lithology. Mainly fragments of "Ostionera rock" of plio-quaternary age coming from the erosion of rocky bottoms and coastal cliffs. They are also present small rolled cobbles of quartzite and sandstone fragments. Other calcareous biota include numerous benthic and planktonic foraminifers, echinoderms, bryozoans, sponge spicules and ostracods.

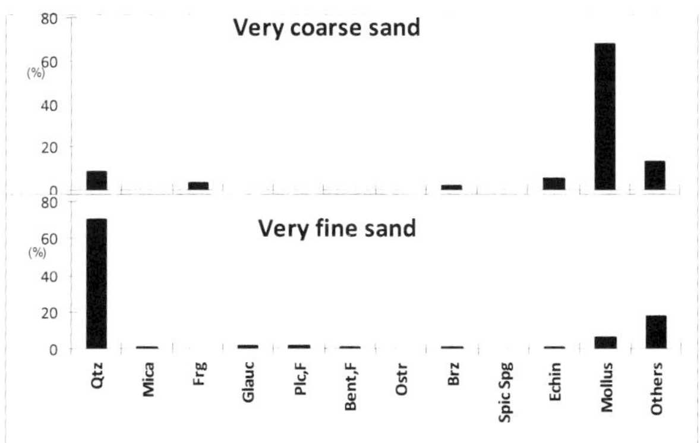

Fig. 8. Sand fraction components of surface sediments in the study zone

4.3 Sediment transport patterns

Sediment transport on marine environments and shelves is mostly a function of wave-current interactions; it depends on surface-wave conditions, bottom-boundary-Layer currents, and bed characteristics, including grain size, density, and surface roughness (Van Rijn, 1984, 1993; Cacchione & Drake, 1990; Wiberg, 1996 and Soulsby, 1997). On the other hand, the distribution of terrigenous sediments in the marine environment can reflect the direction of water mass movement (Mead, 1972; Poulos et al, 1996). Therefore, it is very important to establish a correlation between the transport of sediments and the hydrodynamic system, to understand the processes of dispersion of fine sediments from different supply sources (Gutiérrez Mas et al, 1999; Achab et al, 2000b).Taking into account the different types of data and results obtained in this work, together with knowledge of the sedimentary environments and characteristics of the local hydrodynamic system, a global model of exchange and transport of sediments in the Cadiz bay and the adjoining continental shelf can be elaborated. This model take in consideration the sedimentological and mineralogical data complemented by satellite image and suspended matter analysis, as well as data obtained from recording of side scan sonar.

4.3.1 Transport mode

The distribution of different granulometric facies allows differentiate the fundamental modes of particles transports (Visher, 1969; DeGiovanni, 1970, Komar, 1977). In our case, the limits to distinguish particle transport modes correspond to a storm situation (high energy), in which very fine sand fraction can be transported in suspension (Parrado Roman & Achab 1999). The different types of sediment grains are moved in three possible modes according to its size, forms, density and the hydrodynamic conditions. Bed-load transport (rolling or sliding motion) dominates the higher-energy environments such as coastal zone mainly composed of coarse fractions and shallow areas associated with rocky shoals. The areas affected by this type of transport and its significance increases substantially during storm periods. The suspended transport predominates in the inner part of the Bay (including tidal channels and marshland areas) and to the north of the continental shelf characterized by mud-clay bottoms deposited in low energy conditions. In general, suspension transport is the dominant transport mode on many marine environments including continental shelves (Cacchione & Drake, 1990). The intermediate transport (saltation-suspension) predominates in the outer bay especially the western sector and the central part of the continental shelf. Areas affected by this type of transport coincide with sandy-mud and mud facies bottoms, and may indicate the transport paths of fine sediments and suspended matter in the study area.

4.3.2 Clay minerals and assemblages

Due to their fine grain size, the clays can be transported large distance by rivers, wind and currents, indicating the dominant trajectories of fine sediments and the suspended matter (Maldonado & Stanley, 1981; Gutierrez-Mas et al, 1998; Achab et al, 1998). To establish the transport pathways of suspended sediments, local variation present in the seabed marine sediment was considered. Mineralogical assemblages determined in the clay fraction have been analyzed. Their origin is clearly detrital or inherited, and may reflect the combined effects of the influence of the source areas, the type of weathering on the adjacent continents, and the process of transport and sedimentation (Millot, 1970; Sawheney & Frink, 1978; Stanley & Liyanage, 1986; Weaver,1989; Chamley, 1989; Naidu et al, 1995). In the present study, clay minerals have been used as tracers, since their small size makes them the only particles susceptible of being transported away from the bay, out towards the continental shelf (Gutierrez-Mas et al, 2006; Achab et al, 2008). Other studies indicate that the clay mineral assemblages deposited in front of the rivers mouth and toward deep marine areas can be used as dynamic tracers to deduce transport path and the sources areas of clay minerals (Neiheisel & Weaver, 1967; Bhukhari & Nayak, 1996; Parhan, 1996). Factor analysis was used to establish the relationships between different clay minerals, their associations and the possible sedimentary and dynamic connections among different studied environments (Reyment & Jöreskog, 1993). The Q-mode factor analysis results provide three factors explaining the 100% of the total variance. (table.1). Factor 1 alone explains 96 % of the variance, and it represents the main clay mineral association in the modern marine sediment of the cadiz bay and the adjacent continental shelf. Factor 1 associates illite >> smectite > chlorite > kaolinite > interstratified illite-smectite.

Minerals	Factor 1 (96%)	Factor 2 (3%)	Factor 3 (1%)
Illite	2.15	1.32	0.45
Smectite	0.32	1.7	0.7
Illite-smectite	0.28	0.21	2.03
Chlorite	0.3	0.36	-0.11
Kaolinite	0.29	0.21	0.35

Table 1. Factor scores of the clay minerals in the Cadiz bay and the adjacent continental shelf, Q-mode factor analysis.

The factor loadings distribution shows a range of very pronounced alignment values, as bands perpendicular to the coast (Fig. 9). This association has great significance in the inner continental shelf sediments, which represent the transition zone toward offshore of deposit processes taking place in the proper bay. In this zone two bands are observed: one oriented towards the West and NW and the other one toward the SE and the South. These bands might correspond to sea floor marks generated by flows between the bay and the continental shelf; agreement with the tidal flow pattern established by Alvarez et al (1999) in the Cadiz bay.

The data from clay minerals contents and assemblages have been used to establish the model of transport paths in different area of the study zone. Two flows paths have been differentiated: i) The inflows coming from external marine areas located to the north, in particular the Guadalquivir river mouth and other sources. These flows can transport suspended matter and fine sediments to the Cadiz bay bottoms by the action of marine currents, specially the littoral and the Atlantic Surface water currents of SE direction. Different input flow paths have been established (Fig. 10). The first one (B1) derived from the Atlantic flow, affected by wind and wave of NO, West and SW, and promoted by the coastal configuration of NNW-SSE direction. This flow can reach the Cadiz bay by his northern margin and to mix with the ebb-tidal current, depositing part of its loads in inner zones of the bay. Other flow (B2) oriented E-W, located between the main entrance of the bay and inner shelf and affect bottoms of the transition zone characterized by the exchange between outflows and the inflows to the bay. This is directed towards the SE at depths between 50 and 60m. Parts of these inflows (D) are oriented to the southwest, reaching depths of 100m in the outer shelf. The outer and deeper part of the continental shelf are affected by currents generated by the Southwest storms and the action of deepest flows presents in the study area. ii) The outflows coming from Cadiz bay and littoral zones; can reach the continental shelf by mean of ebb tide currents. The Fig. 10 shows the existence of several flows that appear to move from the innermost zones of the bay to the outer one. These flows (A1) are configured in three main bands: one oriented towards the NNW following the north margin of the bay. Other band oriented towards the West and the third one goes towards the SSW bordering the Cadiz city. To the south of the bay, output flows (A3) are also observed and coming from the Sanctipetri tidal creek mouth that head towards the offshore, are capable to reach depth of 50m (Achab et al, 2000b; Gutierrez-Mas et al, 2006; Achab et al, 2008).

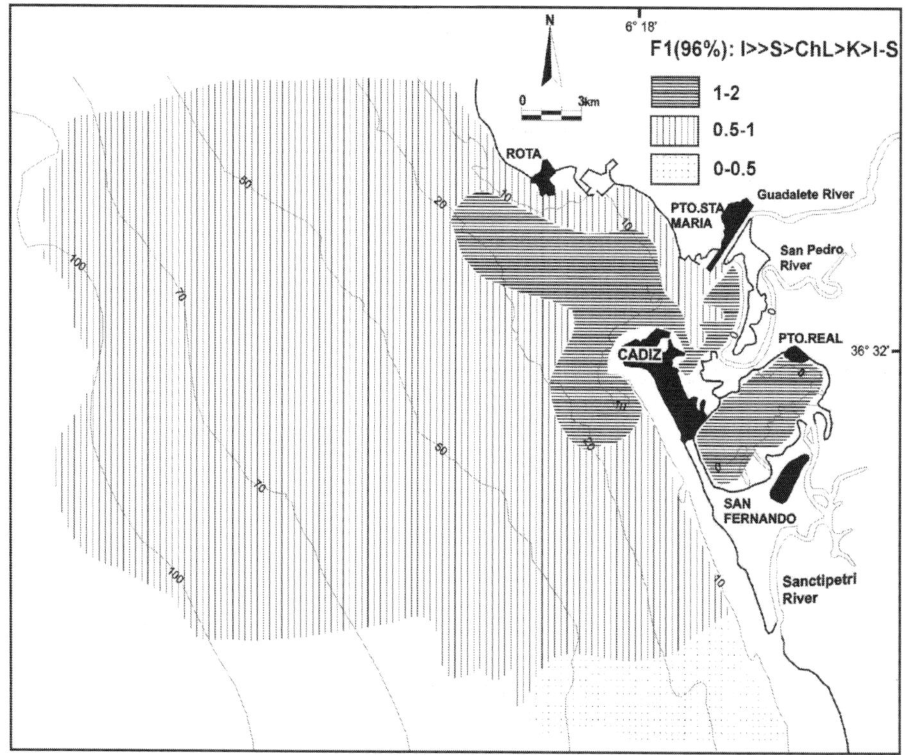

Fig. 9. Factor loading distribution of Factor 1 in the study zone from Q-mode factor analysis

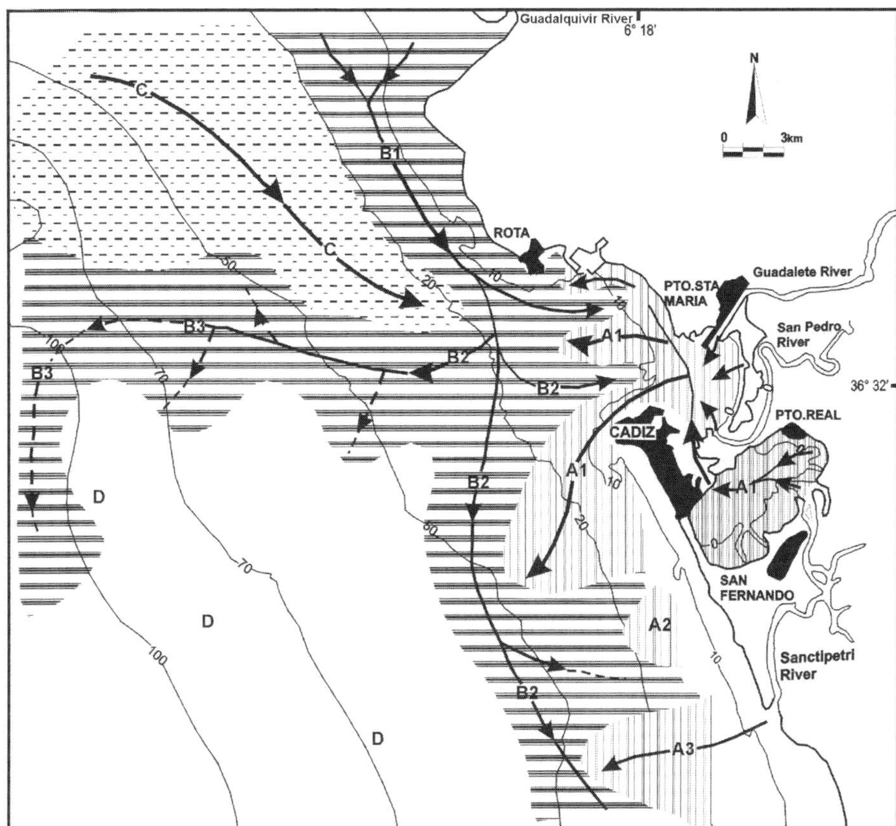

Fig. 10. Diagram of fin sediments transport paths between Cadiz bay and the continental shelf obtained from clay minerals contents and assemblages data. A1: Out flow path coming from Cadiz Bay; A2: Output flow coming from possible old tidal creeks mouth; A3: Output flow coming from Sanctipetri tidal creek mouth; B1: Output flow coming from Guadalquivir river, B2: Inflow path to Cadiz Bay; B3: External flow path to Cadiz Bay; C: Flow path coming from zones located to the north of Cadiz bay ; D: Bottoms affected by external flow path

4.3.3 Land-sat images and suspended matter analysis

The application of remote sensing techniques to the study of suspended matter dynamics allows model for marine and coastal water circulation based on the use of "turbidity patterns" as natural tracers, relating parameters of water quality to satellite images (Balopoulos et al. 1986; Fernández Palacios et al.,1994; Ojeda et al., 1994). The turbidity caused by suspended particles is detectable by the reflective bands of Landsat satellite (Spitzer and Dirks, 1987; Baban, 1995). Therefore, the analysis of these images can provide informations about size and direction of the plumes, deposit area (Lo, 1986) and estimate the concentration of particles in water column (Kleman & Hardisky, 1987, Fielder & Laursen,

1990). In the study area, due to the existence of several sources of fine sediments (Guadalquivir river, Guadalete river, inner bay, tidal channels, etc.), associated to the action of winds, waves and tides regime, is frequently observed turbidity plumes with a high content of suspended matter (Achab et al, 2000a). In the case of the Bay of Cadiz, Bernal (1986) and Guillemot (1987), based on Landsat images showed that depending to hydrodynamic conditions, these plumes follow different directions and can cross the bay area. They reach the inner shelf by the action of tidal ebb, depositing part of its SPM. Once in the open sea, SPM are moved by currents and interact with the general hydrodynamic system affecting coastal areas and the Gulf of Cadiz (Gutiérrez-Mas et al., 2006; Achab et al, 2008).

The concentration of SPM under the hydrodynamic conditions at the sampling time shows values that vary from one area to another, with an average content of 6.5 mg/l. Lower values, between 1.5 mg / l and 5 mg / l are given respectively in the inner shelf and in some areas of the inner bay unaffected by the currents. The highest values occur in the outer bay near the mouth the Guadalete river (25 mg / l). Other high values are given in front of the mouth of the San Pedro River (16 mg / l) and Sanctipetri tidal creek (13 mg / l). The concentrations of SPM are also relatively high in the oriental part of the inner bay reaching 12.87 mg / l. The general transport pattern of the SPM is affected by local processes, which take place in littoral zones, in particular in Cadiz bay and the Guadalquivir estuary (Gutierrez Mas et al. 1999, Achab et al., 2000a). Part of the Atlantic waters rich in SPM coming from the Guadiana and Guadalquivir rivers reaches the bay of Cadiz and can be deposited in lagoons and salt marshes. The resuspension of fine-grained material in the inner zone of the bay during southeast wind and ebb tidal current generate suspended matter outflows towards the outer bay. Considerable quantity of this SPM is injected in the Atlantic waters.

The analyses of satellite images show the existence of water masses of different degrees of turbidity. These, appear as turbid plumes oriented from the inner zone towards the outer bay extending to the continental shelf. These plumes are moved seawards by action of the tidal ebb currents, following the morphology of the coast and the sea bottom (Fig.11). The highest turbidity is observed at the mouths of the Sanctipetri tidal creek and the San Pedro and Guadalete rivers. In particular, these images show that the turbidity pattern coincides generally with the area of the muddiest facies present on the outer bay bottoms and with the geographical locations of the sampling stations providing the highest contents of suspended solids. From the distribution of SPM concentration and the observation of the satellite images, two possible transport paths of turbidity plumes can be deducted: a) one runs preferably by the northern margin of the bay of Cadiz and reaches the Rota city. b) Another flow oriented towards the west, bordering the city of Cadiz, eventually extending to the continental shelf.

4.3.4 Side scan sonar recording analysis

The analysis of bed-forms using Side Scan Sonar recording is considered to be a useful technique in the study of the submarine physiography and to deduce the direction of current and sediment transport (Kenyon, 1970; Belderson et al, 1972, Dalrymple et al, 1978; Fleminig, 1980.). In order to establish the distribution pattern of bottom currents and the flow regime, different bed- form fields have been identified in the bay of Cadiz based on the analysis of Side Scan Sonar recording (Parado Roman et al, 1996, 2000, Gutierrez-Mas et al, 2000). The diferent bed-forms present in the Cadiz bay bottoms (Fig.12) result from the interaction of different hydrodynamic factors (waves, tides, currents, etc.) with seabed sediments. These bed-forms correspond to modern Holocene deposits, coexisting relict forms beside present day and reworked forms. The flow regime has been deducted from typology of bed-forms and grain size of sediments (Rubin and McCulloh, 1980).

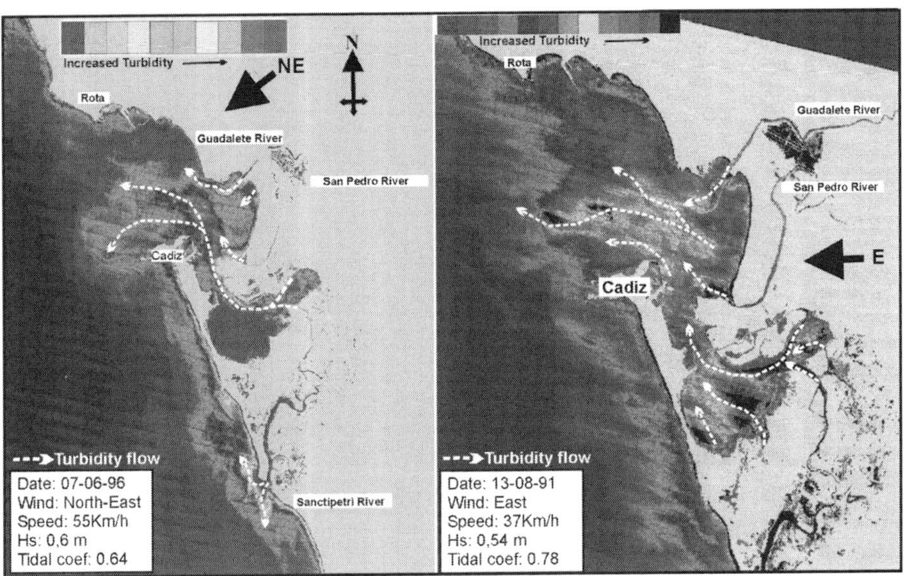

Fig. 11. Turbidity plume paths in the bay of Cadiz during ebb tide from Landsat TM images analysis

In the eastern sector of the bay of Cadiz, especially in Valdelagrana beachfront, rip currents generated by waves are directed to the West and lead to parallel and oblique sand waves of decimeter height. To the south of Santa Catalina tip, the currents are oriented towards the SW and SSW and form sand waves and ripples. North of the main channel, there are small dunes and sand waves formed mainly by effect of tidal currents of NW direction. About the Galera and Diamond shoals, the current is oriented towards the SW and forms large ripples under a high and very high energy, with speeds between 60 and 100cm / s. In the western sector, fields of bed-forms are essentially controlled by the grain size of bottom sediments. In the south part, there are straight and sinuous ripples, some sand waves and low plane bed developed on sandy sediments. To the north and NW, large mud patches appear on muddy bottoms associated to gravitational slides processes. While on sandy bottoms, there are ripples and sand ribbons. These forms indicate an upper-middle energy regime, caused by ebb tidal currents toward the SW, with speed between 30 and 100cm / s. Beside the rocky shoals located near the cities of Cadiz and Rota, we can see straight and sinuous ripples wedged between rocks, which indicate ebb currents to the SW and WSW, with speed between 20and 60 cm / s.

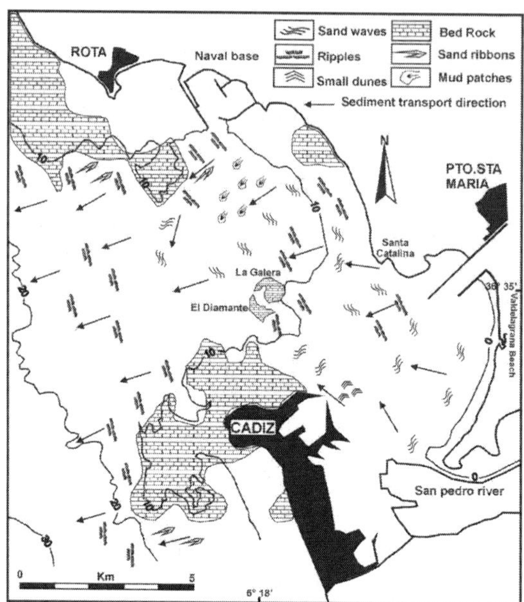

Fig. 12. Bed-forms distribution in the bay of Cadiz and transport direction of prevailing bottom currents (Modified from Gutierrez-Mas et al 2000)

The sonographic data show that different bed-forms fields present in the Cadiz bay bottoms reflect the hydrodynamic regime prevailing in the area and the sedimentary processes. The distributions of these bed-forms indicate that the dominant direction of transport of bottom sediments is toward the west and southwest, and conditioned by the grain size and current velocity. Another factor is the physiography of the seabed, which is particularly evident in sectors with greater presence of rock outcrops.

5. Sedimentary processes and conclusions

The sedimentary dynamics displays a specific behaviour pattern, related with different environments of deposit, and the interaction of these with the hydrodynamic system dominant in the area. The analysis of sediments shows the existence of sedimentary facies with variable disposition and granulometric nature. These facies occupy differents sedimentary environments; sometimes do not correspond with the present oceanographic conditions (Achab and Gutiérrez-Mas, 1999a). The distributions of grain sized and mineralogical facies reflect the action of hydrodynamic agents, which control recent sedimentary dynamics. Cadiz bay bottoms show the presence of muddy-sand and muddy facies covering sediments of coarser nature (sand and gravel), indicating actual processes of transport of fine sediments from the inner bay to be deposited upon sands of the outer bay. This model of transport has been confirmed by means of studies of bed-forms fields carried out in the study area. In situation of strong Easterly winds, suspended matter derived from resuspension of the inner bay bottoms are subsequently transported by ebb tidal currents, with an average speed of 1.02 m / s, which is reduced to 0.77 m / s in the outer bay. This involves the settling of much of the load carried on sandy bottoms. Part of the suspended matter tends to leave the bay being deposited in the inner continental shelf (Gutiérrez Mas et al, 1999 and Achab et al, 2000b). In the continental shelf, the North Atlantic Surface water current transport a large volume of fine sediments from the Guadaiana and Guadalquivir rivers toward the SE. Part of this current can transport fine particles in suspension toward the bay of Cadiz. Satellite images show that water with suspended particulate matter flows out beyond the limits of the bay under condition of easterly winds and ebb tidal current. These flows appear as branch oriented northwards along the eastern edge of the bay and then turn west, influenced by coastline and bottom morphology. A lot of suspended matter no deposited on the outer bay bottoms, can reach the continental shelf. The influence of the hydrodynamic system, the location of terrigenous sources, sea-level changes, and coastal morphology are considered to be the main factors controlling the hydrodynamic model of exchange and transport of sediments between the Cadiz bay and the adjoining continental shelf (Parrado et al., 1996; Gutierrez-Mas et al., 2004; Achab et al., 2005b).

6. Acknowledgements

This paper was supported by projects MAR 98- 0796 and 2002-01142/MAR of the Interministerial Commission of Science and Technology (CICYT) of the Spanish Government; as well as by the project SVT12 of the Ministry of Higher Education of the Moroccan Government. The author would like to give particular thanks to J.M. Gutierrez-Mas (University of Cadiz) for his encouragement during this work and to INTECH Publisher.

7. References

Achab, M. (2000). *Estudio de la tranferencia sedimentaria entre la bahia de Cadiz y plataforma continenetal adyacente: Modelo de transporte mediante el uso de minerales de arcilla como trazadores naturales*. PhD Thesis, Universidad de Cadiz, Spain, ProQuest, ISBN 84-7786-727-5, 538 p.

Achab, M., Gutiérrez Mas, J.M., Parrado Román, J.M., Moral Cardona, J.P, Sánchez Bellón, A., González Caballero, J.L. & López Aguayo, F. (1997). Distribution of recent sedimentary facies in the Cadiz bay bottoms. *Geogaceta*, 21, pp. 155-157.

Achab, M., Gutiérrez Más, J.M., Sánchez Bellón, A. (1998). Transport of fine sediments and clay minerals from the tidal flats and salt marshes in Cadiz bay towards outer marine zones. *European Land-Ocean Interaction Studies*, Huelva-Spain, pp.93-94.

Achab, M. & Gutiérrez Mas, J.M.(1999a). Characteristics and controlling factor of the recent sedimentary infill on the bottom of the Bay of Cadiz. *Geogaceta* 27 pp. 3-6.

Achab, M., Gutiérrez Más, J.M., Moral Cardona, J.P., Parrado Román, J.M., Gonzalez Caballero J.L. & Lopez Aguayo, F. (1999b). Relict and modern facies differentiation in the Cadiz bay sea bottom recent sediments, Geogaceta, 27, pp.187-190.

Achab, M., Gutiérrez Mas, J.M., Luna del Barco, A. (2000a). Concentration and mineralogic composition of suspended matter in the Bay of Cadiz and adjacent continental shelf. *Geotemas*, 1 (14), pp. 81-86.

Achab, M., Gutiérrez Mas, J.M., Sanchez Bellon, A. & Lopez Aguayo F. (2000b). Fine sediments and clay minerals embodiment and transport dynamics between the inner and outer sectors of Cadiz Bay. Geogaceta, 27, pp.3-6.

Achab, M. & Gutiérrez Mas, J.M. (2002). Analysis of terrigenous components present in the sand fraction of Cadiz bay bottoms (SW Spain). *Thalassas,* 18 (1) pp. 9-17.

Achab, M. & Gutiérrez Más, J.M. (2005a). Nature and distribution of the sand fraction components in the Cadiz bay bottoms (SW Spain). *Revista de la Sociedad Geologica de España,* 18 (3-4), pp. 133-143.

Achab, M., El Moumni, B., El Arrim, A. & Gutierrez Mas, J.M. (2005b) Répartition des faciès sédimentaires récents en milieu marin côtier : exemple des baies de Tanger (NW-Maroc) et de Cadix (SW-Espagne). *Bull. Inst. Sci., sect. Sci. Terre,* n°27, pp. 55-63.

Achab, M., Gutiérrez Más, J.M. & López Aguayo, F. (2008). Utility of clay minerals in the determination of sedimentary transport patterns in the bay of Cadiz and the adjoining continental shelf (SW-Spain). *Geodinamica Acta,* 21/5-6, pp. 259-272.

Agi, A. (1987). *Glosary of geology.* (eds, R.L. Bates y J.A.Jackson). 3 De. Amer Geological Institute, Virginia. 788p.

Alvarez, O., Izquierdo, A., Tejedor, B., Mañanes, R., Tejedor, L. & Kagan B.A. (1999). The influence os sediments load on tidal dynamics, a case study: Cadiz Bay. *Estuarine Coastal and shelf Science,* 48 pp.439-450.

Baban, S. M. J. (1995). The use of LandSat imagenery to map fluvial sediment discharge into coastal waters. *Marine Geology,* 1230. pp. 263-270.

Baldy, P. (1977). *Geologie du plateau continental portugaise (au sud du Cap de Sines),* Thése de 3eme cycle, Université de Paris VI, , 113 p.

Balopouls, E.T., Collins, M.B. & James, A.E. (1986). Satelite images and their use in the numerical modeling of coastal processes. *Inter. Jour. of Remote Sensing,* 7. pp. 905-519.

Baringer, M.O. & Price J.F. (1999). A review of the physical oceanography of the Mediterranean out-flow. *Marine Geology,* pp.155 63-82.

Belderson, R.H., Kenyon, N.H., Stride, A.H. & Stubbs, A.R. (1972). *Sonographs of the sea floor, a picture atlas.* Elsevier, Amsterdam. 185 p.

Benavente, J.(2000). *Morfodinámica litoral de la Bahía Externa de Cádiz.* Tesis Doctoral, Univ de Cadiz, 533p.

Benkhelil, J. (1976). *Étude neotectonique de la terminaison des Cordilléres bétique (Espagne).* Thése 3éme cycle. Université de Nice. 180p.

Bernal, Ristori, E. (1986): Aplicaciones de la teledetección espacial al medio marino litoral . 1ª Jornadas de Ingeniería Geográfica. Madrid. *Bol. Inf. Serv. Geográf. Ejerc.*Nª 62. Pp.37-56.

Bhukhari, S.S., Nayak G.N. (1996). Clay minerals in identification of provenance of sediments of Mandovi Estuary, Goa, West Coast of India, Indian. *Journal of Marine. Science,* 25 (4) pp.341-345.

Cacchione, D. A. & Drake, D. E. (1990). Shelf sediment transport: An overview with applications to the northem Califomia continental shelves. In: Mehaute, B. L. & Hanes, D. M. (eds.) The Sea 7b. *John Wiley and Sons,* New York, pp.729-773.

Chamley, H. (1989). *Clay sedimentology,* Springer-Verlag, Berlin, 623p.

Dabrio, C J., Zazo, C., Goy, J.L., Sierro, F.J., Borja, F., Lario, J., Gonzalez, A.& Flores J.A. (2000). Deposicional history of estuarine infill during the last postglacial transgression (Gulf of Cadiz, Southern Spain), Marine Geology, 162, pp.381-404.

Dalrymple, R.W., Knight, R.T. & Lambiase, J.J. (1978). Bedforms and their hydraulic stability relationships in a tidal environment, Bay of Fundi, Canada. *Nature*, 275, pp. 100-104.

Degiovanni, C. (1970). Les concentrations de minJraux lourds sur la plage de Pramousquier (Var), et leurs relations avec les indices d'Jvolution de A. RiviPre. *Compte Rendus de l'AcadJmie des Sciences*, Paris, 217(D), pp. 28-30.

Drake, D. E., Kolpak, R. L. & Fisher, P. J. (1972). Sediment transport on the Santa Barbara - Oxnard shelf, Santa Barbara Channel, California. In: Swift, D. J. P., Duane, D. B. & Pilkey, 0. H. (eds.) Shelf Sediment Transport . *Dowden, Hutchinson and ROSS*, Stroudsburg, pp. 307-331.

Emery, K.O. (1968). Relict sediments on continental shelves of wored. *Am. Ass. of Petrol. Geol. Bull.* pp. 445-464.

Fernández-Palcios, A., Moreira Madueño, J.M., Sánchez Rodriguez, E. & Ojeda Zujar, J. (1994). Evaluation of different methodological approaches for monitoring water quality parameters in the coastal waters of Andalusia (Spain) using Landsat.TM.Data. *Earsel workshop on Remot Sensing and GIS for Coastal zone Managment*. Rijks WaterStaat, Survy Department. Delft, The Netherlands. pp. 114-123.

Fiedler, P. C. & Laurs, R. M. (1990). Variability of the Columbia River plume observed in visible and infrared satellite imagery. Inter. *Jour. Remote Sensing*, Vol. 11(6). pp. 999-1010.

Flemming, B.W. (1980): Sand transport and bedforms patterns on the continental shelf between Durban and Port Elizabeth (Southeast African continental margin). *Sedimentary Geology*, 26, pp. 179-205.

Folk, R.L. & Ward, W.C. (1957) Brazos River bar: a study in the significance of grain size parameters. *Journal of Sedimentary Petrology*, 27, pp. 3-26.

Folk, R.L. (1980). *Petrology of sedimentary roks*. Hemphills.Austin. 182p.

Gao, S & Collins, M.B. (1994). Análysis of grain size trends, for defining sediment transport pathways in marine environments. *Jour. Coast. Res*, Vol 10, nº 1. pp. 70-78.

Gracia, F.J., Rodríguez Vidal, J., Benavente, J., Cáceres, L. & López Aguayo F. (1999). Tectónica cuaternaria en la bahía de Cádiz. *Avances en el estudio del cuaternario español.*, Girona, pp. 67-74.

Greenberg, A.E., Clescer, L.S. & Eaton, A.D. (1992). Standard Methodes for the Examination of Water and Wastewater. 18th Edición 1992. *American Public Heath Association*.

Grousset, F. E., Joron, J. L., Biscaye, P. F., Latouche, C., Treuil, M., Maillet, N., Faugeres, J. C. & Gonthier, E. (1988). Méditerranean Outflow through the Strait of Gibraltar since 18.000 years B. P: Mineralogical and Geochemical Arguments. *Geo- Marine Letters*, 8. pp. 25-35.

Guillemot, E. (1987): *Teledetection des milieux litoraux de la Baie de Cadix*. Thése. Univ. Paris, 146p.

Gutiérrez Mas, J.M., Martín Algarra, A., Domínguez Bella, S., Moral Cardona, J.P. (1990). *Introducción a la Geología de la provincia de Cádiz*. Servicio de Publicaciones. Universidad de Cádiz. 315 p.

Gutiérrez-Mas, J.M., Hernández-Molina, J., Lopez-Aguayo, F. (1996a). Holocene sedimentary dynamics on the Iberian continental shelf of the Gulf of Cadiz (SWSpain). *Continental Shelf Research*, 16 (13), pp. 1635–1653.

Gutiérrez Mas, J.M., Achab, M., Sánchez Bellón, A., Moral Cardona, J.P. & López Aguayo F. (1996b). Clay minerals in recent sediments of the bay and their relationships with the adjacent emerged lands and the continental shelf. *Advances in Clay minerals*, pp. 121-123.

Gutiérrez Más, J.M., Lopez Galindo, A., López Aguayo, F (1997). Clay minerals in recent sediments of the continental shelf and the bay iof cadiz (SW Spain). *Clay Minerals*. 32, pp. 507-515.

Gutiérrez Mas, J. M., Sánchez Bellón, A., Achab, M., Fernández Palacios, A., Sánchez Rodríguez, E. (1998). Influence of the suspended matter exchange between the Cadiz Bay and the continental shelf on the recent marine sedimentation, in: *15 th Inter. Sediment. Congr.* Alicant, Spain, Abstract volume. 404-405.

Gutiérrez Mas, J.M., Sanchez Bellon, A., Achab, A., Ruiz Segura, J., Gonzalez Caballero, J.L., Parrado Román, J.M.& Lopez Aguayo, F. (1999). Continental Shelf zones influenced by the suspended matter flows coming from Cadiz bay. *Bulletin of the Spanish institute of oceanography*, 15 (1-4), pp. 145-152.

Gutiérrez Mas, J.M., Luna del Barco, A., Achab, M., Muñoz Pérez, J.J., González Caballero, J.L., Jódar Tenor, J.M. & Parrado Román J.M. (2000). Controlling factors of sedimentary dynamics in the littoral domain of the Bay of Cadiz. Geotemas, 1 (14), pp. 153-158.

Gutiérrez Más, J.M., Achab, M., Gracia, F.J. (2004). Structural and physiographic control on the Holocene marine sedimentation in the bay of Cadiz (SW-Spain). *Geodinamica Acta*, 17/2, pp. 153-161.

Gutiérrez Más, J.M., López Aguayo, F., Achab, M. (2006). Clay minerals as dynamic tracers of suspended matter dispersal in the Gulf of Cadiz (SW Spain). *Clay minerals*, 41, pp. 727-738.

Hernández-Molina, F.J., Fernández-Puga, M.C., Fernández-Salas, L.M., Llave, E., Lobo, F.J., Vázquez, J.T., Acosta, J. & López-Aguayo F. (1996). Distribución y estructuración sedimentaria de los depósitos del Holoceno Terminal de la Bahía de Cádiz. *Geogaceta*, 20 (2), pp. 424-427.

Imbrie, J. (1963). Factor and vector analysis programs for analysing geological data. *Office Naval Research. Geological. Branch*, Tech. Rep. 6.

Kenyon, N.H. & Belderson, R.H. (1973). Bedforms of the Mediterranean undercurrent observed with side Scan sonar. *Sedim. Geol*, 9, pp. 77-99.

Kenyon, N.H. (1970): Sand ribbons of European tidal seas. *Marine Geology*, 9, pp. 25-39.

Kleman, Y. & Hardisky, M. A. (1987). Remote sensing of estuaries: An overview. Symp. *Remote sensing of envirinment*. (Ann Arbour, Michigan), october. I, pp. 183-204.

Komar, P.D. (1977): Selective longshore transport rates of different grain-size fracction within a beach. *Journ. Sed. Pet.* 47, pp. 1444-1453.

Lo, C.P. (1986). *Applied Remote Sensing*. Longman, London. 393p.

Lobo, F.J., Hernandez-Molina, F.J., Somoza, L., Rodero, J., Maldonado, A. & Barnolas, A. (2000). Patterns of bottom current flow deduced from dune asymetries over the Gulf of Cadiz Shelf (southwest Spain). *Marine Geology*, 164, pp. 91-117

Lopez Galindo, A., Rodero, J., Maldonado, A. (1999). Surface facies and sediment dispersal patterns: southeastern Gulf of Cadiz, Spanish continental margin. *Marine Geology*, 155, pp. 83-98.

Mabesoone, J.M., (1966). Deposicional and provenance of the sediment in the Guadalete estuary (Spain). *Geologie en Minjbouw-Amesterdam*, 45, pp. 25-32.

Madelain, F. (1970). Influence de la topographie du fond sur l´ecoulement Mediterranée entre le detroit de Gibraltar et le cap Saint- Vicente. *Extrait des cahiers Oceanographiques*. 22 (1), pp. 43-61.

Maldonado, A., Stanley D. (1981). Clay mineral distribution patterns as influenced by depositional processes in the southeastern Levantine Sea. *Sedimentology*, 28, pp. 21-32.

Maldonado, A y Nelson, C.H. (1999a). Interaction of tectonic and depositional processes that control the evolution of the Iberian Gulf of Cádiz margin. *Marine Geology*, 155, pp. 217-242.

Maldonado, A., Somoza, L y Pallarés, L. (1999b). The Betic orogen and the Iberian-African boundary in the Golf of Cadiz: Geological evolution (central North Atlantic). *Marine Geology*, 155, pp 9-43.

McCave, I. N. (1975). Vertical flux of particles in the ocean. *Deep-Sea Research*, 22, pp. 491-502.

Meade R.H. (1972). Transport and deposition of sediments in estuaries. *Geological Society of America*, 133, pp. 91-120.

Meliéres., F. (1974). *Recherche sur la dynamique sedimentaire du Golfe de Cadix (Espagne)*. Thése. Sci. Nat. Univ. Paris. 235 pp + ann.

Mezzadri, G. & Saccani, E.(1988). Heavy mineral distribution in late quaternary sediments of the southern Aegean Sea: Implications for provance and sediment dispersal in sedimentary bassins at active margins. *Journal of. Sedimentary Petrology*., 59, pp. 412-422.

MOPT (1992). *Recomendaciones para obras marıtimas 0.3-91. Oleaje. Anejo I. clima marıtimo en el Litoral Español (76 pp.)*. Ministeriode Obras Publicas y Transportes. Direccion General de Puertos.

Moral Cardona, J. P., Achab, M., Domínguez, S., Gutiérrez-Mas, J.M., Morata, D., Parrado, J. M. (1996). Estudio comparativo de los minerales de la fracción pesada en los sedimentos de las terrazas del Río Guadalete y fondos de la Bahía de Cádiz. *Geogaceta*., 20 (7), pp. 14-17.

Millot, G. (1970). *Geology of clay*. Springer. Berlin. HeidelBerg. New York, Masson, Paris, 425p.

Naidu, A S., Han, M.W., Mowatt, T.C. & Wajda, D (1995). Clay minerals as indicators of sources of terrigenous sediments their transportation and deposition: Bering Basin Russian-Alaskan Arctic, *Marine Geology*, 127 (1-4), pp. 87-104.

Neiheisel, J. & Weaver, C.E. (1967).Transport and deposition of clay minerals southeastern United States. *Journal of. Sedimentary Petrology*, 37 (4).1084-1116.

Nelson, C.H., Baraza, J., Maldonado, A., Rodero, J., Escutia, C., Barber, Jr. & John H. (1999). Influence of the Atlantic inflow and Mediterranean outflow currents on late quaternary sedimentary facies of the Gulf of Cadiz continental margin. *Marine Geology*, 155, pp. 99-129.

Nombela, M. A. (1989). *Oceanografía y Sedimentología de la Ría de Vigo*. Tesis doctoral. Universidad de Madrid, 291p.

Ojeda, J., Fernández Palacios, A., Moreina Madueño, J.M, & Sánchez Rodríguez, E. (1994). Programa de seguimiento de la calidad y dinámica del espacio marino y litoral a través de imágenes de satélite (Andalucía, Agencia de Medio Ambiente). *Revista de teledetección*, n° 3, pp. 9-15.

Ojeda, J., Sanchez, E., Fernandez-Palacios, A & Moreira J.M. (1995). Study of the dynamics of estuarine and coastal waters using remote sensing: the Tinto-Odiel estuary, SW Spain. *J.Coastal Conserv.*1, pp.109-118.

Olabarría, C., Urgorri, V. & Troncoso, J.S. (1996). Distribución de los sedimentos de la ensenada do Baño (Ría de Ferrol). *Nova Acta Científica Compostelana* (Bioloxia), 6, pp. 91-105.

Ortega-Huertas, M., Palomo, I., Moresi, M. & Oddone, M. (1991). A mineralogical and geochemical approach to establishing a sedimentary model in a passive continental margin (Subbetic zone, Betic Cordilleras, SE Spain). *Clay Minerals*, 26, pp. 389-407.

Palanques, A., Plana, F. & Maldonado, A. (1987). Estudio de la materia en suspensión en el Golfo de Cádiz. *Acta. Geol. Hisp*, t.21-22, pp. 491-497.

Parhan, W.E. (1996). Lateral Variation of clay mineral assemblages in modern and ancient sediments Proc. *International Clay Conference Jerusalem*, 1, pp. 135-145.

Parrado Román, J.M., Gutiérrez Mas, J.M. & Achab M. (1996). Determination of directions of currents by means the analysis of "bed forms" in the Bay of Cadiz. *Geogaceta*, 20(2), pp. 114-117.

Parrado Roman, J.M. &Achab M. (1999). Grain size trend associated with transport and sedimentary dynamics in the Cadiz bay (SW Spain). *Bulletin of the Spanish institute of oceanography*, 15 (1-4), 269-282.

Parrado Román, J.M., Gutiérrez Mas J. M., Achab M., Luna del Barco A. & Jódar Tenor, J.M. (2000). Flow regime classification in sea bed of Cadiz Bay from bedform field analysis. *Geogaceta*, 27, pp.191-194.

Pevear, D.R. & Mumpton D.R. (1989). Quantitative Mineral Analysis of clays. CMS Workshop Lectures1. *The Clay Mineral Society*, Colorado.

Poulos, S. E., Collins, M B. & Shaw H F. (1996). Deltaic sedimentation, including clay mineral deposition patterns marine embayment of Greece (SE. Alpine Europe). *Journal of Coastal Research*, 12 (4), pp. 940-952.

Ramos, P. (1991). *Climatologia de Càdiz (1961-1990), Instituto Nacional de Meteorologia*. Centro Meteorologico,Territorial de Andalucıa Occidental, 15p.

Reyment, R. &. Jöreskog K.G. (1993). *Applied factor analysis in the natural sciences*. Cambridge University Press, New York, NY 371 p.

Rodero, J., pallares, L. & Maldonado, A. (1999). Late quaternary sequence stratigraphy and continental shelf model controlled by eustatic and paleooceanographic events. Gulf of Cadiz, southwest Iberia. *Marine Geology*. 155, pp. 131-156.

Rubin, D.M. & McCulloch, D.S. (1980). Single and superimposed bedforms: a synthesis of San Francisco Bay and flume observations. *Sedim. Geol.*, 26, pp. 207-231.

Sanz de Galdeano, C. (1990). Geologic evolution of the Betic cordilleras in the western Mediterranean Miocene to the presente. Tectonophysics, 172, pp. 107-119.

Sawheney, B.L. & Frink C.R. (1978). Clay minerals as indicators of sediment source in tidal estuaries of long island sound. *Clays and Clay minerals*, 26 (3), pp. 227-230.

Seibold, E y Berger, W.H. (1982). *The Sea Floor. An introduccion to marine Geology*. Springer-Verlag. 288p.

Shepard, F.P. (1932). Sediments on the continental Shelves. *Geol. Soc.Am. Bull*, 43, pp. 1017-1039.

Soulsby, R.L., (1997). *Dynamics of Marine Sands.* , Thomas Telford Publications, London, UK.

Spitzer, D. & Driks, R. W. J. (1987). Bottom influence on the reflectance of the sea. Int. *Journal. Remote Sensing*, 8, pp. 779-290.

Stanley, D.J. & Liyanage, N.A. (1986). Clay-Mineral variations in the northeastern Nile Delta, as influenced by depositional processes. *Marine Geology*, 73, pp. 263-283.

Van Rijn, L. C. (1984). Sediment transport, part I: bed load transport. *Journal. Hydraul. Eng.*, 110, 10 pp. 1431–1456.

Van Rijn, L. C. (1993). Principles of sediment transport in rivers, estuaries and coastal seas. *Oldemarkt, Aqua*, the Netherlands, 7, 41, pp. 7-43.

Viguier, C. (1974). *Le néogéne de l´andalousie Nord-occidentale (Espagne). Histoir geologique du bassin du bas Guadalquivir*. Thése d´université . Bordeaux. 449p.

Visher, G.S. (1969). Grain-size distribution and depositional processes. *Journal of. Sedimentary Petrology.*, Vol. 39, pp. 1074-1106.

Weaver, C.E. (1989). *Clays, Mud and Shales*. Devlopments in Sedimentology, 44. New York., Elsevier, 589p.

Wegner, C. (2003). Sediment Transport on Arctic Shelves – Seasonal Variations in Suspended Particulate Matter Dynamics on the Laptev Sea Shelf (Siberian Arctic). *Ber. Polarforsch. Meeresforsch.* ISSN 1618 – 3193, 455p.

Wiberg, P.L., Cacchione, D.A., Sternberg, R.W. & Wright, L.D. (1996). Linking sediment transport and stratigraphy on the continental shelf. *Oceanogrphy*, Vo1. 9, No. 3.

Zazo, C. (1980). *El problema del límite Plio-pleistoceno en litoral de Cádiz*. Tesis Doctoral (2T). Univ. Complutense. Madrid. 399p.

Zazo, C., Goy, J.L. & Dabrio C. (1983). Medios marinos y marino-salobres en la bahía de Cádiz durante el pleistoceno. *Rev. Mediterránea. Ser. Geol*, 2, pp. 29-52.

Zazo, C., Goy, J.L., Somoza, L., Dabrio, C.J., Belluomini, G., Improta, S., Lario, J., Bardají,T. & Silva P.G. (1994). Holocene sequence of sea-level fluctuations in relation to climatic trends in the Atlantic- Mediterranean linkage coast. *Journal of Coastal Research*, 10 (4), pp. 933-945.

Zazo, C., Goy, J.L., Lario, J. & Silva, P.G. (1996). Littoral zone and rapid climatic change during the last 20.000 years. The Iberian study case. *Z. Geomorph*, N.F suppl-Bd 102, pp. 119-134.

Zazo, C., Silva, P.G., Goy, J.L., Hillaire-Marcel, C., Ghaleb, B., Lario, J., Bardají, T. & González A. (1999). Coastal uplift in continental collision plate boundaries: data from the Last Interglacial marine terraces of the Gibraltar Strait area (south Spain). *Tectonophysics*, 301, pp. 95-109.

3

The Significance of Suspended Sediment Transport Determination on the Amazonian Hydrological Scenario

Naziano Filizola[1], Jean-Loup Guyot[2], Hella Wittmann[3], Jean-Michel Martinez[2] and Eurides de Oliveira[4]

[1]*Universidade Federal do Amazonas – Department of Geography, Manaus*

[2]*IRD-LMTG – Univeristé de Toulouse*

[3]*Deutsches GeoForschungs Zentrum Potsdam Telegrafenberg, Potsdam*

[4]*Agência Nacional de Águas, Brasília*

[1,4]*Brasil*

[2]*France*

[3]*Germany*

1. Introduction

Rivers play an important role in continental erosion as they are the primary agents of transferring erosion products to the ocean. Understanding rivers and their transport pathways will improve the perception of many processes of global significance, such as biogeochemical cycling of pollutants and nutrients, atmospheric CO_2 drawdown, soil formation and their erosion, crust evolution- in short the interaction between the atmospheric and the lithospheric compartment of the Earth´s system (Allen, 2008). This interaction is characterised by the relative proportions of mechanical degradation vs. chemical weathering, whose products are, in dissolved or solid form, transported by rivers. The sediment load of rivers is thereby controlled by catchment relief, the channel slope and its connectivity to the hill slope, but also by climatic factors such as precipitation. The latter, together with temperature, exert control over chemical weathering that is dependent on physical erosion to a degree that is yet unknown (Anderson et al., 2002; Gaillardet et al., 1999; Riebe et al., 2001). Both mechanical erosion and chemical weathering, are, however, governed by tectonic activity, which drives processes of landscape rejuvenation and preconditions the fluvial transport regime (von Blanckenburg et al., 2004). On the shorter time scale, humans may act as geomorphic agents by constructing dams and reservoirs, and changing land use by deforestation and mining (Hooke, 2000; Syvitski et al., 2005; Wilkinson and McElroy, 2007).

In tropical regions around the globe, large river basins are especially concentrated, and their behaviour plays an important role in river sediment transport (Latrubesse et al., 2005). For example, the tropics represent 25% of the total continental lands and contain 57% of the world's fresh water, and associated catchments contribute 50% and 38% to solids and dissolved solid input into the oceans, respectively (Milliman and Meade,1983, Guyot, 1993 and Latrubesse et al,2005). Large river basins often display mixed river channel forms, as they usually constitute a rapidly eroding sediment source area and an associated depositional area in the lowlands (Latrubesse et al., 2005). Among the highest erosion rates worldwide are found in mountainous areas of the tropics (Pinet & Souriau, 1988; Milliman & Syvitzki, 1992; Summerfield & Hulton, 1994), of which the Amazon basin is an excellent example. Here, most of the sediment that is transported in the main Amazon channel is derived from the Andes, but is also intermittently stored for at least several thousands of years in the floodplain (Meade, 2007). This chapter discuss some factors about suspended sediment transport determination and its importance into the Amazon Basin hydrological scenario. The text is divided in sections. After introduction, section 2 presents the Amazon basin and some topics concerning sediments and hydrology. Section 3 presents a brief review about the works done about Amazon River suspended sediment discharge into the Ocean. Section 4 shows the importance of the suspended sediment discharge values, into the Amazonian hydrological scenario, especially at central portion of the basin. Section 5 presents some information about new techniques contributions to study that scenario. Finally, Section 6 summarizes the chapter and concludes it with some comments about the use of suspended sediment data to help water resource management.

2. The Amazon Basin

The Amazon Basin (Figure 1) is the biggest river basin in the world. It has more than 6 million km^2 in area. This area corresponds to almost 5% of the total global continental land. This basin flows into the Atlantic Ocean at approximately 6.6 10^{12} m^3 yr^{-1}, and this volume is approximately 16% of the world's fresh water (Molinier et al., 1995 and 1996). The Andean mountains take up around 12% of the Amazonian land. Most of the sediment transported by the region's rivers originates from these mountains as a result of rapid erosion processes (Sioli, 1950, 1964; Gibbs, 1967; Meade et al., 1985; Guyot et al., 1994; Filizola, 1999; Wittmann et al., 2011). The most important rivers, in terms of contributing sediment to the main river, are the Rio Solimoes (draining the Peruvian Andes and forming the main tributary of the Amazon in Brazil upstream of the Rio Negro confluence) and the Madeira River that drains in part the Andes, and in part the cratonic Brazilian Shields.

The Rio Amazonas crosses a huge flood plain surrounded by two pre-Cambrian shields (the Guiana shield to the north and the Brazilian shield to the south). Compared to the Rio Madeira and the main channel, rivers with their headwater sources in the shields and those that cross them do not contribute much to the Rio Amazonas in terms of transporting suspended sediment. The flood plain landscape is characterised by areas where deposition and re-suspension occur (Meade et al. 1985). Additionally, extreme climate events, such as El Niño/Southern Oscillation (ENSO), have been identified as important agents for landscape construction. As described by Aalto at al. (2003), episodic sediment accumulation on the Amazonian flood plains is influenced by ENSO. An important correlation was also established between sediment deposition and La Niña occurrences. With a very wet climate, the majority of the Amazon Basin receives 2400 mm yr^{-1}, with extremes that can vary from 100 mm yr^{-1} to 6000 mm yr^{-1}. The Amazon Rain forest covers more than 70% of the basin area, but there are also small areas of tundra (high altitude), desert and savannah.

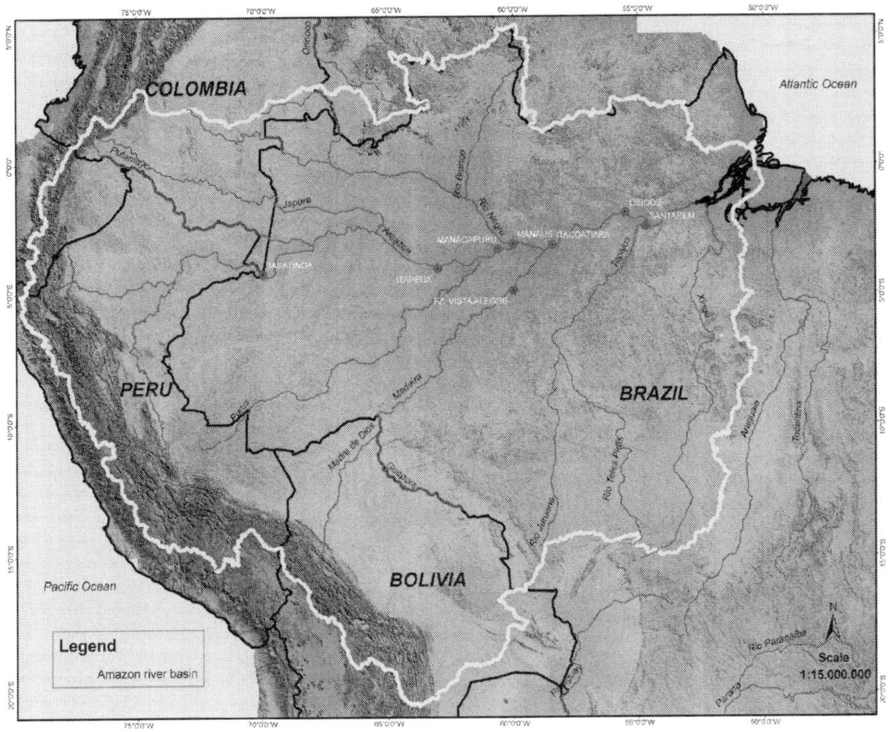

Fig. 1. The Amazon Basin with sites indicated in the text.

3. Amazon Basin suspended sediment discharge into the ocean, a brief review

Since the 1960s, the Amazon ecosystem had been an important subject for systematic science. Initial results, conclusions and also new scientific questions have emerged from those studies, which motivated the emergence of new research groups. Interdisciplinary themes were proposed, and important scientific publications summarised the advancement of this period of knowledge: *The Amazon* (Sioli, 1984); *Key Environments: Amazon* (Prance & Lovejoy, 1985); *Biogeography and Quaternary history in Tropical Americas* (Whitmore & Prance, 1987). The impact of those publications was observed as new environmental policies emerged for sustaining the Amazon region.

Concerning suspended sediment, Sioli (1950, 1957, 1984) classified Amazonian rivers based on their physical and chemical characteristics and partly based on the suspended solid concentrations. Gibbs (1967) investigated the factors that controlled the amount and composition of the suspended load found in the Amazon River. During high and low water periods, Gibbs collected a total of 74 samples. He estimated the Rio Amazonas suspended load to be approximately ~500 10^6 t yr^{-1} or 79 t km^{-2} yr^{-1}. Gibbs also demonstrated that 82% of the total suspended sediment load discharged by the Amazon comes from only 12% of the total area of the basin, which comprises the mountainous environment (the Andes).

For the Óbidos station (~800 km from the Rio Amazonas mouth), Oltman (1968) reported concentration variability from 300 to 340 mg l^{-1} near the stream bed to 50-70 mg l^{-1} near the water surface. He estimated, using a few data points, that the suspended sediment input from the Amazon River into the ocean was 1.5 10^6 t day^{-1} or ~600 10^6 t yr^{-1}. Oltman also indicated the important influence of the black and white water rivers on the proportion of the total water flow contribution at Óbidos.

After that, Schmidt (1972) determined the annual variability of the suspended sediment concentrations. He only reported the suspended sediment concentration (SSC), and no estimations were made as to the suspended sediment discharge. Meade et al. (1979), in the course of the Alpha Helix Project, estimated a mean total suspended sediment (TSS) discharge from Óbidos of approximately 900 10^6 t yr^{-1}. Using that data, Meade (1985) provided the first description of the vertical and lateral variation of suspended sediment in the Amazon River. The result were derived from more than 300 samples collected during two field cruises conducted during the high water periods of 1976 and 1977.

The CAMREX Project, during the 1980s, measured water discharge and collected a series of suspended sediment samples from the Amazon (Richey et al., 1986). In this project, more than 200 new samples were collected during several field cruises, mainly between 1982 and 1985. The samples were collected at a single cross section as depth-integrated composites. As indicated by Meade et al. (1985), a new estimate of 1100 to 1300 10^6 t yr^{-1} suspended sediment discharge from the Amazon into the ocean was made. The CAMREX Project also estimated bank erosion contributions to total suspended sediment discharge (Dunne et al., 1998).

After that, Bordas et al. (1988) and Bordas (1991) used almost 200 samples from both the Brazilian national dataset (managed by the Brazilian National Water Authority) and from several Brazilian companies and obtained results very similar to those reported by Meade et al. (1979). Again similar results were obtained by Guyot et al. (1988 and 1996), especially for the Madeira Basin (250-300 10^6 t yr^{-1}). From this river, Martinelli et al. (1989; 1993) and Ferreira et al. (1988) estimated a somewhat higher suspended sediment discharge of approximately 550 10^6 t yr^{-1}.

The Amazon Shelf SEDiment Study (AmasSeds) group, working with a 190 CTD Probe profiles at the continental shelf, indicated a total flux between 550 and 1,000 10^6 t yr^{-1} from the Amazon River into the ocean (Nittrouer et al., 1986a and 1995).

In the scope of the HiBAm (Hydrology and Geochemistry of the Amazon Basin) Project until 2001 when it was completed and after that until today with the ORE/HYBAM (Environmental Research Observatory on Amazon Basin Hydrology, Geochemistry and Geodynamics) Program, Guyot et al. (1996), Filizola (1999), Filizola (2003), Filizola and Guyot (2004), Guyot et al. (2005) and Filizola & Guyot (2009) calculated the contributions from tributaries to the main stream during several cruises (HiBAm and ORE/HYBAM data) and using data from the Brazilian National Data set (from Brazilian Agência Nacional de Águas). They tested different kinds of sampling, data sources and calculation procedures. The results from their multiple approaches, using more than 5000 samples, were quite similar. They indicated imprecision concerning some data from the Brazilian National Data Set (data at www.ana.gov.br) but also validated it with data from several field cruises made between 1995 and 2001 (with the HiBAM Project) and afterwards with the ORE/HYBAM Programme. With the ORE/HYBAM data set Laraque et al. (2005) tried to characterise the spatial distribution of sediment yields and sediment transfer processes in the Brazilian

Amazon basin between 1998 and 2003. These authors established a relationship between the surface concentration and the mean suspended sediment concentration for the whole cross section at each ORE/HYBAM station. Using the same data set, but beginning from 1995, Guyot et al. (2005) highlighted a marked seasonality in the sediment flux, which did not correspond to the discharge variation at Óbidos. From those ensemble of works the results indicated TSS discharge values between 600 and 900 10^6 t yr^{-1}.

The HiBAm Project and ORE/HYBAM Programme are research initiatives involving several Amazonian countries (Brazil, Peru, Colombia, Equator, Bolivia and Venezuela) and France. Especially ORE/HYBAM is a long-term programme to evaluate the process of matter transport in the Amazon Basin. From this project, data points were acquired daily for hydrology, every ten days for suspended sediment and monthly for geochemistry and physical-chemistry at fifteen gauging stations mainly on the larger rivers of the Amazon basin. The aim of this network was to investigate the piedmont areas, the flood plain tributaries, the tributaries that originate in the Andes and those that drain into the Brazilian and Guiana shields (Cochonneau et al., 2006). The ORE/HYBAM initiative is operated today in cooperation with research institutions and national agencies, creating an independent data set. The main interests are focused on mass transfer within the Amazon basin and towards the Atlantic Ocean, the sensitivity of mass transfer to climatic variability and anthropogenic activities, and the key role of the wetlands in mass transfer. The data acquired with standardised collection and analysis methods are freely available at http://www.ore-hybam.org.

Martinez et al. (2009) used a combined approach of both the HiBAm and ORE/HYBAM data where they also introduced remote sensing data from MODIS spaceborne sensors to estimate the suspended sediment discharge at the Óbidos to be near 800 10^6 t yr^{-1}. They also showed an increase in the suspended sediment discharge of the Amazon River between 1996 and 2007 (see Section 5.1).

Recently, using a completely different approach from those cited above, namely by using cosmogenic nuclide-produced ^{10}Be [1], Wittmann at al. (2011) obtained an estimation of the Rio Amazonas total sediment discharge of approximately 610 10^6 t yr^{-1} (see Section 5.2). This estimation integrates over several thousands of years and thus provides a long-term estimate on the total sediment load.

Finally, Guyot et al. (2011) showed that the whole ORE/HYBAM data set, with more than eight years of data, can be used to give a more complete picture of the Amazon Basin between the Andean region and the portion near Óbidos. The estimated contribution from the basin to the ocean remains within the range of 600 to 900 10^6 t yr^{-1} as found by other authors. With those data the best estimation of sediment discharge at Óbidos is 862 10^6 t yr^{-1}. If other tributaries contributions, downstream Óbidos (Rivers like Tapajós, Xingu, Paru and Jari), are included in the sediment discharge value, the TSS discharge increases to 872 10^6 t yr^{-1}.

4. The significance of suspended sediment discharge values to the actual Amazonian hydrologic scenario

Determining the Rio Amazonas suspended sediment discharge into the Ocean is not a simple task. The presented results (as showed in Table 1) sometimes differ in their methods

[1] This method is described later in the text.

QS (10^6 t yr-1)	Source
500	(Gibbs, 1967)
600	(Oltman, 1968)
900	(Meade et al., 1979)
1,100 - 1,300	(Meade et al., 1985)
550 - 1,000	(Nittrouer et al., 1995 and Nittrouer et al., 1986a)
600 - 700	(Bordas 1988; Filizola,1999)*
600 - 800	(Filizola 2003), Guyot et al., 2005 and Filizola and Guyot ,2009)*
800	(Martinez et al., 2009)*
610	(Wittmann et al., 2011)*
872	(Guyot et al., 2011)

Table 1. Shows a summary (by author) of the determined values of suspended sediment discharged (QS) by the Amazon River (*)at Óbidos or at its mouth.

and approaches used, and different periods of measurement different measurement periods of measurement could explain some of the variability. Wittmann et al. (2011) give a summary of measurement periods for TSS monitoring in their table 2. Further, discrepancies could be attributed to variability in hydrological behaviour, sampling methods, sediment discharge calculation, and others topics discussed further in this section.

In the following, we will review the hydrological behaviour of Amazon Basin tributaries, which will aid understanding sediment transport. When single gauging stations or regions are considered, water discharge appears to behave regular and stable over time, but nevertheless, the system as a whole can be quite complex. Some authors (Meade et al., 1991; Molinier et al., 1996; Dunne et al., 1998) have described some of these complexities (backwater effect, flood effects combinations, river channel morphology, etc.), which must be considered during suspended sediment analysis.

4.1 Water discharge and suspended sediment concentration variability

Molinier et al. (1996) demonstrated the importance of taking into account that the Rio Negro, Rio Madeira and Rio Solimões (see Figure 1) are the most important discharge contributors to the total amount of water discharged from the Amazon Basin into the ocean. The Madeira basin represents 23% of the total Amazon basin area and 15% of total Amazon water discharge into the ocean. The Rio Negro basin represents only 11% of the total Amazon basin area but contributes 14% in terms of water discharge. These values are correlated to areas receiving most rainfall, which are predominantly located in the northwest area of the basin (Espinoza et al. 2008). Despite these water discharge values, the Rio Negro does not transport as much suspended solids compared to the Madeira (Filizola and Guyot, 2009). This difference is attributed to the different sediment source areas (low-erosion cratonic Shields vs. high-erosion Andes) and the lowlands traversed by the rivers.

The high and low water periods of the above cited rivers have special characteristics that display interesting behaviours. The Rio Madeira's high water period takes place between March and April with the average maximum occurring in April, and the minimum value occurring in the end of September. The Rio Solimões' high water period occurs between May and July with the average maximum occurring in June. Additionally, the Rio Amazonas high water period at the Óbidos station takes place between May and June, and the average maximum value occurs at the end of May. Thus, the downstream peak at the

Rio Amazonas at Óbidos occurs before the upstream peak at the Rio Solimões[2] at Manacapuru. This behaviour, as reported by Molinier et al. (1996), is, in fact, the result of the influence of the Rio Madeira high water period, whose maximum occurs two months (on average) in advance of the maximum from the Rio Solimões. This event also causes an advance in the Rio Amazonas discharge peak at the Óbidos gauging station.

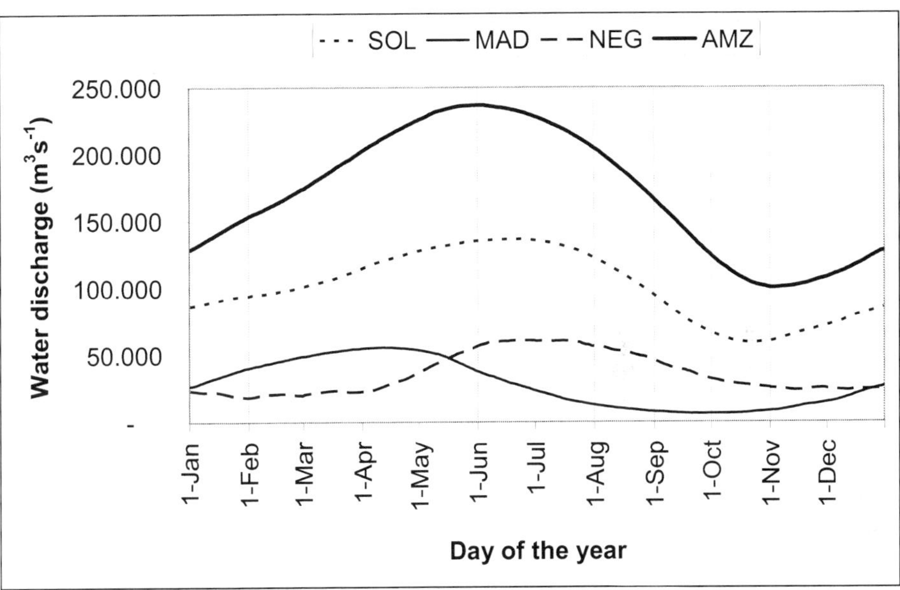

Fig. 2. Average hydrograms of the Rio Amazonas (AMZ) at Óbidos, the Rio Negro (NEG) at Manaus, the Rio Solimões (SOL) at Manacapuru and the Rio Madeira at Fazenda Vista Alegre for the years 1978 to 2008. The data source is the Brazilian National Water Agency dataset (http://www.ana.gov.br). The discharge of the Rio Negro at Manaus was calculated by discharge transposition using upstream to downstream stations correlated with HYBAM Doppler discharge data (http://www.ore-hybam.org) to eliminate the backwater effect.

From Figure 2 and Table 2, we can observe that the Rio Solimões dominates the water flow to the Rio Amazonas at Óbidos; however, from February to May, the Rio Madeira flow controls the variability, and between June and September, the flow of the Rio Negro waters have a greater influence. From October until January, the power of the flow of Rio Solimões is superimposed over the other two large rivers.

Some authors, such as Filizola (2003), Guyot et al. (2005), Bourgoin et al. (2007) and Martinez et al. (2009), showed graphs with the surface and total suspended sediment concentrations at Óbidos as a function of the Amazon River water discharge. The suspended sediment

[2] Rio Solimões and Rio Amazonas are, in fact, the same river; the difference in names comes from the importance of the Rio Negro to local culture and tourism (meeting of the black and white waters respectively from the Negro and the Solimões). Thus, before the mouth of the Rio Negro, the main river is called Solimões, and the name Amazonas (in Brazil) is used downstream of their confluence.

	High Period	Max	Low Period	Min
R. Solimões	May – Jul	mid Jun	Oct - Nov	mid Oct
R. Negro	Jun - Jul	mid Jun	Jan - Feb	end Jan
R. Madeira	Mar - Apr	mid Apr	Sept - Oct	end Sept
R. Amazonas	May - Jun	end May	Oct - Nov	early Nov

Table 2. Summary of the hydrological behaviour of the most important rivers within central Amazonia showing periods for high and low water discharge. Also shown are the months in which the average maximum and minimum values are observed (using data series from 1978 to 2008).

concentration measurements used were collected on a 10-day basis since 1995 at Óbidos and other important sites within the Amazon Basin by the HiBAm project and by ORE/HYBAM after 2003. From this data, Bourgoin et al. (2007) obtained three different equations to calculate QS from water discharge relations over three different periods of time, which were used to develop different scenarios for the temporal dynamics of water and sediment exchanges between the Curuai floodplain (Várzea) and the Rio Amazonas in front of Óbidos. The resulting figures shown by those authors clearly show a non-linear behaviour between water discharge and suspended sediment concentrations as well as three distinct temporal situations that enable distinguishing between the discharge modulation of the Rio Amazonas at Óbidos from that of the Rio Madeira and Rio Negro (see Figure 3). These three characteristic temporal situations are interpreted as follows. The first one, on the left side of the figure, represents a moment in the cycle corresponding to September, October and November and correlates well to the low period of all three rivers (Negro, Madeira and Solimões). The second situation (the highest clockwise) is the result of samples collected between February and April. This period coincides with the highest period of water discharge and sediment input from the Rio Madeira. The third area (on the right) corresponds to the period from May to July, which correlates to the highest period of water discharge from the Rio Negro. The samples collected between December and January (shown on the left side of the figure) are concurrent with the rising period of the Rio Madeira. The same conclusion can be drawn for the Rio Negro when looking at the right side of the figure. It follows that, between March and May, the Rio Negro is rising, with very low SSC, and the flow of Madeira is decreasing. Finally, the base of the figure can be attributed to the period between August and September when the flow of all three rivers is decreasing. Thus, the process described above indicates an SSC multi-control system for the Rio Amazonas at Óbidos. This system is highly dependent on the sediment contributions from the Rio Madeira during a certain period of time and also on absent sediment contributions from the Rio Negro during another period of time.

From the above presented data and Figure 3, it clearly follows that at Óbidos, sediment discharge is not a linear function of discharge.. If linear functions would be applied to estimate suspended sediment discharge from the Rio Amazonas to the Ocean, these would be seriously compromised and probably overestimated. This data provides important information on water and sediment interaction within the Amazon Basin. The water discharge and SSC behaviour demonstrated above shows the key to explaining some of the differences in the suspended sediment values presented at Table 1. It can indicate a kind of "teleconexion" among the upward areas of Rio Madeira and the Rio Amazonas downward region. With dams reservoirs now under constructions at the Rio Madeira this behaviour assumes an important role. If the new Madeira Hydropower Plants reservoirs will reduce sediment discharge it can also will cause impacts at the lower Rio Amazonas region. Stressed here is the significance of the Rio Madeira sediments to support the large Várzea (seasonal humid/flooded plains) areas, situated near of Óbidos, in terms of floodplain fertilization that is of great importance to the local farming and grazing economy.

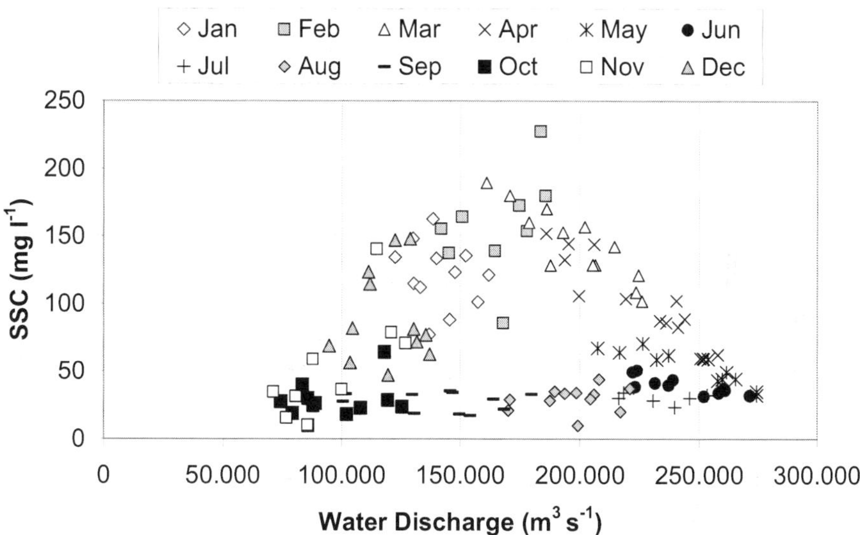

Fig. 3. Suspended Sediment Concentration versus the water discharge data for Rio Amazonas at Óbidos. The graph also shows the distribution of samples collected between 1995 and 1998 according to the month the samples were collected. For details see text. Figure modified from Guyot et al. (2005) using ORE/HYBAM data (http://www.ore-hybam.org).

4.2 Suspended sediment concentration variability (in deep and at the water surface)

Suspended sediment transport and the actual Amazon hydrology scenario can be changed by the new infrastructure construction (e.g., hydropower plants, roads and new agriculture frontier), climate change impacts and land use policies. Monitoring those factors and its variability is an important issue. In the Amazonian Hydrographical Region Book (Brasil, 2006), the Brazilian National Water Resources Management Plan describes the sustainable development of hydropower plant reservoirs and new agriculture frontiers as two of the most important goals to be achieved. Both are affected by sediment transported by rivers within the Amazon region. With a large sediment source such as the Andes to the west and its low general slope (especially from the piedmont to the flood plains) of about 2 cm km^{-1} (Molinier et al. 1996), reservoirs will act as trapping areas, which will increase deposition processes. These trapping areas can reduce, for example, the amount of sediment deposited from Rio Madeira into the main stream (lower Amazon stretch), leading to a diminished fertilisation of the flood plain. However, in other areas, the intensive agricultural land use and deforestation can have the opposite effect. These areas are more susceptible to erosion. Additionally, erosion occurs as a consequence of rain intensification derived from more frequent climatic variations. Thus, monitoring sediment transport in this kind of scenario must consider analyses over both space and time.

Some sampling methods and calculations were evaluated by Filizola and Guyot (2004) for the Óbidos station and for other stations in the basin by Filizola and Guyot in 2009, with the aim to reconcile the different methods and approaches used. For the test conducted at Óbidos, Filizola and Guyot used three types of depth samplers: one 8-liter collapsible bag depth integrator, a Brazilian model of the US P-63 adapted to the Amazonian rivers conditions (deep sections, high water velocities and discharge), and a 12-liter point horizontal oceanographic sampler, also adapted to the Amazonian rivers conditions. Surface sediments were sampled using a simple bucket, and a Doppler device was used for water discharge determination, and to test its use to determine suspended sediment discharge. Sampling surface sediments was carried out to test whether they represent the total suspended sediment discharge. Using the different samplers, the results showed a very small difference between all samplers when using the same QS calculation procedures. The relationship between the surface and total suspended solids was found to be, on average, 28%. It means that, in average, the SSC of a sample obtained in Óbidos at the water surface represents 72% of the total suspended sediment concentration. However, this relationship cannot be extended over the whole hydrological year, because it varies seasonally.

Another important question to be answered is that of the SSC variability with depth. Filizola (2003) showed that increasing SSC with depth does not always correspond to an increase in vertical SSC discharge (qS). Using Doppler data combined with SSC data, Figure 4 was created as an example from the Rio Madeira at Fazenda Vista Alegre. It can be seen that SSC values increase with depth, but qS values do not follow suit. The qS value seems to be influenced by the water velocity component. Filizola (2003) also demonstrated that this heterogeneous behaviour with depth has a seasonal variation, and is smoothed during the low water period.

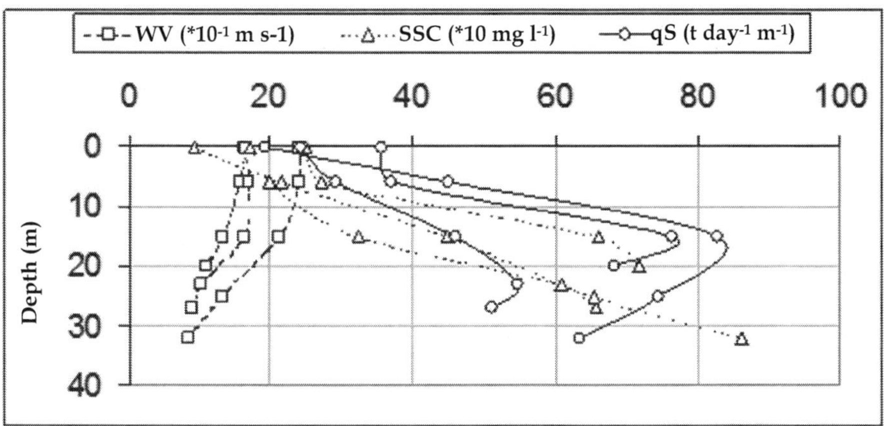

Fig. 4. Rio Madeira at Fazenda Vista Alegre on 22/05/1997 showing the variations with depth of water velocity (WV), suspended sediment concentration (SSC) and the discharge of suspended sediment (qS) for three vertical sampling profiles. Source: Filizola, 2003.

Another important point stems from the idea that stable river sections are good for water discharge determinations are good for TSS discharge determination as well. An interesting case about it has been presented by Filizola et al. (2009) at Manacapuru (Rio Solimões), a station used by Brazilian Agencies as a school site to conduct a yearly training course on large river water discharge measurement methods. Additionally, this site is also an SSC collection point. The authors showed that, at Manacapuru, geological structures, as described by Latrubesse et al. (2002), continue into the river channel and create bottom irregularities that influence the water movement. This movement causes heterogeneous behaviour of the surface SSC compared to that on the bottom. The surface results can be viewed by satellite images with "in situ" data confirming the image impressions (Figure 5). The resulting plume of suspended sediment from the phenomenon varies seasonally (Filizola et al. 2009).

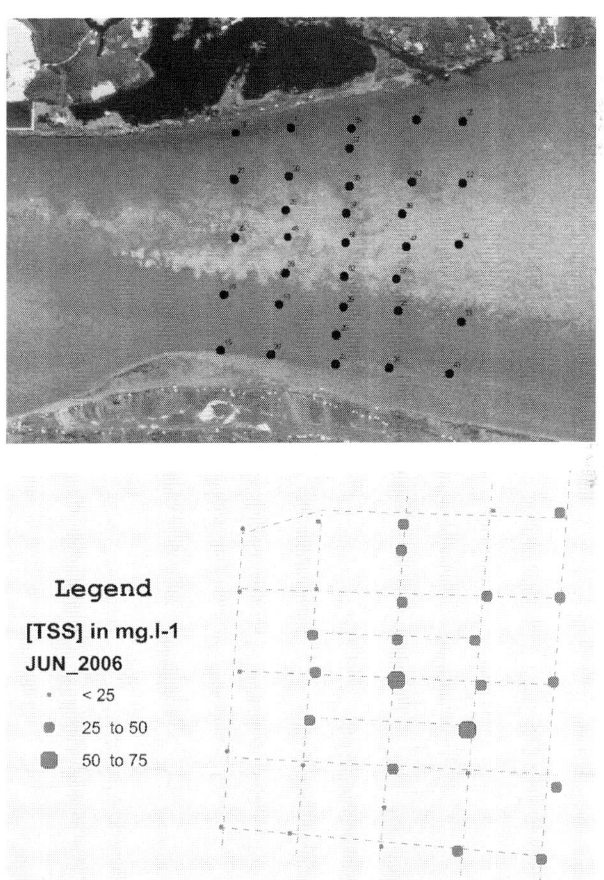

Fig. 5. A satellite image with the plume marked by Filizola et al. (2009) with the "in situ" results from the total suspended sediment collected at the water surface.

Regarding not only the suspended sediment transport, but also the bedload transport, important work was performed by Strasser (2008). He quantified the geometry, sediment transport and dynamics associated with the riverbed configuration of the main course of the Rio Solimões/Amazonas. He reported that, in the Brazilian Amazon territory, dunes are the prevailing bedform along the fluvial reach of the Rio Amazonas with heights that ranged from 0.2 to 12.0 meters and wavelengths of up to 400 meters. The main characteristics of the dune geometry (height and wavelength) remains practically unaltered throughout the year. Using dune geometry and migration measured in different stretches of the river, Strasser (2008) estimated the bedload transport to be 4 10^6 to 5 10^6t yr^{-1} between the Iracema, near the city of Itacoatiara at the Rio Amazonas a few kilometres before the mouth of Rio Madeira, and the Óbidos stations, which represents less than 1% of the total sediment transport of the Rio Amazonas.

All of the numbers produced in terms of suspended sediment concentration and discharge as well as all of the related phenomena seem to indicate that additional research still needs to be conducted with more emphasis on the processes and factors that control the transport of suspended sediment through the Amazonian rivers. Knowledge of the processes will provide a better understanding not only of the Amazonas contributions to the ocean, but will also aid to understanding the importance of tributaries..

5. New technique contributions

In situ networks with stations systematically collecting samples on suspended sediment concentrations and also non-systematic field cruisers performing detailed surveys will continue to be useful and important tools. However, new techniques can contribute, in terms of improving the capacities of remote analysis (in time and space), or estimating a more long-term average of sediment load transportation in the Amazon Basin by using cosmogenic nuclides.

5.1 Satellite approach, an example using MODIS sensors

Using satellite techniques, reflectance of the inland water turbidity can be useful to suspended sediment transport studies. A reflectance wavelength of 700 to 800 nm agrees with SSC in turbid inland waters as described by Martinez et al. (2009). These authors successfully tested this correlation and spaceborne sensors, such as MODIS, which are promising methods to monitor inland water, in the Amazon basin. These sensors offer spatial resolution and spatial coverage that is compatible with the dimensions of Amazonian River system and allow for fine temporal resolution.

The methodology proposed by Martinez et al. (2009) used five atmospherically corrected surface reflectance products from the Terra and Aqua MODIS spaceborne sensors. They used composite images with the following considerations: i) the 8-day composite is compatible with the 10-day field measurement sampling frequency (the ORE/HYBAM data set); ii) the amount of data to be analysed is reduced because a large number of daily images cannot be taken due to persistent cloud cover; and iii) the directional reflectance effects and atmospheric artefacts are significantly reduced. For each date, the composites from Terra and Aqua were automatically scanned, and the image with the lowest cloud coverage was selected. When both composites exhibited low cloud coverage, the composite acquired with the lowest satellite viewing angle was preferred. A pixel-based river mask covering nearly 10,000 pixels was manually outlined over the tested station (Óbidos) to automatically extract the reflectance in each MODIS image.

The surface reflectance data appeared to be robustly linked to the suspended sediment concentration at the river surface over a large range of concentrations and for several consecutive hydrological cycles (Martinez et al. 2009). Thus, MODIS data can be useful as an operational tool to provide more observations in a poorly gauged basin, especially the large rivers, by combining excellent temporal resolution and fine calibration quality. This technique can also be used to map SSC superficial dispersion in the main stream and to observe the impact of tributaries. However, the authors emphasise that the quest for a universal algorithm for suspended sediment retrieval from satellite data is never likely to succeed. This is because the scattering efficiency of suspended particles is very much a function of the average particle size and is quite variable from one catchment to another. Thus, the local calibration of satellite data would have to be developed and tested for each river basin.

5.2 Cosmogenic nuclide-produced ^{10}Be used to derive long-term sediment loads

The cosmogenic nuclide-produced method to derive denudation (or erosion) rates is a useful tool for estimating long-term sediment loads. ^{10}Beryllium ($T_{1/2}$ = 1.39 Myr) is the most widely used cosmogenic nuclide for studying the processes that shape the Earth's surface despite the fact that very low nuclide abundance requires time-consuming chemical ^{10}Be enrichment and costly accelerator mass spectrometer (AMS) analysis. ^{10}Be is produced in situ within mineral lattices (e.g., quartz) through the interaction of cosmic rays, and its production is altitude-dependent (Lal, 1991). In situ-produced ^{10}Be accumulates in near-surface deposits over time such that the concentration of the nuclide is related to both the age and stability of the surface (Lal, 1991). The accumulation rate of ^{10}Be is thereby inversely proportional to the erosion rate of the surface, if the erosion process has been taking place for a period that is long compared to the erosion time scale z^*/ε (see Figure 6A). This time scale is equivalent to the time it takes to erode ~60 cm of silicate rocks, which is a depth where the intensity of cosmic irradiation is reduced by a factor of 1/e through interaction with the material. In this case, the production of nuclides equals the removal of nuclides at the surface by erosion ("cosmogenic steady-state") and a basin-wide rate (i.e., the total rate of chemical and physical removal from the surface) can be calculated. At the same time, this basin-wide rate integrates over all spatially variable erosion rates within a single basin following the concept of "Let nature do the averaging" (see Figure. 6A). More comprehensive synopses of cosmogenic nuclides and their applications are given by Gosse and Phillips (2001), Bierman and Nichols (2004), von Blanckenburg (2005), and Granger and Riebe (2007).

Basin-wide erosion rates from cosmogenic nuclide concentrations are normally derived from actively eroding hillslope settings, where no storage occurs. Long-term sediment storage in large depositional basins might violate the assumption of cosmogenic steady state, and may result in (i) additional nuclide production when exposed to cosmic rays or (ii) the loss of nuclides during decay when sediment is prone to deep burial and shielding from cosmic rays. However, under certain prerequisites, it is possible to correct cosmogenic nuclide-derived erosion rates in lowland basins to yield rates that are not affected by low elevation, depositional areas. This concept is summarised in Figure 6 where production rates of nuclides are altitude-dependent. Thus, the average basin-wide production rate decreases with increasing lowland (low-elevation) area. Additionally, the modeling of cosmogenic nuclide concentrations in lowland areas (Wittmann and von Blanckenburg,

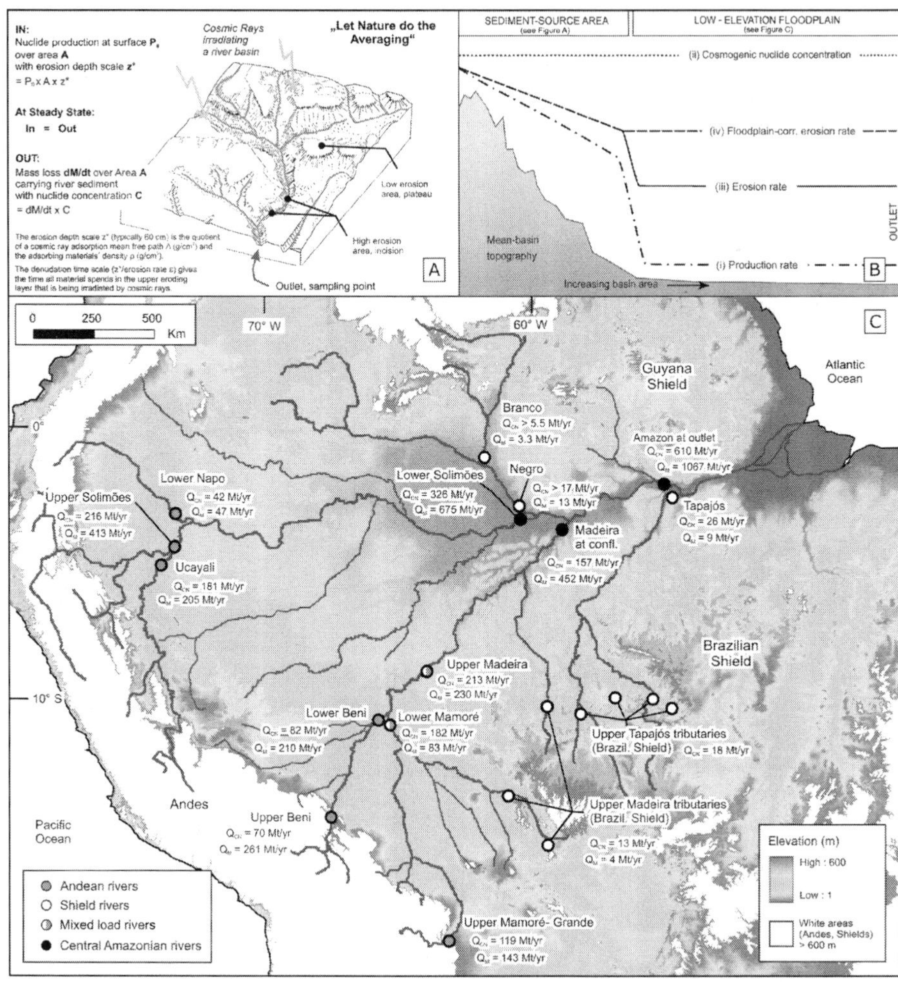

Fig. 6. (A): Illustration of the "Let Nature do the Averaging" concept for cosmogenic nuclide-derived basin-wide erosion rates; modified from von Blanckenburg (2005); (B): Concept of correction applied to cosmogenic nuclide-derived erosion rates in lowland basins; modified from Wittmann et al. (2011); and (C): Sediment loads (Mt/yr) in the Amazon basin calculated from cosmogenic nuclide-derived erosion rates (QCN) compared to published modern loads from gauging (QM); modified from Wittmann et al. (2011). For QM-related references, the reader is referred to Wittmann et al., 2011.

2009) and the testing of the model in a large Amazonian tributary (Wittmann et al., 2009) have shown that the concentration of long-lived nuclides (e.g., ^{10}Be) contained in headwater sediment is not prone to change due to storage in depositional areas of the basin. Thus, uniform nuclide concentrations from the sediment source area to the floodplain are expected. As a result of decreasing production rates, uniform nuclide concentrations translate into decreasing basin-wide erosion rates. As this effect is only caused by decreasing production rates, erosion rates may be corrected using the production rate of the sediment source area only (i.e., to be "floodplain-corrected"). A prerequisite for this correction is that the cosmogenic signal of the sediment-producing area is preserved in the floodplain, which is the case when sediment storage is shorter than the nuclides half-life (Wittmann and von Blanckenburg, 2009).

In the central Amazonian lowlands, cosmogenic nuclide concentrations resemble those measured in the sediment source areas (Andes, cratonic Shields). Thus, erosion rates for the entire basin may be calculated, which integrate over the time it takes the sediment to be transported through the basins, i.e., 5-15 kyr (Dosseto et al., 2006; Mertes and Dunne, 2007). These erosion rates can be converted to loads using the sediment density and can then be compared to decade-scale, gauging-derived sediment loads (see Figure 6). This comparison yields an average factor of ~2 agreements between QCN and QM, which could be due to stable erosion rates in the sediment source areas over the last 5-15 kyr until today, despite changes in climate. Additionally, the slightly better overall agreement between QCN and QM in central Amazonia in comparison to the source areas could be caused by the buffering capability of the large Amazon floodplain (Métivier and Gaudemer, 1999). Changes in Amazon River sediment output fluxes, as measured by long-term cosmogenic nuclides or short-term gauging, could have been stabilised over the last millennia by a persistent, accommodation-dominated floodplain.

6. Conclusions

Sediment transport is part of natural processes undergoing in river basins, and the rivers are the natural ways to transport suspended sediments from the continent to the ocean. This chapter showed that since 1950 the Amazon basin suspended sediment transport has been studied by several scientific groups. Therefore, different results have been presented. These results were obtained from different data sets, sampling and calculation methods. They came not only from several multidisciplinary scientific projects, which operates since the 1970, but also from local hydrometric networks providing long-term data. Now, with the help of all these previous works, it is possible to indicate with more certitude that the suspended sediment flux from the Amazonas to the Ocean is lower than 900 10^6 ton.yr^{-1}.

The value cited above was indicated thanks to the progressive evolution of scientific works and also by an intensive and more frequent sampling procedure, especially done since the year 2000. These data permit a more efficient evaluation about hydrology scenario and its importance to the sediment discharge variability, especially at central Amazonian region. This scenario marks the importance of the biggest Amazon tributaries (Rio Negro and Rio Madeira) to modulate the Rio Amazonas SSC flux at Óbidos (~ 700km to the mouth). New techniques, using satellite data or cosmogenic nuclide, helps to reinforce these results, open other perspectives to the theme and also reinforce that the huge distance between the Amazon River sources in the Andes and its mouth in the Atlantic Ocean makes possible the co-existence of erosion, deposition and re-suspension processes at different scales in time and space.

The transport of suspended sediment is a product of erosion, and erosion and deposition processes within a basin depend heavily on factors such as climate and relief. In Amazonia, the western landscape acts as an important driving mechanism of climate, and the gradient relief has an important effect on the erosion and deposition processes between the highlands and the exit of the basin. However, anthropogenic factors also play an important role in those processes. Within this context, Allen (2008) suggested that the sum of physical, chemical and biological processes that shape the Earth's surface and drive its mass fluxes, affecting suspended sediment transport; need to be investigated over the so-called human timescale. Additionally, integration over larger spatial and temporal scales needs to be performed. More complexity is thereby added by human actions, when aiming at studying the processes that shape our Earth's surface, and thus an increasing need arises for new monitoring systems and new techniques. In today's Amazonia, the environmental scenario reflects challenges especially concerning to water resource management, with integration techniques studies under development. To do so, the suspended sediment transport by rivers is an important issue to be included on that kind of integration.

7. References

Aalto, R.; Bourgoin, L.M. ; Dunne, T. ; Montgomery, D.R.; Nittrouer, C.A. and Guyot, J.L. (2008) Episodic sediment accumulation on Amazonian flood plains influenced by El Niño/Southern Oscillation. Nature, 425:493-497.

Allen, P. A (2008) From landscapes into geological history. Nature, Reprinted from Vol. 451, no. 7176 (Supplement), pp. 274-276, doi:10.1038/nature06586.

Anderson, S. P., Dietrich, W. E., and Brimhall, G. H., (2002) Weathering profiles, mass-balance analysis, and rates of solute loss: Linkages between weathering and erosion in a small, steep catchment: Geological Society of America Bulletin, v. 114, no. 9, p. 1143-1158.

Bierman, P.R., and Nichols, K.K., (2004), Rock to sediment - Slope to sea with Be-10 - Rates of landscape change: Annual Review of Earth and Planetary Sciences, v. 32, p. 215-255.

Bordas M.P. (1991). An outline of hydrosedimentological zones in the Brazilian Amazon basin, 191-203. In Water Management of the Amazon Basin, Braga B.P.F. & Fernandez-Jauregui C. (eds.), Publ. Unesco-Rostlac, Montevideo.

Bordas M.P., Lanna A.E., Semmelmann F.R. (1988). Evaluation des risques d'érosion et de sédimentation au Brésil à partir de bilans sédimentologiques rudimentaires, 359-368. In Sediment Budgets, Bordas M.P. & Walling D.E. (eds.), IAHS Publ. 174.Bourgoin et al. (2007

Bourgoin L.M, Bonnet M-P., Martinez J-M., Kosuth P., Cochonneau G., Moreira-Turcq P., Guyot J-L., Vauchel P., Filizola N., Seyler P. (2007) Temporal dynamics of water and sediment exchanges between the Curuaı´ floodplain and the Amazon River, Brazil. Journal of Hydrology (2007) 335, 140- 156.

Brasil (2006)RASIL, Caderno da Região Hidrográfica Amazônica 2006 Ed. Ministério do Meio Ambiente, Secretaria de Recursos Hídricos. - Brasília: MMA,. 124 p.

Cochonneau, G.; Sondag, F.; Guyot, J-P. ; Boaventura, G.; Filizola, N.; Fraizy, P.; Laraque, A.; Magat, P.; Martinez, J-M.; Noriega, L.; Oliveira, E.; Ordonez, J.; Pombosa, R.; Seyler, F.; Sidgwick, J. & Vauchel, P. (2006). Climate Variability and Change—Hydrological Impacts (Proceedings of the Fifth FRIEND World Conference held at Havana, Cuba, November 2006), IAHS Publ. 308, 2006, 44–50.

Dosseto, A., Bourdon, B., Gaillardet, J., Allegre, C.J., and Filizola, N., (2006), Time scale and conditions of weathering under tropical climate: Study of the Amazon basin with U-series: Geochimica Et Cosmochimica Acta, v. 70, p. 71-89.

Dunne T., Mertes L.A.K., Meade R.H., Richey J.E., Forsberg B.R. (1998). Exchanges of sediment transport between the floodplain and channel of the Amazon river in Brazil. Geological Society of America Bulletin, 110(4): 450-467.Ferreira et al. (1988)

Espinoza, J. C., Guyot, J.L., Ronchail, J., Cochonneau, G., Filizola, N., Fraizy, P., Labat, D., Noriega, L., de Oliveira, E., Ordoñez, J.J., Vouchel, P., (2008) Contrasting regional runoff evolution in the Amazon Basin (1974-2004) Journal of Hydrology, Volume 375, Issues 3-4, Pages 297-311.

Ferreira, J. R.; Devol, A.H.; Martinelli, L.A.; Forsberg, B.R.; Victoria, R.L.; Richey, J..E. & Mortatti, J. (1988) Chemical composition of the Madeira river: seasonal trends and total transport. Mitt. Geol. Paläont. Inst. Univ. Hamburg, Scope/Unep Sonderband 66, 63-75.

Filizola N. (1999). O fluxo de sedimentos em suspensão nos rios da Bacia Amazônica brasileira. Publ. ANEEL, Brasilia, 63 p.

Filizola, N. & Guyot, J-L. (2009) Suspended sediment yields in the Amazon basin: an assessment using the Brazilian national data set. Hydrol. Process. 23, 3207–3215. DOI: 10.1002/hyp.7394.

Filizola, N. (2003). Transfert sédimentaire actuel par les fleuves amazoniens. Thése, UPS, Toulouse III. 273pp.

Filizola, N. and J.L. Guyot, (2004). The use of Doppler technology for suspended sediment discharge determination in the River Amazon. Hydrological Sciences–Journal des Sciences Hydrologiques: 49(1)143-153.

Filizola, N. e Guyot, J-L (2007) Balanço do fluxo de sedimentos em suspensão da Bacia Amazônica. 2007 In: Workshop Geotecnologias Aplicadas às Áreas de Várzea da Amazônia: trabalhos apresentados no workshop realizado em Manaus, de 17 a 18 de julho de 2007. - Manaus: Ibama. 104 p.

Filizola, N., Seyler, F., Mourão, M. H., Arruda, W., Spínola, N., and Guyot, J-L. Study of the variability in suspended sediment discharge at Manacapuru, Amazon River, Brazil. (2009) Latin American Journal of sedimentology and basin analysis, Vol. 16 (2) 2009, 93-99 Ed. Asociación Argentina de Sedimentología - ISSN 1669 7316

Gaillardet, J., Dupré, B., Louvat, P., and Allègre, C. J., (1999) Global silicate weathering and CO2 consumption rates deduced from the chemistry of large rivers: Chemical Geology, v. 159, no. 1-4, p. 3-30.

Gibbs R.J. (1967). The Geochemistry of the Amazon River System. Part I. The factors that control the salinity and the composition and concentration of the suspended solids. Geological Society of America Bulletin 78 : 1203-1232.Gosse and Phillips (2001),

Gosse, J.C., and Phillips, F.M., (2001), Terrestrial in situ cosmogenic nuclides: theory and application: Quaternary Science Reviews, v. 20, p. 1475-1560.

Granger, D.E., and Riebe, C.S., (2007), Cosmogenic nuclides in weathering and erosion, in Drever, J.I., ed., Treatise on Geochemistry, Volume 5: Surface and Ground Water, Weathering, and Soils: London, Elsevier.

Guyot J.L. (1993) Hydrogéochimie des fleuves de l'Amazonie bolivienne. Collection Etudes & Thèses, ORSTOM, Paris, 261 p.

Guyot J.L., Bourges J., Hoorelbecke R., Roche M.A., Calle H., Cortes J., Barragan M.C. (1988). Exportation de matières en suspension des Andes vers l'Amazonie par le Rio Béni, Bolivie, 443-451. In Sediment Budgets, M.P. Bordas & D.E. Walling (eds.), IAHS Publ. 174.

Guyot J.L., Filizola N., Quintanilla J., Cortez J. (1996). Dissolved solids and suspended sediment yields in the Rio Madeira basin, from the Bolivian Andes to the Amazon, 55-63. In Erosion and Sediment yield : Global and Regional Perspectives, Exeter, July 1996. IAHS Publ. 236.Guyot et al. (2011

Guyot, J.L. et al, .(2011) Hydrology and sediment transport in the Amazon basin, from the Andes to the ocean. In: Proceedings of the World's Large Rivers Conference, Vienna, 2011.

Guyot, J.L., Filizola, N., Laraque, A., (2005). Régimes et bilan du flux sedimentaire à Óbidos (Pará, Brésil) de 1995 à 2003. In: Walling, D.E., Horowitz, A.J. (Eds.), Sediment Budgets 1, vol. 291. IAHS. Publ., pp. 347–354.Nittrouer et al., 1995

Hooke, R. L., (2000) On the history of humans as geomorphic agents: Geology, v. 28, no. 9, p. 843-846.

Julien P. (1998). Erosion and sedimentation. Cambridge University Press. Cambridge, UK. 280pp.

Lal, D., (1991), Cosmic ray labeling of erosion surfaces: in situ nuclide production rates and erosion models: Earth and Planetary Science Letters, v. 104, p. 424-439.

Laraque, A., Filizola, N., Guyot, J.L., (2005). Variations spatiotemporelles du bilan sédimentaire dans le bassin amazonien brésilien, à partir d'un échantillonnage décadaire. In: Walling, D.E., Horowitz, A.J. (Eds.), Sediment Budgets 1, vol. 291. IAHS Publ., pp. 250–258.

Latrubesse, E. and E. Franzinelli, (2002). The Holocene alluvial plain of the middle Amazon River, Brazil. Geomorphology 44: 241-257.

Latrubesse, E., Stevaux, J.C., Sinha, R., (2005). Tropical rivers. Geomorphology 70,187–206.

Martinelli L.A., Forsberg B.R., Victoria R.L., Devol A.H., Mortatti J., Ferreira J.R., Bonassi J., Oliveira E. 1993. Suspended sediment load in the Madeira River. Mitt. Geol.-Paänont. Inst. Univ. Hamburg, Sonderband 74, 41-54.Martinez et al (2009

Martinelli, L. A., Devol, A. H., Forsberg, B. R., Victoria, R. L., Richey, J. E. & Ribeiro, M. N. (1989) Descarga de sólidos dissolvidos totais do Rio Amazonas e seus principais tributários. Geochim. Brasil. 3(2), 141-148.

Martinez J.M., Guyot J. L., Filizola N. and Sondag F. (2009) Increase in suspended sediment discharge of the Amazon River assessed by monitoring network and satellite data. CATENA, Vol. 79, Issue 3, 15 December 2009, Pages 257-264.

Meade R.H. (1985). Suspended Sediment in the Amazon River and its Tributaries in Brazil during 1982-1984. U.S. Geological Survey Open File Report 85-492, Denver, 39 p.

Meade R.H., Nordin C.F., Curtis W.F., Costa Rodrigues F.M., Do Vale C.M., Edmond J.M. (1979). Sediment loads in the Amazon River. Nature 278 : 161-163.

Meade, R. H., (2007) Transcontinental moving and storage: The Orinoco and Amazon Rivers tranfer from the Andes to the Atlantic, in Gupta, A., ed., Large Rivers: Geomorphology and Management, John Wiley & Sons Ltd, p. 45-63.

Meade, R. H., Dunne, T., Richey, J. E., Santos, U. de M., and Salati, E., (1985), Storage and remobilization of suspended sediment in the lower Amazon River of Brazil: Science, v. 228, p. 488–490.

Mertes, L.A.K., and Dunne, T., (2007), Effects of tectonism, climate change, sea-level change on the form and behaviour of the Amazon River and its floodplain, in Gupta, A., ed., Large Rivers: Geomorphology and Management, John Wiley & Sons Ltd, p. 115-144.

Métivier, F., and Gaudemer, Y., (1999), Stability of output fluxes of large rivers in South and East Asia during the last 2 million years: implications on floodplain processes: Basin Research, v. 11, p. 293-303.

Milliman J.D., Meade R.H. (1983) World wide delivery of river sediment to the ocean. Journal of Geology, 91(1): 1-21.

Milliman J.D., Syvitzki J.P.M. (1992). Geomorphic / tectonic control of sediment discharge to the ocean: the importance of small mountainous rivers. Journal of Geology, 100: 525-544.

Molinier M., Guyot J.L., Oliveira E., Guimarães V. (1996). Les régimes hydrologiques de l'Amazone et de ses affluents, 209-222. In L'hydrologie tropicale : géoscience et outil pour le développement, Paris, Mai 1995. IAHS Publ. 238

Molinier, M., J.L. Guyot, E. De Oliveira, V. Guimarães and A. Chaves, (1995). Hydrologie du bassin de l'Amazone. In Proc. Grands Bassins Fluviaux Péri-atlantiques, vol. 1, 335-344. PEGI, Paris, France.

Nittrouer, C.A., Kuehl, S.A. Sternberg, R.W. Figueiredo, A.G. and Faria, L.E.C. (1995) An introduction to the geological significance of sediment transport and accumulation on the Amazon continental shelf, Mar. Geol. 125, 177-192 .

Nittrouer, C.A., T.B. Curtin and D.J. De Master ,, (1986a): Concentration and flux of suspended sediment on the Amazon continental shelf. Cont. Shelf Res. 6. 151-174.

Oltman R.E. (1968). Reconnaissance investigations of the discharge and water quality of the Amazon. U.S. Geological Survey Circ., 552, 16 p.

Pinet P. and Souriau M. (1988). Continental erosion and large scale relief. Tectonics, 7: 563-582.

Prance G.T. and Lovejoy T. (1985). Key environments : Amazonia. Pergamon Press, Oxford. 442 p.

Richey, E.J., Meade, R. H., Salatti, E., Devol, A. H., Nordin Jr., C. F., and dos Santos U. 1986 Water Resources Research, Vol. 22, No. 5, 756-764.

Riebe, C. S., Kirchner, J. W., Granger, D. E., and Finkel, R. C., (2001) Minimal climatic control on erosion rates in the Sierra Nevada, California: Geology, v. 29, no. 5, p. 447-450.

Schmidt G.W. (1972). Amonts of suspended solids and dissolved substances in the middle reaches of the Amazon over the course of one year (August, 1969 - July 1970). Amazoniana 3(2) : 208-223.

Sioli H. (1984). The Amazon and its main affluents: hydrography, morphology of river courses, and river types, 127-165. In The Amazon, Sioli H. (ed.), Junk.

Sioli, 1957

Sioli, H. (1950). Das Wasser im Amazonasgebiet. Forsch.u.Fortschr. vol. 26, p. 274-280.

Sioli, H. (1957). Valores de pH de águas Amazônicas. Boletim do museu paraense Emilio Goeldi. Geologia 1: 1-35.

Strasser, M. (2008) Dunas fluviais no rio Solimões - Amazonas - Dinâmica e transporte de sedimentos. Tese de doutorado, UFRJ - Programa de Engenharia Civil (PEC /COPPE), 147p.

Summerfield M.A. and Hulton N.J. (1994). Natural controls of fluvial denudation rates in major world drainage basins. Journal of Geophysical Research, 99: 13871-13883.

Syvitski, J. P. M., Vorosmarty, C. J., Kettner, A. J., and Green, P., (2005) Impact of humans on the flux of terrestrial sediment to the global coastal ocean: Science, v. 308, no. 5720, p. 376-380.

von Blanckenburg, F., (2005), The control mechanisms of erosion and weathering at basin scale from cosmogenic nuclides in river sediment: Earth and Planetary Science Letters, v. 237, p. 462-479.

von Blanckenburg, F., Hewawasam, T., and Kubik, P. W., (2004) Cosmogenic nuclide evidence for low weathering and denudation in the wet, tropical highlands of Sri Lanka: Journal of Geophysical Research-Earth Surface, v. 109, no. F3.

Whitmore T.C. and Prance G.T. (1987). Biogeography and Quaternary History in Tropical America. Claredon Press, Oxford. 214 p.

Wilkinson, B. H., and McElroy, B. J., (2007) The impact of humans on continental erosion and sedimentation: Geological Society of America Bulletin, v. 119, no. 1-2, p. 140-156.

Wittmann, H., and von Blanckenburg, F., (2009), Cosmogenic nuclide budgeting of floodplain sediment transfer: Geomorphology, v. 109, p. 246-256.

Wittmann, H., von Blanckenburg, F., Guyot, J.L., Maurice, L., and Kubik, P.W., (2009), From source to sink: Preserving the cosmogenic 10Be-derived denudation rate signal of the Bolivian Andes in sediment of the Beni and Mamoré foreland basins: EPSL, v. 288, p. 463-474.

Wittmann, H., von Blanckenburg, F., Maurice, L., Guyot, J.L., Filizola, N., and Kubik, P.W., (2011), Sediment production and delivery in the Amazon River basin quantified by in situ-produced cosmogenic nuclides and recent river loads: Geological Society of America Bulletin, v. 123, p. 934-950.

4

Sediment Transport in Rainwater Tanks and Implications for Water Quality

Mirela I. Magyar, Anthony R. Ladson and Clare Diaper
Monash University/Atura Pty Ltd
Australia

1. Introduction

Prolonged drought and increased water demands because of population growth have led to water storages. In Australia, the introduction of permanent water restrictions in urban areas and water conservation education programs have resulted in increased uptake of rainwater tanks as an alternative water source at the household scale. While there is in-depth understanding of the water savings that can be attributed to the substitution of mains water by water from rainwater tanks, there is limited understanding of the quality of the water and sediment collected in the tanks. This chapter provides information on tank water and sediment quality gained through field work, a laboratory investigation and development of a mathematical model.

The layout of this chapter is as follows. Section 1.1 describes the rainwater tanks design while Section 1.2 provides background information for this work on water quality in rainwater tanks in residential areas. Section 2 explains the series of methods employed to collect and analyse the sediment in rainwater tanks, while Section 3 presents the series of results for sediment quality in the tanks, including a summary of the experimental program implemented to understand the factors affecting sediment re-suspension from rainwater tanks, the results of leaching tests and development of a model. The implications of these results on rainwater tanks sediment quality are brought together in Section 4, discussion and 5, conclusions.

1.1 Background – The design of an urban rainwater tank

A typical urban rainwater tank system layout in Australia consists of the following components: rainwater is collected from the roof and conveyed directly to the rainwater tank through an inlet pipe positioned at the top of the tank. Water from the tank is supplied to the end use through an outlet situated close to the base of the tank. The outlet can be located anywhere between 50 mm to 600 mm above the tank base. The inlet/outlet configuration focuses on maximising the storage capacity in the tank, without considering water quality implications. The tank overflow is connected to the urban stormwater system.

The rainwater tank acts as a sedimentation tank between the rain events, being reported to improve water quality from inlet to outlet (Coombes et al. 2000). There are, however, significant differences between storage and sedimentation tanks (Magyar et al. 2006) because of the unsteady flow effects (Vaes 1995) into a rainwater tank. A sedimentation tank needs to satisfy certain conditions, such as: a low velocity inflow, ideally horizontal, is required to enhance settling of particles, while careful selection of inlet and outlet positions is required to avoid turbulence (Tchobanoglous and Burton 1991). None of these factors traditionally considered in a sedimentation tank design match the rainwater tank design described in this study, therefore requires further investigation.

To maximise utilisable volume of water from the tank, the outlet pipe is positioned as close as possible to the base of the tank. At the same time, for effective water savings, rainwater tanks are connected to end uses that constantly withdraw water from the tank, so that storage capacity is available to capture the next rain event. This, combined with the use of smaller volume rainwater tanks in the urban environment, has the potential to reduce the depth of water above any accumulated sediment. As the inflow is located at the top of the rain tank, there is potential that accumulated sediment is resuspended during inflow (rain events) and that the outflow water will be contaminated by the re-suspended sediment. Therefore, characterisation of sediment in rainwater tanks is important, as metals from roof materials and environmental pollution accumulate over time in the tank. The metals are of interest because they can be toxic to humans when water containing those metals is drunk or used for high contact end uses, such as showering.

There are limited studies investigating the sediment processes taking place in an urban rainwater tank, such as: sedimentation rate, accumulation rate, the potential of sediment and attached heavy metals being mixed and re-suspended and ultimately delivered to the end use. There are no studies investigating the implication of the current type of tank design and position of inlet and outlet, nor investigation of the potential for metals to leach from sediment. Recent studies highlighted knowledge gaps in our understanding of processes occurring in rain tanks (Evans et al. 2006) and knowledge gaps in sediment re-suspension and precipitation processes in rainwater tanks (Spinks et al. 2003).

There are no guidelines recommended values for water or sediment quality from rainwater tanks. Therefore, in order to compare quality of water or sediment from the urban tanks, a number of other relevant guidelines were considered. These guidelines were the Australian Drinking Water Guideline (NHMRC & NRMMC 2004), the Recreational Water Guidelines (ANZECC and ARMCANZ 2000), the Agricultural Irrigation Guidelines (ANZECC and ARMCANZ 2000) and for sediment comparison, the EPA Victoria guidelines (EPA Victoria 2007). The ADWG have the same recommended values for metals as the WHO international guidelines (WHO 2004).

1.2 Background - Water quality in rainwater tanks in suburban areas of Melbourne

To understand the variability of water quality, nine rainwater tanks were investigated, with three sampling rounds over a two year period (Magyar et al. 2006; 2007; 2008). The tanks were located in suburban areas of Melbourne (Fig. 1). Characteristics of the nine tanks are described in Table 1 and in more detail elsewhere (Magyar et al. 2006; 2007; 2008). Water from the tanks was used for residential outdoor uses (including irrigation and car washing).

A summary of the water quality from the nine tanks for the three sampling rounds (2006, 2007, 2008) is compared to the Australian Drinking Water Guideline (ADWG) (NHMRC & NRMMC 2004); the Recreational Water Guideline (RWG) (ANZECC and ARMCANZ 2000); the Agricultural Irrigation Guidelines for long term trigger values (up to 100 years) (AIG-LTV) and for short term trigger (up to 20 years) (AIG-STV) (ANZECC and ARMCANZ 2000) (Table 2). Both the AIG-LTV and AIG-STV assume that the annual application of irrigation water is 1000 mm, that contaminants are retained in the top 150 mm of soil and that soil bulk density is 1300 kg/m^3.

ID	Location	Roof material	Tank material	Size, m^3	Outlet height from base, mm	Distance to road, m	Tank age, years	Maintenance
S1	Doveton	PT, PPS, ZnAL	PVC	4.5	100	15	3	Once a year
S2	Brunswick	PVC	PVC	2.25	90	50	>3	Very rarely
S3	Brunswick	GI	PVC	2.75	50	50	>3	Very rarely
S4	Brunswick	GI	ZnAl	2	40	50	>3	Very rarely
S5	Brunswick	Tiles	FC	23	150	50	20	Very rarely
S6	Brunswick	Tiles	PVC	2.25	60	50	>5	Very rarely
S7	Northcote	Tiles	PVC	2.27	40	15	3.4	Once/2 years
S8	Northcote	ZnAl	ZnAl	2	35	15	4 months	New tank
S9	Northcote	PM	St.St.	0.23	30	15	7	Once/1.5 year

PT: painted tiles; PPS: pre-painted steel; GI- galvanised iron; ZnAl: 55% Aluminium zinc coated steel; St.St: stainless steel.

Table 1. Rainwater tanks characteristics

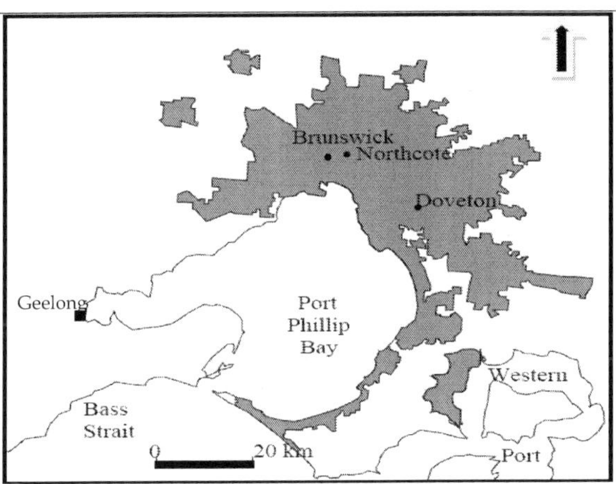

Fig. 1. Location of the rainwater tanks (Brunswick, Northcote and Doveton) in metropolitan Melbourne (denoted by the grey shaded area), Australia

Water from the nine rainwater tanks was found contaminated with several heavy metals: aluminium (Al), cadmium (Cd), iron (Fe), nickel (Ni), lead (Pb) and zinc (Zn). None of the nine tanks was found to meet the ADWG at all times, with concentrations of Al, Cd, Fe, Ni, Pb and Zn exceeding recommended levels for drinking or recreational guideline values. Of particular concern were the high concentrations of Pb in the tank water which were consistently above the acceptable limits, therefore the discussion in this paper will focus on Pb. Metals concentrations did not necessarily increase over time, suggesting that several factors may have contributed to metals behaviour, including: the water level in the tank at

the time of sampling, pH of the water, oxygen content and ionic composition in the water column and sediment level and quality in the tank.
Although analysis was undertaken to identify environmental or local conditions leading to water quality variations in the tank, no specific directions were found.

		Al	Cd	Cr	Cu	Fe	Mn	Ni	Pb	Zn
		mg/L	mg/L	mg/L	mg/L	mg/L	mg/L	mg/L	mg/L	mg/L
Detection limits and guidelines values										
DL		0.002	0.0007	0.0006	0.0003	0.0003	0.0002	0.0004	0.0006	0.0006
ADWG		0.2	0.002	0.05	1	0.3	0.1	0.02	0.01	3
RWG		0.2	0.005	0.05	1	0.3	0.1	0.1	0.05	5
AIG-LTV		5	0.01	0.1	0.2	0.2	0.2	0.2	2	2
Water quality results										
2006 sampling	Min	<DL	<DL	<DL	0.005	0.069	0.002	0.001	0.003	0.374
	Mean	0.339	0.004	0.002	0.085	0.191	0.009	0.004	0.047	1.139
	Max	0.616	0.004	0.003	0.596	0.805	0.018	0.016	0.348	3.123
2007 sampling	Min	<DL	<DL	<DL	0.001	0.025	0.004	0.002	0.080	0.204
	Mean	0.106	0.020	0.012	0.083	0.141	0.024	0.006	0.147	1.935
	Max	0.376	0.027	0.012	0.294	0.496	0.060	0.013	0.251	6.495
2008 sampling	Min	0.072	<DL	<DL	0.005	0.057	0.008	<DL	0.005	0.212
	Mean	0.228	0.035	0.070	0.086	0.366	0.035	0.030	0.041	0.838
	Max	0.876	0.103	0.139	0.298	1.606	0.092	0.138	0.114	1.614

DL: detection limit

Table 2. Water quality in nine suburban rainwater tanks

Fractionation (>63 µm, 0.45-63 µm and <0.45 µm) of the water samples revealed that metals were still found in particulate form in suspension, although sampling took place during steady conditions of no flows in the tanks (sampling was at least one day after a rain event). An example of Pb fractionation is shown in Fig. 2, where it can be observed that 0 to 80% was found in a dissolved form and the remainder (20-100%) was attached to particles in suspension.

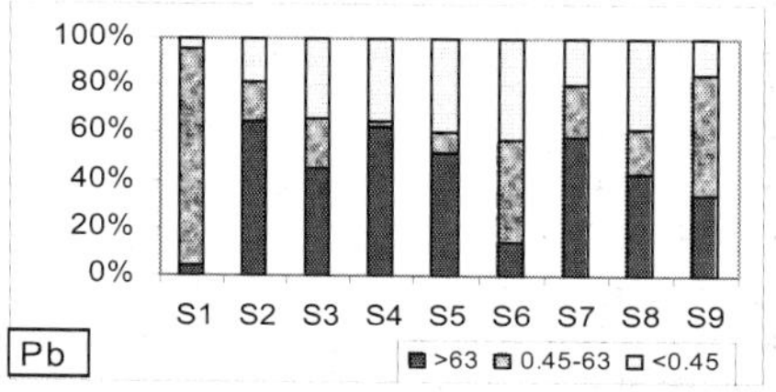

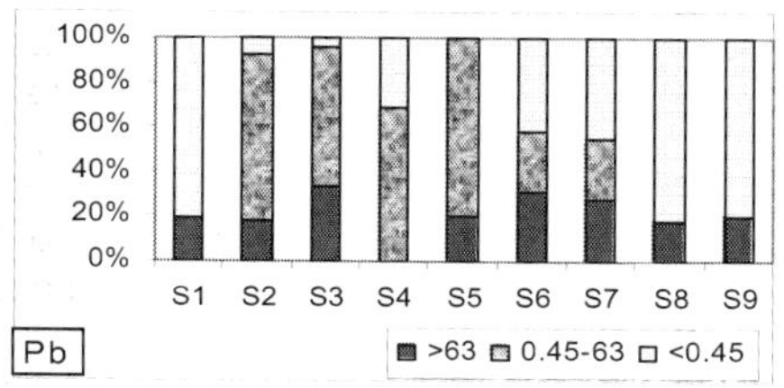

Fig. 2. Fractionation of Pb in the water samples in 2006 (left) and 2007 (right) runs (Magyar 2010); (>63 and 0.45-63 µm represent the particulate fraction while <0.45 µm represents the dissolved fraction)

Investigation of the pH in the tanks (Table 3) revealed that with the exception of one sample (pH= 8.1), water in the tanks was always acidic, with pH values ranging from 3.6 to 6.7. The pH measured values were similar to those observed in other studies where acidic pH in PVC rainwater tanks has been reported (Simmons et al. 2001; Coombes et al. 2002). The acidic pH in tanks was probably due to acidic rain events, as this was not unusual for Melbourne (Siriwardene et al. 2008) or in other places (Dean et al. 2005; Sabina et al. 2008).

Site	2006	2007	2008
S1	5.0	5.5	6.7
S2	4.9	5.9	6.5
S3	4.8	5.8	6.4
S4	4.6	6.2	6.7
S5	4.8	6.0	8.1
S6	4.5	5.6	6.4
S7	4.9	3.6	5.1
S8	4.4	4.4	6.4
S9	4.3	4.4	5.2

Table 3. Measurement of pH in the tanks

The variation of pH was found to contribute to the variation of dissolved concentrations of metals in the tanks' water, with a higher dissolved metal concentration observed when pH was lower. The higher pH measured in the 2008 samples (Table 3) may explain the lower concentrations of Pb in water in the 2008 sampling compared with the 2007 sampling (Table 2). This is because at higher pH (pH > 6), the fraction of dissolved Pb (Pb^{2+} ions) decreases and particle bound Pb is predominant with the settlement of Pb particles enhanced.
An example of the distribution of Pb fractions in the nine tanks as a function of pH in the 2007 sampling run is shown in Fig. 3, left. The dissolved and particulate (consisting of 0.45-63 µm + > 63 µm) fractions of Pb were plotted as a function of the pH measured in the tank water. It can be seen that this is in good agreement with the theoretical calculated distribution of Pb fractions (Fig. 3, right) (Bodek et al. 1988). The measured results show that as the pH decreases below 5.5, the dissolved Pb phase is predominant. Fifty-six percent of water samples had a pH less than 5.5.

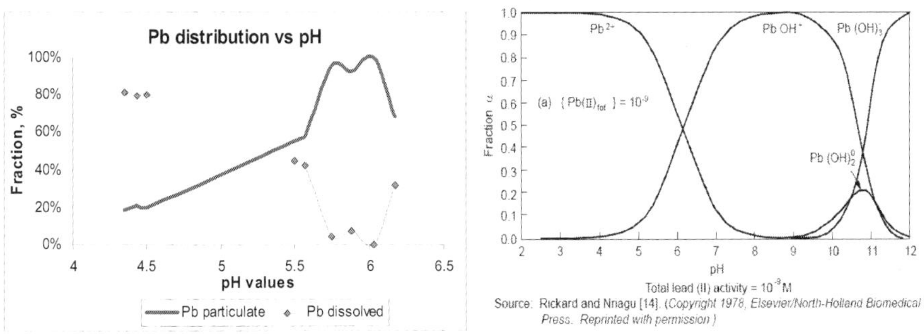

Fig. 3. Distribution of Pb fractions as a function of pH measured in this study (left) vs calculated distribution of Pb(II) hydroxy species as a function of pH and fraction of total dissolved lead (II) (Bodek et al. 1988) (right)

To better understand the water quality variations in the tank (as shown above), the investigation proceeded to sediment quality analyses, which are described below.

2. Methodology

In this section the method used for sediment sampling and analysis is described. First, grab samples from rainwater tanks are discussed; second, the continuous measurements of sediment accumulation; third, the chemical analysis used to test for metals in sediment; fourth, discussion of the particles size distribution in the tanks; fifth, the methods used for leaching tests are described; and sixth, the experimental procedures used to test sediment re-suspension are detailed.

2.1 Rainwater tanks sediment grab sampling

To determine the variability of metals in the sediment in rainwater tanks, grab samples of sediment from the nine urban tanks described above, were obtained three times over the two years period (2006 to 2008).

Several challenges have been identified for sampling sediment in a rainwater tank, including (i) limited access to the base of the tank, from the top opening; (ii) a dark and confined sampling space that is under several meters of water; (iii) due to drought conditions and water restrictions imposed in Melbourne, tank owners were not keen to waste tank water for the purpose of sediment sampling; (iv) it was unclear whether sediment is evenly distributed over the base of the tank; (v) analyses required gentle handling of the sample, to avoid break-up of aggregates of particles or flocs and a minimum exposure to air (APHA/AWWA/WEF 1995) (vi) to avoid introducing contaminants from the sampler itself into the sediments sample, there were restrictions on the materials of construction (Murdoch and MacKnight 1994; APHA/AWWA/WEF 1995).

An extensive literature review identified that there was no suitable sampling device available on the market; therefore the Magyar sediment sampler was designed for this project (Fig. 4). The Magyar sediment sampler is based on the Conbar telescopic dipper design (Forestry Suppliers Inc. 2005), but with a ladle designed to scrape sediment at the base of a flat tank and a lid designed to avoid losing sediment when the sample is retracted.

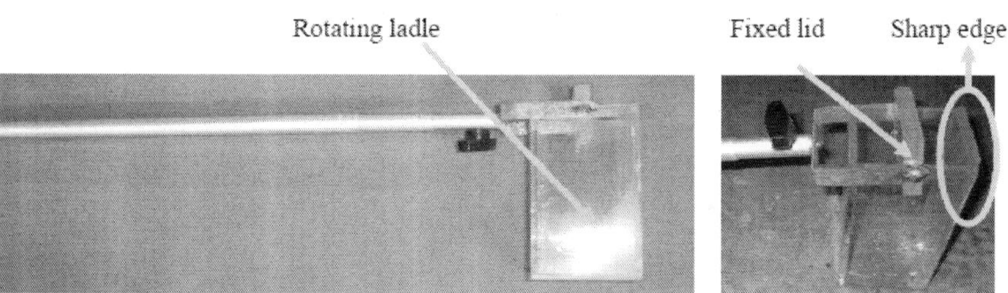

Fig. 4. Magyar sediment sampler (Magyar 2010)

2.2 Measuring sediment spatial distribution and accumulation in tanks

The experiments were intended to provide information on the areas of the tank that act as accumulation zones and the ones that act as sediment transportation zones. Sediment spatial distribution over the base of the tank was measured by locating four sediment traps (Fig. 5, left) at the base of the tanks, in locations as shown in Fig. 5, right. The sediment traps were designed with dimensions of 90 mm diameter and 50 mm height and were made of Perspex pipe. The traps were inserted in tanks and sediment was collected every three months. During sediment collection, a lid was used (Fig. 5, left) to avoid losing sediment from the trap on retraction from the base of the tank. The volume of accumulated sediment was measured with Imhoff cones.

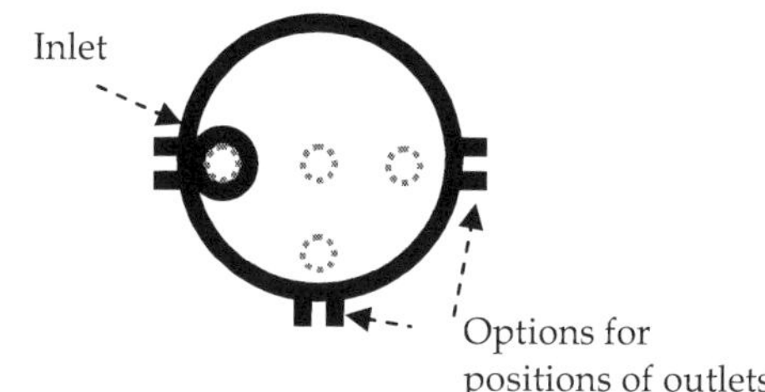

Fig. 5. Sediment traps (left); location of sediment traps at the base of the tanks (right); Traps are shown as dotted circles, inlet as a solid black circle and outlets as short parallel line segments

2.3 Chemical analyses

Sediment samples were taken immediately after collection to the Water Studies Centre laboratory, at Monash University, Australia. The sediment samples were fractionated in three wet fractions: >63 μm (very fine sand), between 0.45-63 μm (clay and silt) and <0.45 μm (dissolved). The fractionated sediment samples were analysed for metals on a dry weight basis and were oven dried (at 105°C) and a pre-weighed sub-sample was acid digested with Suprapur nitric acid. Each fraction was acid digested for analysis of metals ions and for analyses of total concentrations of Al, Cd, Cr, Cu, Fe, Mn, Ni, Pb and Zn according to Standard Methods 3030D and 3120B (APHA/AWWA/WEF 1995). The method is a verified laboratory in-house method with a recovery between 80-120%. The samples were analysed for metals with an Inductively Coupled Plasma Optical Emission Spectrometer (ICP OES).

For quality control, tests were also performed for blanks, duplicates, spikes and standard reference materials (SRMs) which represented at least 10% of the total number of samples per batch. The acceptable recovery for duplicates was 90-110 % and for SRMs and spikes was 80-120%. For the purposes of statistical and graphical presentation, results below ICP-OES detection limit were taken as half of the detection limit.

2.4 Particle size distribution

The sediment samples were tested for particle size distribution (PSD) by the light scattering method, standard method 2560D (APHA/AWWA/WEF 1995), using a Mastersizer 2000 instrument at a commercial laboratory (HRL Technology Pty Ltd) and a Beckman Coulter LS32 version 3.01 at CSIRO in Clayton, Victoria, Australia. In order to work within the sensor's accuracy, both instruments required an obscuration rate of at least 20% in the sample mixing unit. If the obscuration was less than 20%, more sample volume was added in the mixing unit and if the obscuration was too high, the sample was diluted with deionised water. Prior to each test, the particle size characterization instruments were calibrated to give a linear alignment. Each sample was tested in triplicate, to avoid bias due to small volumes taken from the sampling bottle. The results show the mean values of these three tests.

2.5 Leaching tests

Two analytical methods were used to test the potential of heavy metals to dissociate from sediment and contaminate the water column, as well as to contaminate ground water if they were disposed to land fill. These methods were: (i) a simplified sequential extraction of metals and (ii) the leaching of metals by Australian Standard Leaching Procedure (ASLP) (AS 4439.2 1997).

The sequential extraction method was developed by Sahuquillo (1999) and Marqui (2004) who simplified the method developed by Tessier (1979). The method selectively extracts particulate trace metals into chemical forms likely to be released in solution under various environmental conditions. The simplified method included the analysis the three fractions shown in Table 4.

Step	Fraction		Procedure
1	EX	Acid extractable (exchangeable + bound to carbonates	25 mL of 0.17 M acetic acid (pH 3), shaken for 16h at 20°C
2	OM	Reducible (bound to Fe/Mn hydroxides)	40 mL 0.1 M hydroxylamine-hydrochloride solution (pH1.5), shaken for 16 h at 20°C
3	OX	Oxidisable (bound to organic matter)	5 mL of 30% H2O2 (pH 1.5), shaken for 1 h at 20°C; heat lightly covered to 85°C for 1 h; added 5 mL of 30% H2O2 (pH 1.5) and reduced volume of solution at 85°C over 4-5 h; added 40 mL of 1 M ammonium acetate solution (pH 1.6), shaken for 16 h at 20°C.
Note: All pH adjustments were done with concentrated HNO3			

Table 4. Chemical reagents and analytical conditions in the modified three-step sequential extraction procedure adopted after Sahuquillo et al. (1999)

The sequential extraction method was undertaken for sediment in the 2007 sampling round. The prepared samples were then analysed for metals by using the ICP-OES with the quality control protocols as explained in Section 3.3. A more detailed description of the method is included in Magyar (2010). The following approach was adopted:

Total particulate metal = EX + OX + OM + RES

where, EX- fraction containing the Exchangeable + bound to carbonates fraction; OX- fraction bound to iron and manganese oxides; OM- fraction bound to organic matter; RES- fraction bound to the residual fraction.

The EX, OX and OM fractions can be affected by changes in environmental conditions of the water in the tanks leading to release of the metals in a soluble form. These conditions include changes in: pH, the ionic composition, oxygen levels in the tank and temperature. The metal attached to the residual fraction is not expected to be released in solution for a longer (undetermined) period of time.

The leaching tests by ASLP- Australian Standard Leaching Procedure were based on the EPA method 1311 (EPA Victoria 2007) and Australian Standards methods (AS 4439.2 1997; AS 4439.3 1997). The leaching tests were undertaken for sediment samples in the 2008 round.

2.6 Sediment re-suspension tests

A series of experiments were undertaken to investigate the effects of rain tank design and variation of water level in the tank on sediment transport. A laboratory tank (Fig. 6) was used for 16 experiments and sediment re-suspension was measured as total suspended solids in the water flowing from the tank.

The analyses for total suspended solids (TSS) were performed according to standard method 2540D (APHA/AWWA/WEF 1995) and the method detection limit was adopted as 0.5 mg/L. This high level of accuracy was obtained by using a 0.0001 g scale which was periodically calibrated and by employing additional quality assurance and control measures (Magyar 2010).

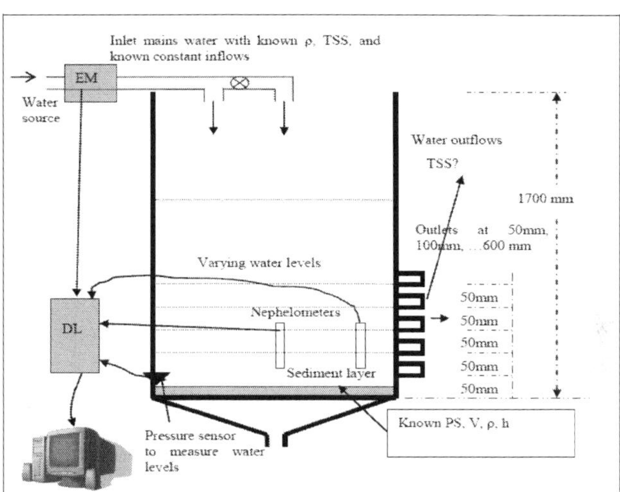

Fig. 6. Schematic of the laboratory tank

3. Results

3.1 Total and fractionated sediment

Results of total and fractionated metals concentrations are shown in Fig. 7, Fig. 8 and Fig. 9, for 2006, 2007 and 2008 sampling rounds respectively.

For these figures, total metal concentration= metal attached to particles of 0.45-63 μm and >63 μm; Fill material, Soil Category C and Soil category B are allowable concentration limits as per local EPA standards (EPA Victoria 2007); no guideline values are set for Al, Fe, Mn and total Cr.

The results show that concentrations of Cd, Cu, Ni, Pb and Zn exceeded the maximum concentration allowed in sediment to be disposed of as fill material, classifying the tank sediment as 'prescribed waste- contaminated soil category C' (EPA Victoria 2007). In some of the tanks, the high concentrations of Pb (sites S5, S6, S7, S9) and Zn (site S2) exceeded these levels, being comparable with 'prescribed waste-contaminated soil category B'. This meant that sediment from these tanks would have needed to be chemically immobilised (a process whereby the solubility, leachability, availability or reactivity of a waste and its components is reduced by chemical reaction and/or physical encapsulation in a solid matrix) (EPA Victoria 2007) prior to disposal to a licensed site for contaminated waste.

The results of the 2008 sampling round (Fig. 9) show the total metal concentration in the sediment, as in this round sediment was not fractionated. Sites S5 and S8 were not accessible for sampling, therefore no results are shown for these two tanks.

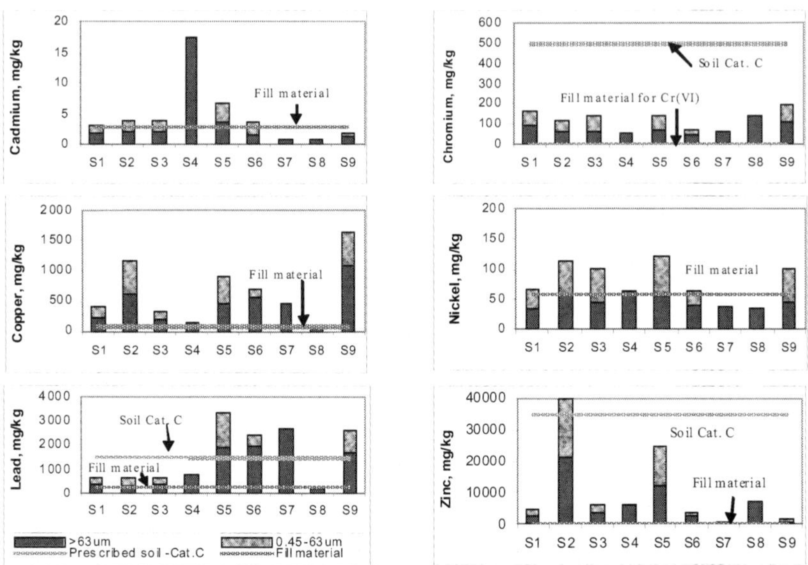

Fig. 7. Total and fractionated metal concentrations in sediment samples (2006 sampling run)

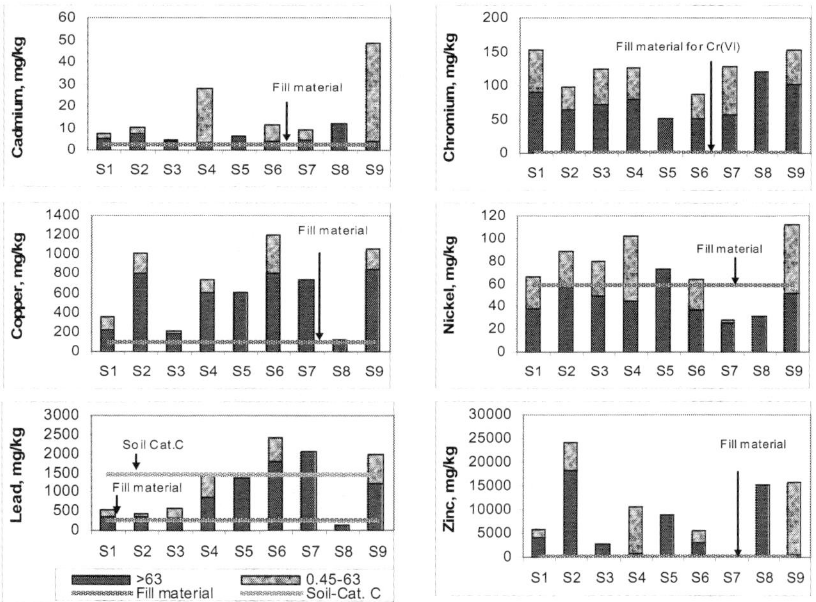

Fig. 8. Total and fractionated metal concentrations in sediment samples from the 2007 sampling run

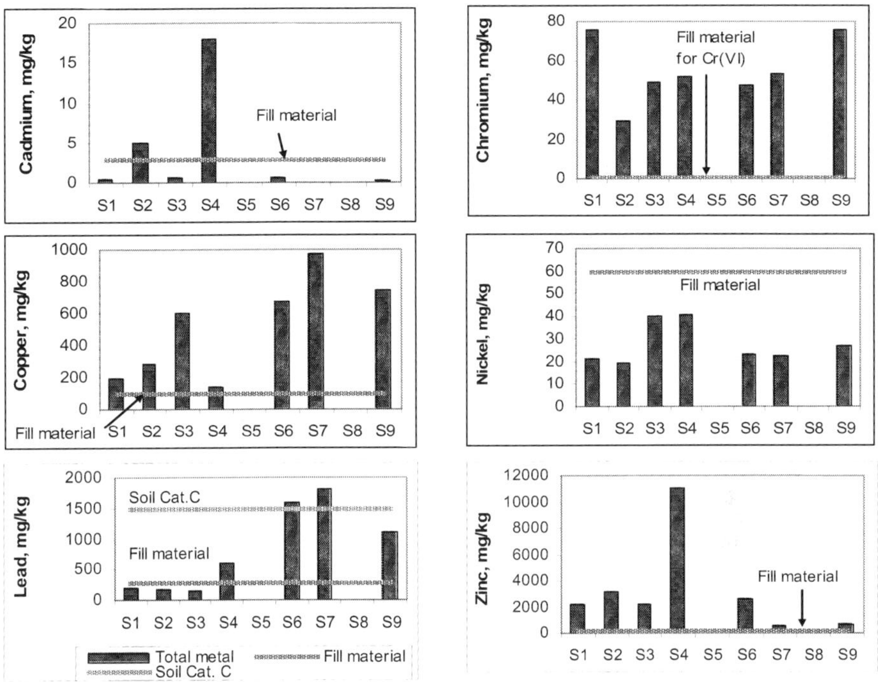

Fig. 9. Total metal concentrations in sediment samples from the 2008 sampling run

Although the samples were taken at 10 to 14 month intervals over a period of two years, the results from all three sampling rounds demonstrate that the sediment in rainwater tanks was contaminated with heavy metals. Since this sediment is not regularly removed through cleaning, it can become a source of pollution within the tanks.

The contamination from sediment can occur in two ways: by chemical interactions, influenced by changes in pH (see Section 3.4) and by ways of physical movements which include sediment re-suspension during inflows (see Section 3.5).

3.2 Sediment spatial distribution and accumulation

Sediment accumulation was measured in five of the tanks (labelled S1, S2, S3, S5 and S6). Measurements were made in four periods over 1 year. The dates and length of each period are shown in Table 5 and an example of sediment appearance is shown in Fig. 10.

Period	Dates	Number of days
1	27 Sep 2006 - 20 Dec 2006	85
2	21 Dec 2006 - 23 Mar 2007	93
3	24 Mar 2007 - 21 Jun 2007	90
4	22 Jun 2007 - 25 Sep 2007	96

Table 5. Sediment accumulation measuring periods

For each period of sediment accumulation, sediment depth was measured in the sediment traps and converted to an annual rate in mm/year. The average measurements for each tank and for each period are shown in Fig. 11, along with the overall average for each tank (summarised in Table 5). The horizontal lines in Fig. 11 represent the average annual accumulation rate for each tank based on all four periods.

Fig. 10. Example of sediment appearance in the traps and measurement in the Imhoff cones

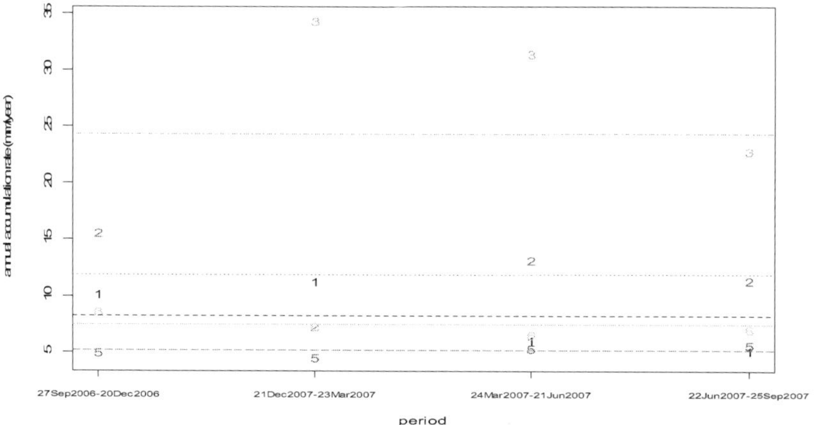

Fig. 11. Annual accumulation rate for each tank for each period (also see Table 6)

Tank	Average annual accumulation rate (mm/year)
S1	8.2
S2	11.9
S3	24.3
S5	5.2
S6	7.4
Average	11.4

Table 6. Average sediment accumulation rates

The average annual sediment accumulation rate for all samples was around 10 mm and this value was used in the laboratory tests presented in Section 3.5 for sediment thickness.

3.3 Sediment particle size distribution

The cumulative distribution of sediment particles from the nine tank is presented in Fig. 12, where some variation can be observed between the tank sediment. This is likely to be due to different local environmental conditions.

Although the PSD of sediment presented here is only related to the nine tanks investigated, it gives valuable information about sediment characteristics in rainwater tanks, especially that no other information was available. These sediment characteristics were used in the laboratory experiments.

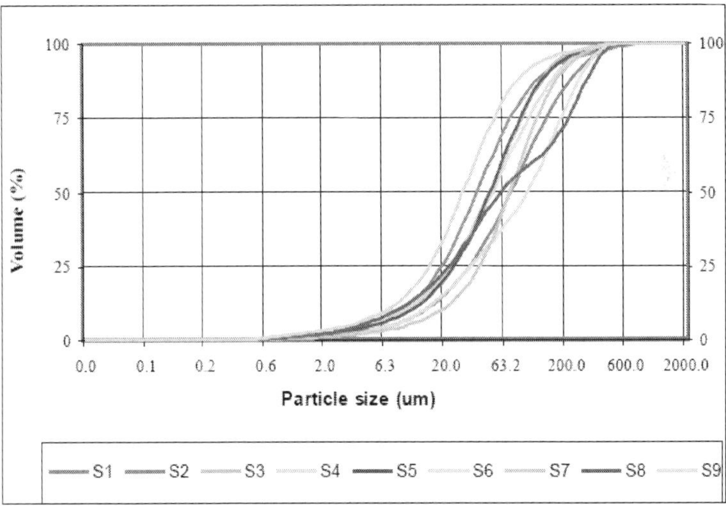

Fig. 12. Sediment particle size distribution (S1, S2,...S9 as per Table1)

3.4 Potential of metals to dissociate from sediment

The chemical characterisation of metals attached to particles can give insight into the potential for these metals to detach from the particles in certain conditions and to become a free soluble ion. The results of the mean distribution of metals in the RES, EX, OM and OX fractions in tanks sediment is included in Table 7, where RES- residual fraction; OM- organic matter fraction; OX- Fe/Mn hydroxides fraction; EX- exchangeable and attached to carbonates fraction. The analysis was based on results from six tanks (sites S1, S2, S3, S4, S6 and S9).

Results in Table 7 demonstrate that although metals were attached to particles in sediment, a significant fraction (determined by the EX, OM and OX fractions) of the total was likely to become available and released in the water column due to changes in pH, ionic composition and oxygen levels in the water column.

An example of Pb distribution along the studied fractions is shown in Fig. 13 and the distribution of all other metals can be found in Magyar (2010).

Overall, the mean values for Pb showed that Pb in the particle range >63 µm was mostly

	Fraction	Al	Cd	Cr	Cu	Fe	Mn	Ni	Pb	Zn
RES	0.45-63 µm	55%	67%	30%	22%	63%	55%	32%	43%	43%
	>63 µm	66%	42%	61%	41%	60%	41%	51%	29%	18%
EX	0.45-63 µm	2%	10%	1%	10%	1%	11%	10%	8%	24%
	>63 µm	1%	17%	1%	5%	1%	16%	6%	3%	30%
OM	0.45-63 µm	25%	7%	56%	48%	15%	14%	31%	15%	8%
	>63 µm	12%	9%	29%	24%	10%	12%	18%	17%	12%
OX	0.45-63 µm	18%	16%	13%	20%	22%	20%	28%	34%	25%
	>63 µm	21%	32%	10%	30%	30%	32%	25%	51%	41%

Table 7. Mean distribution of metals in the RES, EX, OM, OX fractions

attached to the OX fraction followed by RES, OM and EX fractions. In the particle range 0.45-63 µm, Pb was distributed in the order RES, OX, OM and EX.

As shown in Section 3.1, Pb concentration exceeded the EPA fill material recommended value in eight tanks (2007 sampling round) and Pb was predominantly found in the particle range >63 µm in all tanks.

Sequential extraction of Pb in the fraction >63 µm (Fig. 14), established that more than half of the Pb was attached to the RES fraction in tanks from sites S1 (52%) and S7 (78%), but in all other tanks there was potential for Pb to dissociate from sediment in the order: S2 (51%), S4 (53%), S3 (77%), S5 (91%), S6 (95%) and S9 (99%). Although the 0.45-63 µm particle range was found in a smaller percentage in the sediment, fractionation predicted that there was potential for Pb to dissociate from this fraction as well, between 4% (site S3) to 98% (site S9).

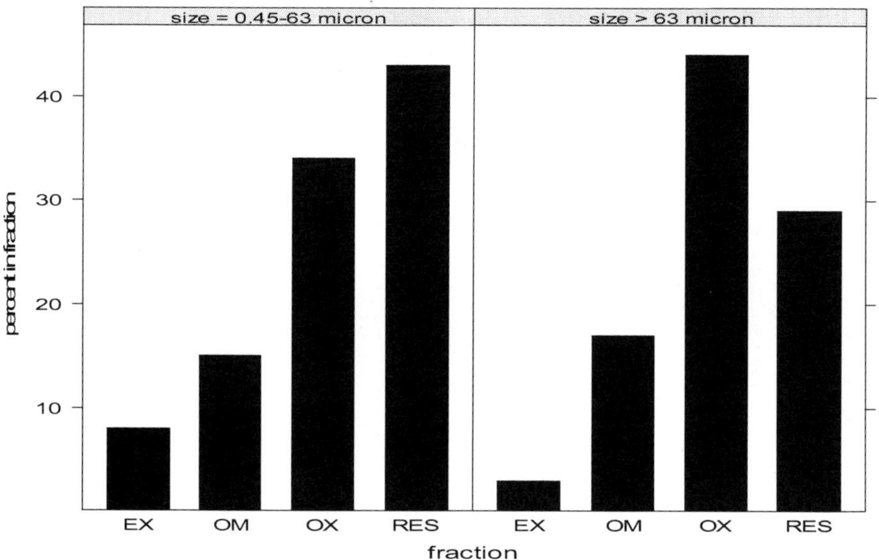

Fig. 13. Mean distribution of Pb in the RES, EX, OM, OX fractions (2007 sampling round)

The highest percentages of Pb associated with the EX fraction were found in tanks from sites S7, S6 and S9. Interestingly, these three tanks presented the highest total Pb concentration in the tank water in the same order, which now could be argued that it was due to dissociation of Pb from sediment in the tanks, as these tanks had also measured the lowest pH in the tank water (see Magyar 2010).

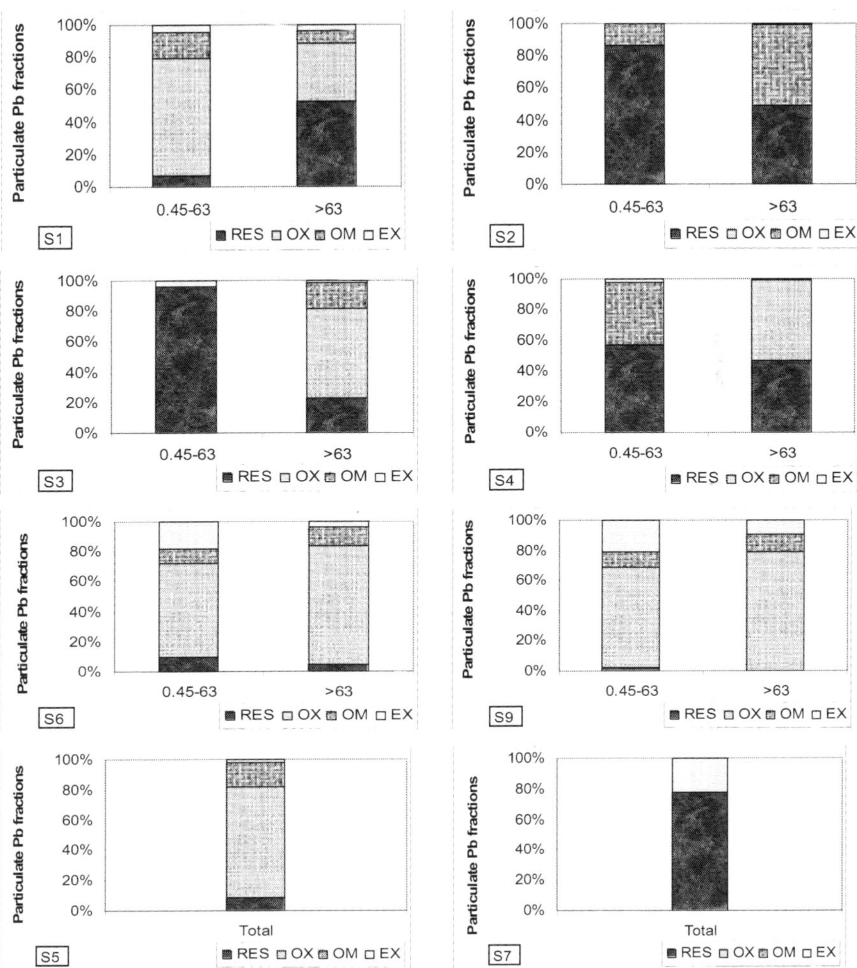

Tank from site S8 – not enough sediment to perform this analysis (new tank); Tanks from sites S5 and S7- not enough sediment for the fraction 0.45-63 μm, thus it was combined with the >63 μm fraction

Fig. 14. Fractionation of Pb in sediment

Tests for leaching by ASLP (see Section 2.5) for sediment in 2008 sampling round found that if pH would have decreased in tanks, metals from sediment would have had the potential of leaching into the water column (Table 8). Of particular concern were the high concentrations of Pb and Zn leached from sediment, which were up to 110 times above the ADWG for Pb (from S9), by up to 11 times above the ADWG for Zn (from S6). Although some dilution with water in the tank would have occurred, depending on the volume of water contained in the tank at different times, metals from sediment would have still significantly contaminated the water column.

Relevant standards		Units	Cd	Cu	Pb	Zn
Fill material		mg/kg	<3	<100	<300	<200
EPA -Category C		mg/kg	<100	<5000	<1500	<35000
EPA Category B		mg/kg	<400	<20000	<6000	<140000
EPA-leachable concentration Category C		mg/L	<0.2	<20	<1	<300
EPA-leachable concentration Category B		mg/L	<0.8	<800	<4	<1200
Tank sites	Tests performed					
S1	Total concentration	mg/kg	0.3	190	200	2200
	Leachable concentration	mg/L	<0.005	0.11	0.17	24
S2	Total concentration	mg/kg	5	280	180	3100
	Leachable concentration	mg/L	NP	NP	NP	NP
S3	Total concentration	mg/kg	0.6	600	140	2200
	Leachable concentration	mg/L	<0.005	0.24	0.07	18
S4	Total concentration	mg/kg	18	140	610	11000
	Leachable concentration	mg/L	NP	NP	NP	NP
S6	Total concentration	mg/kg	0.5	670	1600	2600
	Leachable concentration	mg/L	0.01	1.1	1.1	32
S7	Total concentration	mg/kg	<0.08	970	1800	580
	Leachable concentration	mg/L	NP	NP	NP	NPNP
S9	Total concentration	mg/kg	0.2	750	1100	660
	Leachable concentration	mg/L	<0.005	1.2	1.1	6.6

Table 8. Summary of the leaching by ASLP tests (NP- not performed as not enough sediment was available)

3.5 Sediment re-suspension

Given that the tank sediment is highly contaminated with metals, any re-suspension of that sediment will be an issue for the end use of the tank water. Re-suspension was investigated as described in Section 2.6.

A series of experiments were undertaken to determine the effect of tank, sediment and inflow characteristics on ΔTSS at the outlet. These experiments are described in Magyar (in press) and are summarised here.

The response variable was ΔTSS which is the difference in TSS value of the water as measured at the outlet (positioned 50 mm above the tank base) and the TSS of the inflowing water (mains water). Therefore ΔTSS provides a measure of the amount of sediment mobilised by inflow of water to the tank. The predictor variables were: sand size, water depth, type of inlet, and sediment depth. The 16 different experimental setups and range of predictor variables are shown in Table 9. All the results are plotted in Fig. 15, where ΔTSS is plotted on a log scale.

Several observations can be made from Figure 15. Experiments 2, 6 and 14 (symbols, 2, 6 and E) resulted in the highest sediment remobilisation. All of these had high flow rates (1 L/s) and a central inlet. The combination of a central inlet and a conical base (experiment 6) produced the highest sediment remobilisation across most water depths. This is where a high inflow rate interacted with a thick sediment layer as the conical base was filled up to be level. Low sediment remobilisation was associated with low flow rates, a side inlet and low sediment depth (experiments, 7 and 11).

Experiment	Symbol on graph	Inlet	Flow (L/s)	Sand Depth (mm)	Base	Sand size (micron)	Range of water depths (mm)
1	1	Central	0.5	10	Flat	63-106	50-1000
2	2	Central	1	10	Flat	63-106	50-1000
3	3	Side	0.5	10	Flat	63-106	50-1000
4	4	Side	1	10	Flat	63-106	50-1000
5	5	Central	0.5	10	Conical	63-106	50-1000
6	6	Central	1	10	Conical	63-106	50-1000
7	7	Side	0.5	10	Conical	63-106	50-1000
8	8	Side	1	10	Conical	63-106	50-1000
9	9	Central	0.5	10	Conical	106-129	50-1000
10	A	Central	1	10	Flat	106-129	50-600
11	B	Side	0.5	10	Flat	106-129	50-600
12	C	Side	1	10	Flat	106-129	50-600
13	D	Central	0.5	20	Flat	106-129	50-600
14	E	Central	1	20	Flat	106-129	50-600
15	F	Side	0.5	20	Flat	106-129	50-600
16	G	Side	1	20	Flat	106-129	50-600

Table 9. Range of variables used in each of the 16 experiments

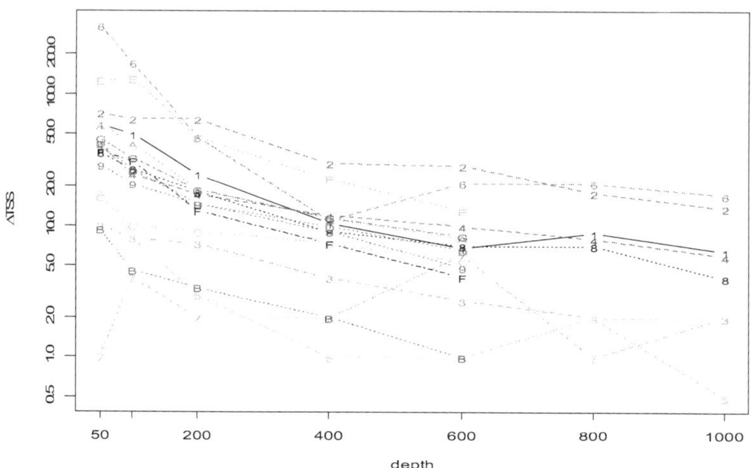

Fig. 15. Results of sediment re-suspension from the 16 experiments (symbols are defined in Table 9)

3.6 Modelling sediment mobilisation

A linear modelling approach was used to explore and quantify the dependence of ΔTSS on experimental variables. Following a Box Cox analysis, ΔTSS was log transformed to bring it closer to normality. Indicator variables were used to characterise base type, sediment size, sediment depth, and flow rate (Table 10). This approach was appropriate because only two values were used for each of these variables during the experiments.

Variable	Indicator variable = 0	Indicator variable = 1
Inlet	Side	Central
Inflow rate	0.5 L/s	1 L/s
Sediment depth	10 mm	20 mm
Base type	Conical	Flat
Sediment size	106-129 micron	63-106 micron

Table 10. Specification of dummy variables

The model was of the form:

$$\log(\Delta TSS) = \beta_0 + \beta_1(\text{water level}) + \beta_2(\text{i\_inlet}) + \beta_3(\text{d\_inflow rate}) + \\ + \beta_4(\text{i\_sediment depth}) + \beta_5(\text{i\_base}) + \beta_6(\text{i\_sediment size}) \quad (1)$$

Where, i = 'indicator' variable as defined in Table 11 and Log = natural log.

Model coefficients and standard errors are shown in Table 11 and Fig. 16. All coefficients were significant ($p < 0.0003$). Fitted versus measured values in Fig. 17, along with the lowess fit suggest the model is a reasonable fit to the data. This was also confirmed when examining the standard diagnostics: residuals versus fitted values, QQ plot of residuals, scale-location plot and Cooks distance plot.

Variable	Coefficient	Std. Error
(Intercept)	1.12	0.24
Water level	-0.0022	0.00022
i_inlet	0.88	0.13
i_flow	1.26	0.13
i_sediment depth	0.81	0.20
i_base type	0.65	0.17
i_sediment size	0.76	0.19
Residual standard error 0.64 on 89 degrees of freedom		
Adjusted R-squared: 0.73	F-statistic: 45.45 on 6 and 89 DF, p-value: < 2.2e-16	

Table 11. Coefficients, standard errors and regression diagnostics

Subsets of predictor variables were explored, starting with just the water level and adding each of the variables in equation 1 in turn. For each of these subsets the adjusted R squared, Akaike's Information Criterion (AIC), the corrected AIC and Bayesian Information Criterion (Sheather 2010) were calculated. All these criteria confirmed that the complete model (equation 1) was appropriate.

The implications of the sediment mobilisation model are shown in Fig. 18. Equation 1 is thought of representing a family of curves where each specific curve depends on the value of the indicator variables and the water depth. A high value of ΔTSS will occur where there is shallow water depth, central inlet, high flow rate, high sediment depth, flat base and fine sediment. Conversely, low sediment mobilisation will occur at higher water levels, side inlet, low inflow rate, little sediment on the base and when the sediment is coarse. The two extreme curves, those that represent the greatest and least sediment mobilisation are shown in Fig. 18. Sediment mobilisation is approximately 80 times higher under high sediment mobilisation conditions compared with low sediment mobilisation conditions.

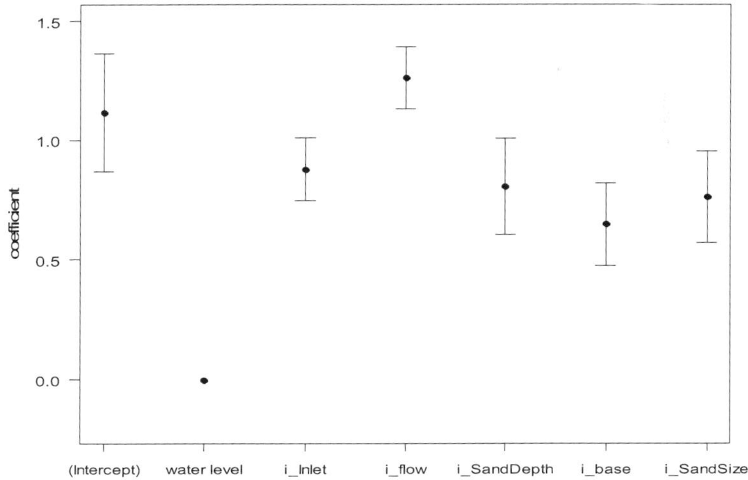

Fig. 16. Model coefficients and standard errors. Note: the standard error for the water depth coefficient is too small to see at this scale.

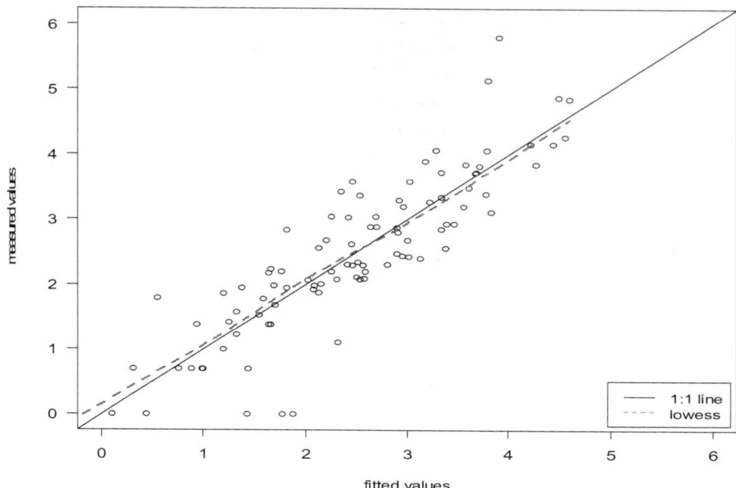

Fig. 17. Fitted versus measured values

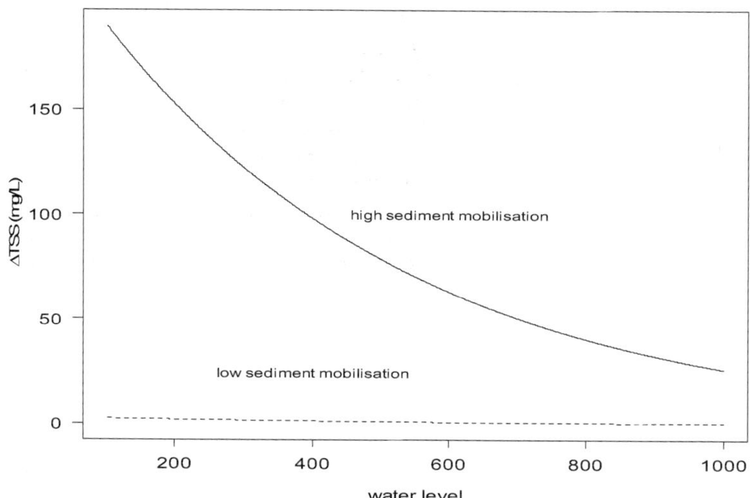

Fig. 18. Sediment mobilisation as predicted by equation 1.

4. Discussion

Nine urban rainwater tanks in Melbourne were investigated for water and sediment quality over a two year period. The tank water was often found with high concentrations of metals and field and laboratory tests demonstrated that the contaminated tank sediment was potentially a significant source of pollution.

The particle size distribution of tank sediment varied between the tanks and was attributed to different local environmental conditions for each tank. The sediment layer in five of the tanks was not evenly distributed over the tanks' base, but no clear trend was observed to determine which areas act as accumulation and which as transportation areas.

Sediment from the nine rainwater tanks was highly contaminated with heavy metals, at levels comparable with the Victorian EPA regulation of prescribed contaminated soil which would require special disposal to a contaminated licensed site and sometimes chemical treatment prior to this disposal. Based on this classification, none of the tanks sediment was suitable for disposal in landfill due to potential for metals leaching into the ground water. At the same time, since sediment is not commonly removed from the tanks, it can potentially contaminate the water column. This contamination from sediment can occur by chemical dissociations which are influenced by changes in pH and temperature in the tank and by physical sediment re-suspension during inflows to the tank.

The sequential extraction tests revealed that a significant percentage of the total particulate metal can be associated with fractions (EX, OX and OM) that are relatively weakly bound to the particle and that can break with changes in pH, oxygen level, temperature and ionic composition in the tanks water. This suggests that analyses for the total metal concentrations of the sediments may not appropriately determine sediments' potential to contaminate the tank water. The leaching tests confirmed this finding as well.

Part of the total metal in sediment remained attached to the residual fraction (RES) regardless of changes in environmental conditions within the tank. This information can help with determining the efficiency of post tank treatment (e.g. filtration). While a filter installed as a post tank treatment will remove certain size particles, it may only remove the metals attached to the residual fraction, as all other fractions can still dissociate if pH of water flowing through the filter is acidic.

The laboratory tests determined that regardless of the tank configuration, sediment at the base of the tank re-suspends during inflows and contaminates the out-flowing water. A top side inlet located opposite the bottom outlet led to the lowest TSS in outlet and thicker sediment led to higher contamination at outlet. For low flows, a conical base tank could reduce sediment re-suspension by up to 60% when compared with a flat base tank, but re-suspension could occur during intense rain events (Magyar, et al. in press).

Metal contamination in the water column of urban rainwater tanks is often reported: Coombes et al., 2000, Simmons et al., 2001, Spinks et al., 2003, but there has been limited success in determining the source of pollution or in explaining the variation in quality. We have undertaken work in this area, but so far results are inconclusive, with Pb contamination being attributed to several sources (Magyar et al. 2008; O'Connor et al. 2009). We have also confirmed that metal contamination of tank water and sediment is widespread, as details from these nine tanks were confirmed with a survey of 54 tanks in Melbourne (Magyar et al., 2008). High concentrations of metals in water can have a health impact if water is used for uses that involve human contact (e.g. showering), or ingestion.

5. Conclusion

Our research suggests that metal contamination of tank water and sediment is common in urban areas. Prior to the set of investigations undertaken by the authors and presented in this paper, sediment from urban rainwater tanks did not capture much research attention as it was considered an insignificant source of tanks' water contamination. However, the results presented above establish that contaminated sediment in rainwater tanks can become a source of pollution to the water in the tanks due to physical transport (re-suspension) and chemical interactions (leaching).

We found that pH in the tanks was usually acid and that metals in sediment were commonly attached to fractions that can be easily broken by changes in pH and temperature. Experimental results showed that sediment re-suspension was common under conditions that usually occur in urban rain tanks.

Currently there are no management protocols for sediment from rainwater tanks from individual urban households, but this current work demonstrates that they should be considered for future. The management protocol could include not only guidelines advising acceptable metal concentrations in tank sediment, but also disposal methods and maintenance periods.

6. Acknowledgement

The authors would like to thank Monash University and CSIRO Water for a Healthy Country for their financial support.

7. References

Anzecc and Armcanz (2000) Australian and New Zealand guidelines for fresh and marine water quality. Paper No. 4, Volume 1, Australian and New Zealand Environment and Conservation Council (ANZECC) and Agriculture and Resource Management Council of Australia and New Zealand (ARMCANZ)

APHA/AWWA/WEF, Ed. (1995) Standard Methods for the Examination of Water and Wastewater. Washington DC, USA, American Public Health Association/American Water Works Association/Water Environment Federation.

AS 4439.2 (1997). Wastes, sediments and contaminated soils. Part 2: Preparation of leachates-Zero headspace procedure. Standards Australia.

Bodek, I., Lyman, W. J., Reehl, W. F. and Rosenblatt, D. H., Eds. (1988) Environmental Inorganic Chemistry. Properties, Processes and Estimation Methods, Pergamon Press. A special publication by SETAC

Coombes, P. J., Argue, J. R. and Kuczera, G. (2000) "Figtree Place: A case study in Water Sensitive Urban Development (WSUD)." Urban Water 1 (4): 335-343.

Coombes, P. J., Kuczera, G. and Kalma, J. D. (2002). Economic, water quantity and quality impacts from the use of a rainwater tank in the inner city. 27th Hydrology and Water Resources Conference, Melbourne, Australia, Engineers Australia.

Dean, C. M., Sansalone, J. J., Cartledge, F. K. and Pardue, J. H. (2005) "Influence of hydrology on rainfall-runoff metal element speciation." Journal of Environmental Engineering 131 (131): 632-642.

EPA Victoria (2007). EPA Publication 448.3. Classification of Wastes. EPA Victoria.

Evans, C. A., Coombes, P. J. and Dunstan, R. H. (2006) "Wind, Rain and Bacteria: The Effect of Weather on the Microbial Composition of Roof-Harvested Rainwater." Water Research 40 (1): 37-44.

Forestry Suppliers Inc. (2005) "Conbar tesecopic dipper - Catalog 59." Retrieved May 2005, from http://www.forestry-suppliers.com/product_pages/View_Catalog_Page.asp?mi=6570.

Magyar, M. I. (2010). Investigating factors affecting outlet water quality from residential rainwater tanks in Australia. PhD, Civil Engineering, Monash University, Melbourne. 377pp.

Magyar, M. I., Ladson, A. R., Mitchell, V. G. and Diaper, C. (in press) "The effect of rainwater tank design on sediment re-suspension and subsequent outlet water quality." Australian Journal of Water Resources.

Magyar, M. I., Mitchell, V. G., Ladson, A. and Diaper, C. (2006). Investigating how tank configuration of residential rainwater tanks affects outlet water quality. 30th Hydrology & Water Resources Symposium Launceston, Tasmania, 4-7 December.

Magyar, M. I., Mitchell, V. G., Ladson, A. R. and Diaper, C. (2007) "An investigation of rainwater tanks quality and sediment dynamics." Water Science and Technology 56 (9): 21-28.

Magyar, M. I., Mitchell, V. G., Ladson, A. R. and Diaper, C. (2008) Lead and other heavy metals: common contaminants of rainwater tanks in Melbourne. Water Down Under 2008 Conference, incorporating the Hydrology and Water Resources Symposium and the 4th International Conference on Water Resources and Environmental Research, Adelaide, Australia.

Margui, E., Salvado, V., Queralt, I. and Hidalgo, M. (2004) "Comparison of three stage sequential extraction and toxicity characteristic leaching tests to evaluate metal mobility in mining wastes " Analytica Chimica Acta 524: 151-159.

Murdoch, A. and MacKnight, S. D. (1994) Handbook of Techniques for Aquatic Sediments Sampling, Lewis Publishers. 236.

NHMRC & NRMMC (2004). Australian Drinking Water Guidelines. National Health and Medical Research Council and Natural Resource Management Ministerial Council. Australian Government, Canberra.

O'Connor, J. B., Mitchell, V. G., Magyar, M. I., Ladson, A. R. and Diaper, C. (2009). Where is the lead in rainwater tanks coming from? . H2009, 32nd Hydrology and Water Resources Symposium, Newcastle, Australia, Engineers Australia. 30 November- 3 December.

Sabina, L., Kus, B., Shon, H.-K. and Kandasamy, J. (2008) "Membrane fouling propensity after adsorption as pretreatment in rainwater: a detailed organic characterisation." Water Science and Technology 58 (8): 1535-1539.

Sahuquillo, A., Lopez-Sanchez, J. F., Rubio, R., Rauret, G., Thomas, R. P., Davidson, C. M. and Ure, A. M. (1999) "Use of certified reference material for extractable trace metals to assess sources of uncertainty in the BCR three-stage sequential extraction procedure " Analytica Chimica Acta 382: 317-327.

Sheather (2010) A modern approach to regression. New York, R. Springer.

Simmons, G., Hope, V., Lewis, G., Whitmore, J. and Gao, W. (2001) "Contamination of potable roof-collected rainwater in Auckland, New Zealand." Water Research 35 (6): 1518-1524.

Siriwardene, N., Duncan, H., Mitchell, V. G., Taylor, G. and Magyar, M. I. (2008) Rainfall water quality sampling in Melbourne, eWater CRC, E1 project - Technical Note

Spinks, A. T., Coombes, P., Dunstan, R. H. and Kuczera, G. (2003) Water quality treatment processes in domestic rainwater harvesting systems. 28th International Hydrology and Water resources Symposium, 10-14 November 2003, Wollongong, Australia, Institute of Engineers, Australia.

Tchobanoglous, G. and Burton, F. L., Eds. (1991) Wastewater Engineering. Treatment, Disposal and Reuse. Third Edition, Civil Engineering Series., McGraw-Hill, Inc.
Tessier, A., Campbell, P. G. C. and Bisson, M. (1979) "Sequential extraction procedure for the speciation of particulate trace metals." Analytical Chemistry 51 (7): 844-851.
Vaes, G. (1995) Design of storage sedimentation basins. NOVATECH'95, Lyon, France.
Who (2004) Guidelines for Drinking-water Quality, Volume 1: Recommendations, World Health Organization, Geneva, 2004. 515.

5

Fine Sediment Deposition at Forest Road Crossings: An Overview and Effective Monitoring Protocol

John F. Rex and Ellen L. Petticrew
University of Northern British Columbia
Canada

1. Introduction

Fine sediment (< 2mm) is an integral component of naturally functioning streams but can become a pollutant when development activities increase stream concentrations beyond that of the natural regime or when the sediments carry contaminants. Watershed development activities including urbanization, agriculture, and forestry can influence fine sediment quantity, quality, and its transport and storage regime by altering the natural timing and volume of water and sediment delivered to the stream channel. These development activities increase water and fine sediment delivery to streams by removing vegetation cover, disturbing soils, and connecting these disturbed areas to streams through roads, ditch lines, and/or simplified ground surfaces that enhance runoff (Bilby *et al.*, 1989; Corner *et al.*, 1996; Keutzweiser and Capell, 2001). Although development activities can affect the transport of larger particle sizes (e.g. gravels and cobbles), fine sediments from sand to clay are emphasized here as they have a significant impact on instream biota but also because they can impair the effectiveness of potable water supply treatment and increase its costs (Gadgil, 1998). In addition, fine sediments are a transport vector for hydrophobic contaminants (Bábek *et al.*, 2008; Taylor and Owens, 2009).

This chapter provides 1) an overview of the effect that forestry generated fine sediment has on receiving stream biota and 2) an effective protocol for measuring fine sediment levels at forest road stream crossings. The routing and downstream accumulation of sediment from these stream crossing point sources is a management concern because it will affect stream biota and streambed composition at each of its temporary in-stream storage areas. This sedimentary cumulative watershed effect (CWE) is one of the most detrimental consequences of forest harvesting activities on a watershed. However, the CWE is difficult to assess because its effect is dependent upon the grain size being introduced, the number and size of streams that transport it, and the original sedimentary state of the streambed it encounters (Bunte and MacDonald, 1999). Further, while it may be possible to determine the change in fine sediment levels at a single point in the stream, it is difficult to determine which upstream land use activity instigated the change. Bunte and MacDonald (1999) suggest that to manage a watershed for cumulative sediment effects it is necessary to monitor for a minimum of 5 - 10 years pre-and-post harvesting because sediment transport is highly variable. However, this type of program is often considered cost prohibitive and of a longer reporting timeline than that required by many resource management programs.

The sampling protocol presented here can bridge the gap between a long-term study and the need for immediate information to address management needs. Application of the sampling protocol presented here will make it possible to designate sediment load increases to an identified activity on a site-specific basis. That is, by monitoring a representative number of sites for a specific type or set of disturbances (e.g. stream crossings or riparian harvest areas) the sediment contribution from those disturbances can be estimated within an affected watershed and modeled. Further, these data can be used to justify the modification of those practices found to contribute significant amounts of fine sediment. Although developed for the quantification of fine sediments generated by forest road use and crossing construction and/or maintenance, this protocol can be used at other types of stream crossings as well as for the collection of fine sediments for contaminant assessment.

The first section of this chapter provides an overview of forest road effects on stream sediment transport and storage regimes as well as its associated biological effects. Section 2 outlines the monitoring protocol, suggested here to be an effective tool for monitoring fine sediment deposition at forest road crossings. Fine sediment is collected using three fish habitat sampling techniques, namely the McNeil corer, gravel bucket, and infiltration bag. These techniques are not quantitatively compared but instead their effectiveness in a sampling protocol is verified through a field assessment using eight case studies from the central interior of British Columbia, Canada (Section 3). The number of samples required for each technique to estimate the magnitude of the road crossing disturbance was determined using field data as well as standard statistical formulas. The resultant protocol includes an outline of procedures that can be adapted to address objectives for a range of sampling programs from trend monitoring to impact assessment studies. Factors to consider before applying these procedures to other development activities and areas are summarized in section 4.

1.1 Forest roads and the sediment load

Forestry, the focus of this chapter, influences water and sediment delivery to streams through tree and understory removal, silviculture activities, as well as road construction, maintenance, and decommissioning. Vegetation removal decreases evapotranspiration, which can increase delivery of precipitation to the stream and thereby flow levels. Clearing vegetation can disturb soils, enhance soil erosion, and increase fine sediment delivery to streams if clearing locations are connected to streams and riparian buffers are absent (Corner *et al.*, 1996; Kreutzweiser and Capell, 2001; Litschert and MacDonald, 2009). Forest roads are also associated with direct stream contribution of sediment due to road surface erosion and sediment delivery at stream crossings (Bilby *et al.*, 1989; Fransen *et al.*, 2001; Lane and Sheridan, 2002) as well as increasing the frequency of landslides (Reid and Dunne, 1984; Fransen *et al.*, 2001; Wemple *et al.*, 2001). Forest road generated sediment can be transported along the road's surface, in rivulets, or its ditches, where it can be delivered at stream crossings. Stream crossings were selected for study in this project because of their consistent occurrence in the literature and the ease in designating them as a point source for increased sediment levels downstream.

Roads can contribute fine sediment to streams during construction, use and maintenance as well as decommissioning. A range of literature from around the world provides insight into the magnitude of the disturbance. For example, Burns (1970) indicated that sediment loads in a harvested California basin were greatest during the road construction period although they were sustained for several years with continued harvesting. Beschta (1978) identified a 150% increase in the sediment load after road construction while Bilby *et al.*, (1989) identified that 680 of 2000 road drainage points directly contributed sediment to streams, and that most of those sites were found on first or second-order channels. Sediment loads can be highest during road construction and can contribute as much sediment to streams as landslides (Cederholm *et al.*, 1981). Close to 80% of the sediment eroded from tropical forestry landings and road surfaces was delivered to stream channels within 16 months of construction (Sidle *et al.*, 2004). Sediment generation can be highest the year after road construction (Megahan *et al.*, 2001) but roads will continue to deliver sediment throughout their active usage period. Reid and Dunne (1984) found that a highly active road contributed 130 times more sediment than lower usage roads. Road maintenance, including grading and ditchline vegetation removal, can increase soil erosion potential. Luce and Black (1999) identified a seven-fold increase in erosion from recently de-vegetated ditches compared to those where vegetation was left intact. Decommissioning of roads reduces the contribution of fine sediment to streams and the amount of fine sediment in streambeds downstream of previous stream crossings (McCaffrey *et al.*, 2007).

Once fine sediment enters the stream at road crossings it contributes to the sediment load, which refers to the amount of sediment passing one point within a stream over a given time period (Leopold, 1997). The sediment load is not equivalent to the rate of upstream erosion. Depending upon watershed characteristics including size, geology, and terrestrial-aquatic connectivity, 75% or more of the sediment eroded in a watershed can be stored in transitional areas such as the base of hillslopes, floodplains, and pools (Leopold, 1997). The sediment load generally consists of two components, namely suspended sediment that moves downstream through the water column as well as bedload sediment that moves downstream through streambed migration or saltation along the surface of the streambed (Leopold, 1997). Although these sediment forms are separated for ease of classification, there is transition between forms in streams as a response to hydrologic conditions. Suspended sediments can settle to the bedload when their settling rate exceeds the force of flow and bedload can be suspended when shear stress is sufficient to lift them off the bottom and they can stay suspended if turbulent flow is sufficient (Knighton, 1998). Accordingly, both sediment forms should be discussed when reviewing the effect of fine sediment contribution from forest roads to streams.

1.2 Biotic effects of increased fine sediment

Fine sediment delivered to streams will have an effect on all stream trophic levels, the scale of which is dependent upon the amount of sediment delivered and its retention period. Primary producers are influenced by fine sediments in suspension as well as settled fines that blanket the streambed surface. Increased water column turbidity or the blanketing action of settled fines will reduce light penetration and productivity. Benthic macroinvertebrates will also be affected by suspended and settled sediment. Invertebrate populations and community structure will respond to increases in suspended and streambed fine sediment concentrations. Fine suspended sediments can abrade invertebrates initiating a downstream migration, while increased streambed fines can decrease intergravel oxygen levels prompting their migration. Fish can be similarly affected by increased fine sediment concentration in streams and the streambed. Increased turbidity can reduce predatory abilities and physically damage the fish, for example by gill abrasion.

An increase in streambed fine sediment composition can reduce intergravel oxygen levels and negatively affect incubating eggs and young residing in the gravels.

1.2.1 Primary producers

Aquatic primary producers range in size from the easily visible macrophytes such as Canadian pond weed, *Elodea canadensis,* to the microscopic periphyton such as the diatom *Navicula*. While macrophytes often adhere to the streambed via roots, periphyton may attach to rocks, sand, or plants with gelatinous stalks (South and Whittick, 1987). Regardless of their size, all aquatic plants can be affected by increases in fluvial sediment loads. Sediment can affect plants by reducing light penetration via light reflection and absorption in the water column or by settling atop benthic forms. The decrease in ambient light as turbidity increases can decrease algal biomass and productivity unless the community is able to adjust efficiency to compensate for lower irradiance (Parkhill and Gulliver, 2002). Sediment can also physically damage plants through direct contact and if it deposits in high concentrations it can prevent attachment or may smother them (Newcombe and MacDonald, 1991; Waters, 1995; Wood and Armitage, 1997).

Davies-Colley *et al.* (1992) identified that clay additions from placer mining operations reduced the photosynthetic active radiation (PAR) depth in streams, which in turn reduced periphyton productivity. Further, they found that periphyton biomass decreased upon exposure to placer runoff and that remaining biomass had a high clay content, which made it a poor food source for stream invertebrates. King and Ball (1967) noted that road construction activities and sediment additions resulted in a 68% decrease in the stream's periphyton community. Brookes (1988) found that stands of the macrophyte *Ranunculus sp.* were smothered downstream of a channelization project during low flow because these species could not alter their rooting depth.

1.2.2 Benthic macroinvertebrates

Benthic macroinvertebrates form the next trophic levels above primary producers and their functional feeding groups range from the herbivorous scrapers to the carnivorous piercers (Peckarsky *et al.*, 1990). While herbivorous invertebrates may be negatively affected by the reduced food quality of clay laden periphyton (Suren, 2005), the following discussion focuses on direct effects experienced by all invertebrates. These include the alteration of substrate composition, instigation of drift due to sediment deposition or saltation, decreased respiratory rates due to sediment depositing on respiratory structures, feeding behaviour alterations, and direct mortality of immobile life stages (Rutherford and MacKay 1986; Wood and Armitage, 1997; Gibbins *et al.*, 2007).

A stream's benthic invertebrate community structure and density is strongly associated with the streambed substrate. Initially, it was believed that invertebrate diversity increased with increasing substrate size. This has been shown to be only true for the surface dwelling Ephemeroptera, Plecoptera, and Trichoptera (EPT) groups (Waters, 1995). Invertebrate community structure is positively affected by increased concentrations of stream detritus, which can increase oxygen exchange and act as a food source (Culp *et al.*, 1983). So, attempts to define community structure must consider streambed substrate composition as well as hydrology. Fine sediment deposition on the streambed can clog interstitial streambed spaces, which may reduce interstitial oxygen levels. Further, it can restrict the size of depositing detritus (Culp *et al.*, 1986). This alteration of the benthic environment can induce an escape or drift response from those organisms unable to cope with the change. Bedload movements and saltating sediments can also increase drift upon contact with surface dwelling invertebrates (Quinn *et al.*, 1992; Gibbins *et al.*, 2007). Culp *et al.* (1986) noted that during their controlled addition of sands to a surveyed stream channel, the invertebrate population was reduced by more than 50% within 24 hours of sand exposure as a result of catastrophic drift. Similarly, Larsen and Ormerod (2010) found that low-level and short term increases in particles less than 2 mm initiated drift and Gomi *et al.*, (2010) suggest that sedimentation of particles greater than 4 mm increases the rate of invertebrate drift.

Invertebrate feeding behaviour alteration and direct mortalities can also occur if sediment concentrations are sufficiently high. Filter feeders will not be able to effectively capture prey items in high concentrations of sediment (Waters, 1995). Immobile life stages such as pupae obviously cannot drift yet they require flowing water for oxygen exchange, so where sediment deposits are thick, exposed pupae may suffocate (Rutherford and MacKay, 1986).

Although invertebrate communities can be affected in several manners, it is important to recognize that exposure duration is equally important as the concentration of fine sediment (Rosenberg and Wiens, 1978). Most of the aforementioned studies determined that community structure and density often returned to pre-disturbance levels once the sediment wave had passed through the sample area. So, extreme but temporally short events such as a road washout may be less damaging than chronic sediment sources that are not as visibly extreme such as increased erosion from riparian or fire areas (Minshall *et al.*, 2001).

1.2.3 Fish

The majority of published studies have focused on sediment effects on salmonids because these fish are sensitive to increases in suspended fine sediment and sedimentation. Generally, fish can be affected at the behavioural and physiological levels (Waters, 1995). Behavioural responses are the first observable reactions to increased sediment and are also the most transitory. They are often a response to increased suspended sediments and include avoidance and increased cough frequency (Anderson *et al.*, 1996). Physiological responses are dependent upon life stage and the type of sediment encountered, suspended or depositing. As this chapter focuses on sedimentation those effects are discussed here.

Excess sedimentation, can affect fish populations by reducing habitat and directly affect individuals through increased egg mortalities and reduced fry emergence. Habitat alteration through increased sedimentation can result in a reduction of fish food resource and over-wintering sites due to in-filling of pools, as well as the alteration of spawning gravels (Waters, 1995; Anderson *et al.*, 1996; Wood and Armitage, 1997;). Further, increased bedload transport may result in deep scour or fill which can remove or bury eggs and fry (Montgomery *et al.*, 1996).

Scrivner and Brownlee (1989) documented a 50% decrease in coho (*Oncorhynchus kisutch*) and chum salmon (*O. keta*) populations of Carnation Creek, British Columbia, Canada following forest harvesting. They attribute this to high levels of fine gravel and sand transport resulting from increased streambank erosion and removal of large organic debris dams. Specifically, they noted that sands formed an impermeable layer within the streambed at varying depths depending upon previous storm flows. They postulated that these layers of sand isolated salmon redds and prevented efficient oxygen exchange or fry emergence.

Excess sedimentation has consistently been shown to affect fish communities but the biologically active grain size varies between studies. McNeil and Ahnell (1964) determined that spawning success of pink salmon (*O. gorbuscha)* was inversely proportional to streambed permeability and the concentration of sediment smaller than coarse to medium sands (less than 0.833 mm). Others have reported similar findings but focussed on grain sizes ranging between 0.25 and 6.4 mm (Chapman 1988; Reiser and White 1988; Lisle, 1989; Platts *et al.*, 1989).

2. Fine sediment monitoring protocol

Fine sediment concentration in the streambed is not commonly monitored possibly due to the lack of a standard sampling protocol and/or potentially onerous bulk sample volumes required to describe the substrate (Church *et al.*, 1987; Zimmerman *et al.*, 2005). The goal of this project was to develop a straightforward field-verified monitoring protocol for assessing road generated fine sediment input and corresponding streambed change that can be used in remote regions. Study areas near stream crossings are selected, which depending on the sampling design have one or more study sites. Within each study site multiple sample locations are identified. Prior to collecting sediment samples, study sites must be

Fig. 1. Laboratory photo of the McNeil corer, gravel bucket, and infiltration bag sampler.

described using site establishment procedures to verify site similarity for their future comparison. Site establishment includes the measurement of stream width and slope, streambed surface conditions using pebble count, discharge, and shear stress at each sampling location. The sampling design will vary depending on the program objectives but can include the before-after-control-impact design (BACI). A short discussion on pseudoreplication issues is also provided to ensure statistically valid results are generated. The sampling techniques used to assess sedimentation were the McNeil core which collects bulk samples to a depth of 30 cm in the streambed, the gravel bucket which captures settling sediment, and the infiltration bag which captures settling sediment and fine sediment moving through the streambed (Figure 1).

2.1 Site establishment

To assess fine sediment contributions from forest road crossings, study sites were established within the same stream reach to reduce environmental variability. A reach was defined here as two repeating units where a unit is composed of two habitat features such as riffle and run, or pool and riffle. Site establishment data should be collected at all sites and include measurements of active and bankfull channel width, discharge, mean depth, habitat units, gradient, pebble count, and sampler placement depth and overlying velocity at the time of sampling.

The active channel width of a stream is the horizontal distance over the stream channel between stream banks that is covered by water (Fig. 2a). Bankfull width is the channel width where water would just begin to spill into the active floodplain (Fig. 2b, Platts *et al.*, 1983). Bankfull indicators include changes in streamside vegetation, slope, bank material, undercuts and stain lines (Harrelson *et al.*, 1994). Measurements of these two parameters should be collected at a minimum of five points along the sample reach.

a) b)

Fig. 2. Identifying the a) active channel width and b) bankfull channel width.

Discharge data was collected at each site using the midsection velocity-area method which measures velocity for 40 seconds at 10-20 evenly spaced locations along a relatively flat channel cross section selected downstream from the sample area. Care was taken to ensure there were no obstructions to interfere with flow measurement. When water depths were less than 1 m, a single velocity reading was taken at 60% of the depth but when depth exceeded 1 m, readings were taken at 20% and 80% (Mosley and McKerchar, 1993). Discharge was calculated for each section and then summed for the channel. The mean stream depth value can be determined from the average of the depths collected during the velocity readings.

The study area was sketched in field notes with specific attention to fish habitat features (e.g. pool, riffle, woody debris, dams, etc.) and streambed sample locations. The sketch is recommended because it may be required at a later date to determine if the study area consisted of one or more reaches and to evaluate the similarity of sample replicate sample locations between and within sites. Channel gradient was measured with a clinometer but an Abney level would also be appropriate. To measure gradient, field staff positioned themselves at a distance greater than the channel width apart along the stream's edge. The clinometer was sighted from one staff to the other at the same distance from the ground. A minimum of five measurements were collected and then averaged to determine the mean channel gradient.

The streambed surface sediment was characterized using a pebble count (Leopold *et al.*, 1992). Counts were conducted at several cross-sections in the sample reach to ensure that representative portions of each habitat unit were sampled. Starting at bankfull elevation, the sampler blindly reached to their left or right foot. The first particle that was touched was removed and the intermediate axis, or width, of the particle was measured and recorded by the second staff member on a tally sheet that was divided into grain classes defined by the Wentworth Scale. The sampler then moved a standard step distance and selected another pebble at the top of the same foot used in selecting the first pebble. This continued until a minimum of 100 pebbles was counted. The pebble count procedure is widely used to assess surface sediment composition by regulatory agencies and researchers alike (Bevenger and King, 1995; Bunte *et al.*, 2009)

Overlying water depth and velocity was measured at each sample location for each technique applied. Flow measurements associated with infiltration bag and gravel bucket locations should be collected when the sampler is installed in the streambed and then again before it is removed. This information can be used to assess shear stress conditions at each sample location and to ensure that sample replicate sites are comparable within and between sites both during and over the sampling interval.

2.2 Bulk sampling: McNeil corer

Since its development in 1964, the McNeil corer has become a commonly applied technique for assessing spawning gravel composition in streams because it was a significant improvement over the previously applied techniques of visual observation or shovel sampling (McNeil and Ahnell, 1964; Schuett-Hames *et al.*, 1994). The McNeil core provides a quantitative and repeatable sampling method (McNeil and Ahnell, 1964). McNeil core samples are measures of bulk streambed composition that are collected by inserting the core tube into the streambed and removing all sediments within the tube (Fig. 3). The tube is inserted into the streambed by torquing the corer while keeping it level using the handle on top of the basin or on the sides of the basin. These handles also help staff to keep the corer from rocking during the sampling process, which would disturb the fine sediments. Once the core tube has been fully inserted, the sample is removed from the tube by hand and transferred to a sample bucket until the end of the core tube is reached. The McNeil core has been used to quantify increased fine sediment loading downstream from industrial activities such as coal mining operations (MacDonald and McDonald, 1987) and roads (Hedrick *et al.*, 2007).

The corer design recommended here differs from the original design identified in Figure 3. The original design was modified because it was both heavy and expensive to construct. The modified version was made with heavy gauge aluminum rather than stainless steel, which made it considerably lighter and inexpensive to construct. In addition, it is larger than the original, standing 0.9 m (vs. 0.45-0.6 m) tall with an outer basin diameter of 0.6 m, and the coring tube which is equipped with a replaceable ring of steel teeth, is 0.2 m in diameter and can penetrate the streambed to a depth of 0.25 m (Fig. 1). Finally, the core handles were placed on the sides of the outer basin and not along the top. Another modificiation recently presented by Watschke and McMahon (2005) is lighter still because it uses a 19 L plastic bucket as the basin instead of aluminum.

The sample procedure was also altered. The original technique required sampled sediments to be brought up through the core tube and placed in the basin. Remaining fine sediments in the tube that were kept in suspension by infiltrating water were removed by a single valve pump or the tube was capped which created a vacuum that allowed trapped water to be lifted from the stream and placed in a bucket. For this program, sampled sediment was transferred directly to a clean 4 L bucket. Also, rather than pumping out sediment laden water, the water level within the tube was measured for later volume calculation, it was then mixed to suspend settling fine sediment and a 1 L water sample was taken. This 1 L sample was analyzed for suspended solids using standard techniques (APHA, 1995) that also allowed measurement of organic and inorganic fine sediment fractions. The mass of suspended solids (SS) was calculated using the concentration of SS and the volume of water in the tube. This data was then added to silt/clay fraction measured during screening of the bulk sample.

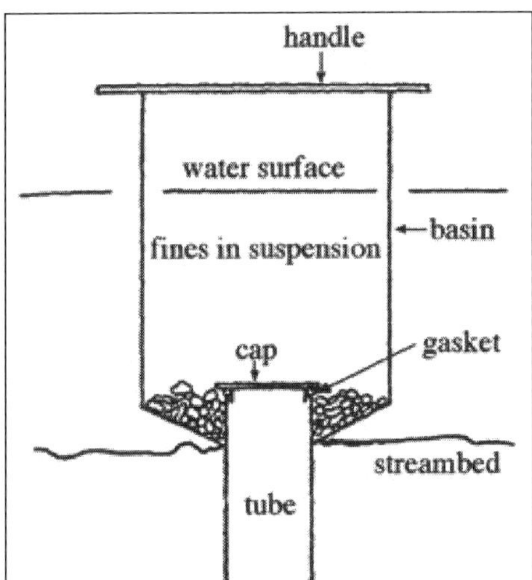

Fig. 3. The original McNeil-Ahnell corer design (McNeil and Ahnell, 1964).

McNeil core samples were collected from riffle areas near pool tail-outs as follows:

1. Sample locations were approached from a downstream direction so as to not step over the sample area prior to coring.

2. Field staff faced upstream and positioned their body over the corer and placed their hands on the handles (Fig. 4).
3. The corer was kept perpendicular to the streambed as field staff turned the corer into the streambed being careful not to use a rocking motion.
4. Once the corer was fully driven into the streambed staff checked to ensure the basin was flush to the streambed.
5. The core sample was removed by a hand to a standard depth, the top of the ring of teeth.
6. Sediment was rinsed off the sampler's hand into the bucket and a 1 L water sample from a core tube was collected to determine the mass of fine sediments.

Fig. 4. McNeil core sampling location approached from the downstream direction. The sample is taken by leaning over the core and forcing it into the streambed until it is flush.

2.3 Sediment traps: Gravel buckets and infiltration bags

2.3.1 Gravel buckets

Gravel buckets are sediment traps used to measure deposition onto and infiltration into the streambed (Fig. 5). It is not a commonly referenced sampling technique but the bucket size used and recommended here is consistent with that of Lisle and Eads (1991) as well as Larkin *et al.* (1998). These samplers consisted of a 4 L hard plastic bucket filled with washed and screened gravel from the hole dug for its placement. The screen used to remove fine sediment is the choice of the program manager but we recommend a minimum of 2 mm screen to remove sand and finer sediment from the cleaned gravel. If field collected gravels are not used and instead the sampler wants to use a standard experimental gravel, we recommend using a cleaned angular gravel with a 1.8 cm intermediate axis because it more effectively traps fine sediment than circular gravels at velocities greater than 0.4 m/s (Meehan and Swanston, 1977).

Gravel buckets can be placed in McNeil core sample locations once the core is extracted or they can be installed in runs with a stream depth less than 30 cm. Once sites are chosen, install and collect gravel bucket samples as follows:

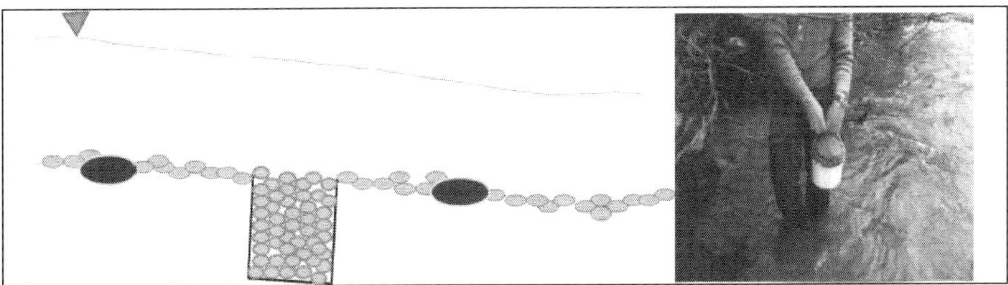

Fig. 5. Gravel bucket schematic showing flush placement with streambed (modeled after Lisle and Eads, 1991). Inset photo is a field image of an overfilled gravel bucket.

1. Identify sample locations by measuring water depth and velocity to ensure comparability between sites and locations.
2. Starting upstream, dig a hole to the approximate depth of the gravel bucket (~20 cm for 4 L buckets). Larger material should be placed to the side to refill vacant areas around the bucket once installed. If using stream sediment, screen materials to remove smaller sizes (e.g. the < 2 mm or if preferred a larger sieve size i.e. 6.3 mm).
3. Place the sealed and filled bucket level and flush to the streambed.
4. Re-measure velocity and depth to ensure replicate site similarity.
5. Move downstream to install the next gravel bucket sampler.
6. Once all upstream sampling is complete and suspended sediment generated by bucket installation has been flushed downstream or settled out, the gravel bucket lids can be removed by staff in a downstream direction. Once the final lid was removed field staff exited the channel below the last bucket.
7. During the retrieval visit, staff should enter the stream below the last bucket and replace gravel bucket lids in an upstream direction.
8. Once lids are replaced, measure the overlying depth and velocity to determine change since installation.
9. Remove buckets in an upstream direction.

2.3.2 Infiltration bags

Infiltration bags measure the amount of sediment moving vertically and horizontally through a streambed. The bags are a modified form of the wire basket retrieval system presented by Sear (1993). To prevent the loss of fine sediments when removing openwork wire baskets, Sear placed them in a collapsed polyethylene bag that was forced open with a foam collar. The bag was lifted up over the basket prior to basket removal and it prevented the loss of 26 to 40% of the collected sample. This technique has been successfully used by many including Heywood and Walling (2007) who documented sedimentation in Atlantic salmon (*Salmo salar*) redds and Petticrew *et al.*, (2007) who identified that this technique captured more fine sediment than natural gravels following a reservoir flood wave release. The infiltration bag is conceptually similar to these techniques but does not have the wire mesh or foam collars.

The infiltration bag is a waterproof fabric bag that is approximately 20 cm in diameter and 35 cm long (Lisle and Eads, 1991). It is attached by hose clamp to a brightly coloured steel ring that is also 20 cm in diameter. The bag is collapsed into the ring and is buried to a depth of 30 cm in the streambed (Fig. 6). The bag can be removed from the streambed by hand or a winch/pulley that hooks onto lines extending from the buried steel ring.

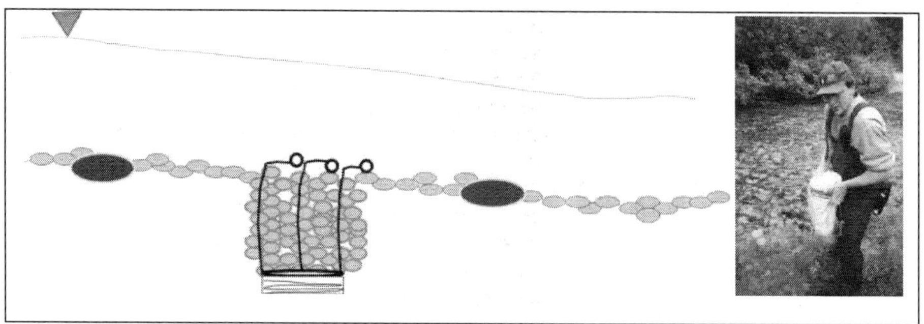

Fig. 6. Infiltration bag is shown collapsed within the steel ring and recovery lines are shown extended to the streambed surface. Inset photo shows a bag after removal from streambed.

Infiltration bags were placed in shallow runs that were less than 30 cm deep. Following the site selection process, infiltration bag samples were deployed and collected as follows:

1. Infiltration bag sites were excavated as staff moved in a downstream direction so that any suspended sediment generated by this disturbance would move downstream of the sample area.
2. Holes were dug to a depth of 35 cm and a width of 30 cm. This provided ample room for the steel ring and incorporates the 30 cm depth typically referred to in the literature for salmonid redds.
3. The collapsed bag was placed into the bottom of the hole and the reference gravel was poured into the hole until it was level with the surrounding streambed. When backfilling was a problem, a sheet metal sleeve was used to support the streambed walls during placement of the bag and reference gravel.
4. The recovery lines were held by hand so they remained on the surface after the reference gravel was poured.
5. Staff then moved downstream to the next bag.
6. To retrieve bags the sites were approached from downstream. The lines were located and attached to the winch/pulley system.
7. The bag was brought to the streambed surface and then capped with a 4 L gravel bucket lid so that upon removing it from the stream bed the overlying water was not sampled.
8. The sample was transferred to a 4 L bucket for transport. The bag was rinsed and re-deployed.

2.4 Grain size analysis

Samples can be pre-sieved in the field to remove larger grain sizes such as reference gravel or transported to the lab whole for gravimetric analysis. Case study samples were sieved using a 16, 9, 6.3, 4.0, 2.8, 2.0 mm and 500, 250, 125, 63 μm sieves. Particles above 9 mm were not included in the analysis because the focus of the study was fine sediment < 6.3 mm. Lab procedures included the following:

1. Sediment was removed from the sample container by inverting it onto a drying tray lined with a pre-weighed plastic sheet. A wash bottle was used to rinse fine sediments at the bottom of the pail and wash them onto the tray. The sample was spread in a thin layer to promote drying.
2. After the sample was air-dried to a constant weight, the weight of the air-dried sample was taken. It was corrected for the weight of the plastic sheet.
3. The sample was placed in portions in the top sieve of a stack consisting of 6.3, 4.0, 2.8 and 2.0 mm pre-weighed sieves and a bottom pan. Dry sieves were shaken by hand until particles no longer pass through to the next sieve. Each sieve was then removed and weights were recorded (corrected for sieve weight).
4. The sample collected in the bottom pan from step 3 (the 'minus 2 mm fraction') was then moved in portions to a stack containing a 500, 250, 125 and 63 μm cleaned sieves. These samples were wet sieved. The portions were limited to a maximum of 50 g because any larger may have caused the sieves to become overloaded or clogged. Sieves were often checked to ensure they were not being clogged.
5. Once the wash water ran clear the sieves were removed one at a time (i.e. from coarse to fine) and the captured sample was transferred to a pre-weighed aluminum dish. The contents were then oven-dried at 105°C. The sample weight was corrected for aluminum tray weight and recorded.
6. Because the < 63 μm (silt/clay) fraction is lost during washing it was determined by subtraction of the larger fraction weights from the total "minus 2 mm" sample weight.
7. Once tabulated, grain size classes were named according to the Wentworth size class system (Bevenger and King, 1995)

These data can be tabulated as percent less than, percent retained on sieve, and sample weight retained on each sieve. The percent retained on the sieve and those sample weights should be used in the analysis. Percent retained on sieve data can be renamed percent composition and was used for the analysis of McNeil core data because it provides a measure of streambed composition. Weight data was used for the traps because it provides a measure of sediment loading.

2.5 Study design, pseudoreplication, and data analysis

The study design selected by individual users will vary based upon the intent of the monitoring program. The case study program presented here was a compliance monitoring program (MacDonald *et al.*, 1991) to assess the effect of stream crossings on water quality. The impact-control study design was used with paired sample sites above the crossing for control and downstream of the crossing for treatment. The impact control design is a response-based study design used when the activity has already occurred but the BACI approach is recommended if temporal control samples can be collected (Manly, 2001). Findings generated from either design can be strengthened if more than one site is sampled in the control and treatment areas. Single sites were used here, but samples were collected more than once which enhances the impact-control site comparison because it can show the magnitude of the impact (Manly, 2001).

Pseudoreplication refers to the use of data drawn from studies where treatments are not replicated or replicates are not statistically independent (Hurlbert, 1984). Using replicates that are not independent artificially increases the sample size and enhances the potential for Type 1 error. Accordingly, each set of McNeil core, gravel bucket, and infiltration bag samples taken in each time interval may be considered correlated and therefore not independent or true replicates. To address this concern, we recommend that an analysis of variance (ANOVA) of mean values for each sample set be conducted. This approach was applied for case study findings where more than one data set was collected for a given technique. For example, mean values for each grain size collected with McNeil cores during each of two trips were computed. The mean values for each grain size over the two trips were grouped by site, yielding two sets of values for each grain size and location. An ANOVA of these values provides results free of pseudoreplication effects (Manly, 2001). However, this analysis was not applied across all case studies because some did not have more than one sample set.

Generally, there are two approaches for interpreting sediment data, the first is to use the raw data and the second is to generate central tendency measures. Raw data measures incorporate each grain size's weight or percent composition while central tendency measures attempt to reduce all grain size information to one number that best describes the entire particle size range. Central tendency measures include the Fredle Index, geometric mean diameter, and median particle size D_{50} (Waters, 1995; Platts *et al.*, 1983). The goal of the case study program was to quantify inputs of fine sediment from crossings so raw data was used, however when the study goal is to describe streambed composition, a wider range of raw data and/or central tendency measures may be more appropriate.

Prior to conducting a statistical analysis, the data were viewed graphically to allow for the determination of normality, designation of outliers, and to assess the potential for significant differences. Normality is a standard assumption of the parametric statistics applied and required confirmation. Data outliers were viewed in light of site establishment data to see if stream conditions at the sample location could explain data values. For example, did an outlying sample have higher or lower overlying water velocity than the other samples? Finally, by plotting sample means and their 95% confidence intervals the potential for significant differences was assessed.

Percent composition data for the McNeil core are not normally distributed. To normalize percent data arc-sin transformation was used (Sokal and Rohlf, 1995). Raw weight data for gravel bucket and infiltration bags were normally distributed and did not require transformation. To determine the presence of significant differences between sites a two-way ANOVA was applied using site and grain size as factors. Here site has two categories, identified by namely up and downstream while grain size has seven categories ranging from fine gravel to silt/clay. Tukey's post-hoc comparison or *honestly significant difference* (HSD) procedure was used to identify grain sizes that were significantly different between sites when a main effect (site difference) was observed (Sokal and Rohlf, 1995). If there was a significant interaction effect, that indicated a relationship between the two factors, i.e. grain size composition is influenced by site. Differences were identified by reviewing graphs and conducting individual t-tests, however Bonferroni adjustment should be applied to lower the risk of committing a Type 1 error if several t-tests are applied.

2.6 Sample numbers

Bulk sample estimates such as the ISO approved standards (Church *et al.*, 1987) can be used for the McNeil corer but these standards are not applicable to the trap techniques or the gravel bucket and infiltration bag. To determine the effective sample size for these techniques a short sampling program was conducted in three creeks of increasing size (5, 9, and 11 m) after the case studies were complete. Effective sample size is defined here as the number of samples after which there is limited gain in precision.

Twelve McNeil core and gravel bucket samples and 10 infiltration bag samples were collected from each stream. These sample numbers, 12 and 10, were selected because they exceeded the numbers collected during the eight case studies and also exceeded those numbers observed in the literature for McNeil coring (MacDonald and McDonald, 1987; Schuett-Hames *et al.*, 1994) and infiltration baskets (Heywood and Walling, 2007; Petticrew *et al.*, 2007). Also, the impetus of this project was to develop a protocol that will be widely used and sample numbers higher than these would likely hamper the application of this protocol in remote areas because sample weights would be too heavy to return to the lab.

2.6.1 Coefficient of variation approach

The data were subdivided into clusters ranging from 4 to 12 samples based upon similar site depth and overlying velocity. These two parameters were chosen because they are easily measured in the field and they have a substantial effect on settling environment (Knighton, 1998). Coefficients of variation (CV) were then calculated for each grouping. The CV expresses the standard deviation as a proportion of the mean (Sokal and Rohlf, 1995). As sample numbers increase, the mean and standard deviation change as does the CV. The CV is used here to represent changes in precision. Specifically, this exercise focuses on finding the sample number where the CV stabilizes. The effective sample number, defined here as that number of samples after which precision gains are small (i.e. < 5%), was determined for the three streams. The weight of sediment deposited at each site as measured by the given technique was then included in these ranked tables (e.g. Table 1) and clusters were formed starting with the largest number of similar values. For example, in Table 1 six samples have a water depth of 8 cm to represent the first cluster (CV=18.9%). The second cluster is identified by including the maximum number of samples of the depth that is most similar to the original cluster. In this case the two samples with depths of 9 cm were included (CV=16.3%). The third cluster now incorporates the single sample at 7 cm because it is more similar to the original cluster than the samples at 10 cm (CV=17.1%). The final cluster includes all samples and it has a CV of 18.4%. Note that the change in CV from six to twelve samples is less than 3%. Very little precision appears to be gained by increasing sample numbers but eight is chosen as the effective sample number because it has the lowest CV.

The results of the clustering CV analysis show the importance of maintaining similarity in site selection. Increased sample numbers are expected to improve the accuracy of the mean and the variation around the mean. However, in some cases the smallest CV was found with the lower sample sizes, which also had the narrowest range of water depth and velocity. This potentially reflects the magnitude of the change of the controlling variable used for clustering (i.e. 6 samples from the same water depth versus 9 samples that incorporate 3 depths). As depth and velocity are important controlling variables for sediment deposition it is best to maintain equivalent conditions for all replicates. This is clearly not always possible and therefore results in natural variability. The range sampled here was not expected to generate large differences but may be affecting the variation. Generally, most sample sets returned data with low variability (CV of 8-30%) with the exception of the largest creek (Young's at 11.6 m wide) where CV values were generally above 30%. Trapping techniques generally showed higher variability as measured with the CV than the McNeil core (Table 2).

Sample Identifier	Sample Depth (cm)	Sample Weight (grams)	Clusters of Samples
10	7	70.6	
9	8	74.4	
4	8	124.8	
11	8	89.7	6
3	8	81.3	8
5	8	85.2	9
1	8	78.1	12
2	9	90.4	
6	9	92.8	
12	10	76.9	
7	10	94.7	
8	10	122	
		Cluster CV	18.9(6) 16.3(8) 17.1(9) 18.4 (12)

Table 1. Gravel bucket data grouped by sample depth for the 5 m wide creek.

Sample Technique	5 m	9 m	11 m
Gravel Bucket	8 (12)	9 (13)	10 (25)
Infiltration Bag	4 (23)	8 (36)	10 (38)
McNeil Core	6 (18)	8 (24)	10 (24)

Table 2. Recommended sample number for each sample technique by stream width (CV in brackets).

2.6.2 Formula-based sample size estimates

Another approach to determine sample size was used for comparison to the sample estimates from CV alone. Sample number estimates were calculated using the following formula from Sokal and Rohlf (1995):

$$N \geq 2\,(\sigma/\delta)^2\,\{t_{\alpha\,[v]} + t_{2(1-\rho),\,[v]}\}^2 \qquad (1)$$

Where: N = sample number

σ = true standard deviation (approximated)

δ = smallest true difference desired to detect

t = t-distribution

v = degrees of freedom of the sample $\sigma_{approx.}$

α = significance level

ρ = desired power (i.e. probability a difference is found if it exists)

Example Calculation for Gravel Buckets in a 4 m wide stream:

Gravel Buckets

CV = 7.6% for 9 replicates

We want to detect a 20% difference 90% of the time.

$v = 2(9-1) = 16$, $\sigma_{approx} = 7.6\,Y/100$, 20% difference is $\delta = 20Y/100$

$N \geq 2\,(7.6Y/100\ /\ 20/100)^2\,\{t_{.05\,16} + t_{.2\,16}\}^2$

$N \geq 2\,(7.6/20)^2\{1.746+1.34\}^2$

$N \geq 2.74 \sim 3$ samples

To confirm 3 is correct, re-calculate equation 1 using 3 rather than 9 replicates, which gives an answer of 5.35. When 5 is used the answer is 4, so 4 is a good approximation.
The formula based sample number requirements were typically much larger than those generated using CV alone (Table 3). As with most equation based sample estimates, these numbers ensure statistical requirements are met, i.e. in our example the ability to detect a 5, 10, or 20% difference 90% of the time (Equation 1). Note that the statistical formula approach does not consider environmental limitations. For example, can and should you collect 250 gravel bucket samples in an 11 m wide creek? The availability of similar sampling sites is implicit within the CV analysis.
The difference between the CV and formula based sample estimates can be explained. First, the CV analysis looks at the decrease in variability with increased sample numbers (based on clusters) with an upper limit of 12 samples. So, if the lowest CV is 25% at 8 replicates, this is the best sample number within the possible sample size of 12 replicates despite the high variability. Although this seems a shortfall of the method, seven of the eight case studies saw a significant difference between sites when using less samples than suggested by the CV analysis for those people interested in using the formula based approach we suggest that the 20% detection limit is most relevant as other studies have focussed on quantifying this level of difference between locations (Rood and Church, 1994).

	5 meter			9 meter			11 meter		
Detectable Difference	**20 %**	**10%**	**5%**	**20 %**	**10%**	**5%**	**20 %**	**10%**	**5%**
Gravel Bucket	4	14	56	9	40	75	250	986	1972
Infiltration Bag	30	115	450	46	176	684	83	306	1227
McNeil Core	12	43	146	11	39	143	28	108	421

Table 3. Calculated sample numbers for each technique assuming a 90% chance of finding 20%, 10%, or 5% differences between sample sites.

2.7 Case study site description

The eight sites selected for case study were located within the central interior of British Columbia in the Prince George, Vanderhoof, and Mackenzie Forest Districts (Fig. 7). All streams were fish-bearing systems with active channel widths ranging between 4 and 12 m (Table 4).

Stream	Channel Width (m)	Activity
Spruce Creek	9.0	Ditch Erosion
Government Creek	8.0	Bridge Construction
Youngs Creek	11.6	Historical Crossing
Nithi River	4.5	Bridge Construction
Big Bend Creek	7.0	Bridge Construction
Cluculz Creek	5.0	Culvert Replacement
Greer Creek	7.0	Bridge Construction
Mugaha Creek	10.0	Bridge Washout

Table 4. Summary information of the eight case studies.

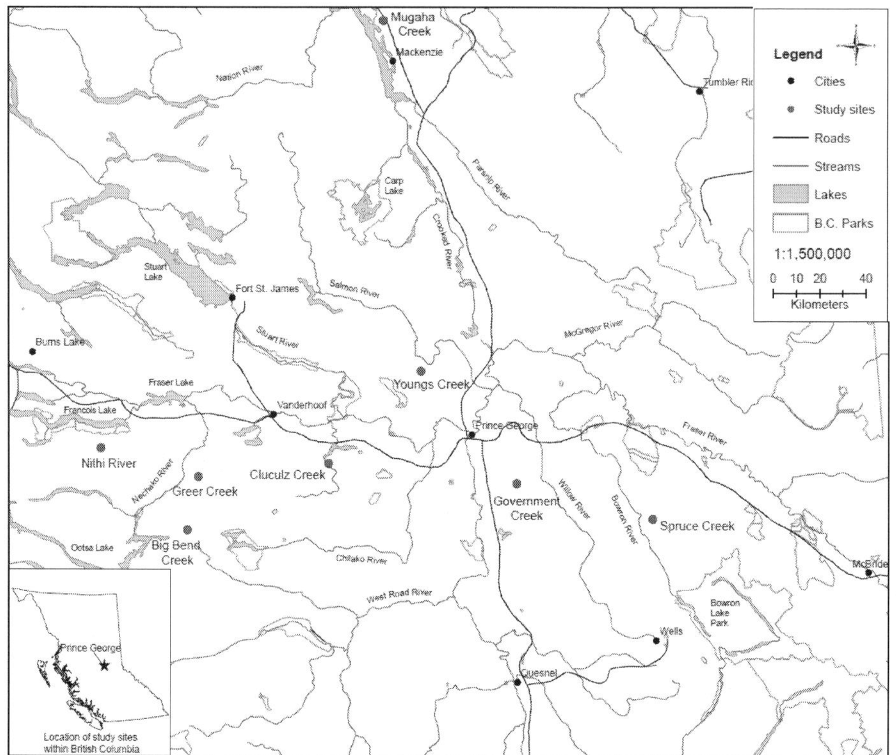

Fig. 7. Location of the eight case study sites in the British Columbia central interior.

3. Case study results

Seven of the eight case studies had significantly higher levels of fine sediment depositing downstream of the selected forest road activity for one or all of the techniques used (Table 5). The Greer Creek study did not show a significant increase in fine sediment downstream of bridge construction. This is likely the result of a decrease in discharge during our study period and the application of sediment control measures by the construction crew, which consisted of hay bales and geo-textile. For brevity the complete findings from each of the creeks will not be presented, instead details will be presented for one study site, Cluculz Creek.

3.1 Case study example: Cluculz Creek

Cluculz Creek is a 5 m wide fish bearing stream in the Vanderhoof Forest District. Its fisheries population includes the Kokanee salmon (*Oncorhynchus nerka)* and rainbow trout (*O. mykiss*). This site was selected because the crossing was undergoing road improvements consisting of culvert replacement and channel bank revetment. The culverts had repeatedly failed to accommodate spring flows often resulting in a road washout. So, the culverts were

being replaced and the channel bank was being reinforced with boulders to direct flow through the new pipe arch (Fig. 8).
The culvert replacement was a large scale disturbance to Cluculz Creek below the crossing. Baseline samples were collected on July 25 and the construction activities occurred between August 15-21. The creek was redirected through a temporary channel for several hours on August 18 while the culverts were pulled out and the pipe arch was installed. Construction period samples were collected with gravel buckets. Post-construction samples were collected with some or all techniques on September 23, October 22, and November 16. Construction activities were found to cause a significant increase in fine sediment depositing downstream of the road crossing for the total period.

Study Area	Technique	Sample Number per Visit	Results
Big Bend Creek	McNeil Core Gravel Buckets	4 and 6 4 and 6	Higher sand downstream Higher sand and clay downstream
Cluculz Creek	McNeil Core Gravel Bucket Infiltration Bag	3, 4, 6, and 6 4, 6, and 6 4 and 4	Higher sand and clay downstream Higher sand and clay downstream Higher very fine gravel upstream
Government Creek	McNeil Core	3	Higher sand downstream
Greer Creek	McNeil Core Gravel Bucket	4 and 6 4	No site differences No site differences
Mugaha Creek	McNeil Core Gravel Bucket Infiltration Bags	6, 6, and 6 6 and 6 4 and 4	Higher sand downstream Higher sand downstream Higher sand downstream
Nithi River	McNeil Core Gravel Bucket	3 and 4 4	Higher sand downstream No site differences
Spruce Creek	McNeil Core Gravel Bucket	6, 6, and 6 6 and 6	No site differences Higher sand downstream
Young's Creek	McNeil Core Gravel Bucket Infiltration Bag	3, 4, and 6 4 and 6 3	Higher sand downstream Higher sand downstream No site differences

Table 5. Sampling technique, sample number per visit, and a result summary for each of the eight case studies.

The most dramatic increase at the downstream site was observed for the construction period bucket samples, which were retrieved on August 21 three days after the creek was redirected. Bucket sample weights for each grain size were up to threefold greater at the

a)

b)

Fig. 8. Upstream view of Cluculz Creek showing a) two culverts and channel before construction and b) post-construction pipe arch and bank revetment.

downstream site (Fig. 9). Although this signal response is clearly shown in the gravel bucket samples, it is interesting to note that a similar response was not identified for the McNeil core samples. An important aspect of the sampling protocol is identified in these results because the lack of correspondence between techniques may have nothing to do with their sensitivity but instead be a function of operator bias and/or the specific sample location. That is, the first gravel buckets were installed in McNeil core sampling locations during the July baseline 25 visit (pre-construction sampling date). Upon our return on August 21, the streambed in this area had changed from its original charcoal grey colour to tan as a result of the high amount of sediment deposited in the area. Despite this obvious increase in deposited sediment, McNeil core samples were not collected there because that area was sampled during the previous visit. Instead, samples were collected downstream of the buckets, outside of the high deposition area.

McNeil core data showed a significant difference in the sand and silt/clay fractions between the up and downstream locations four and ten weeks after the culvert replacement (Fig. 10 and 11) as the sand moved along the bed downstream from the originally effected area due to increasing fall flow volumes.

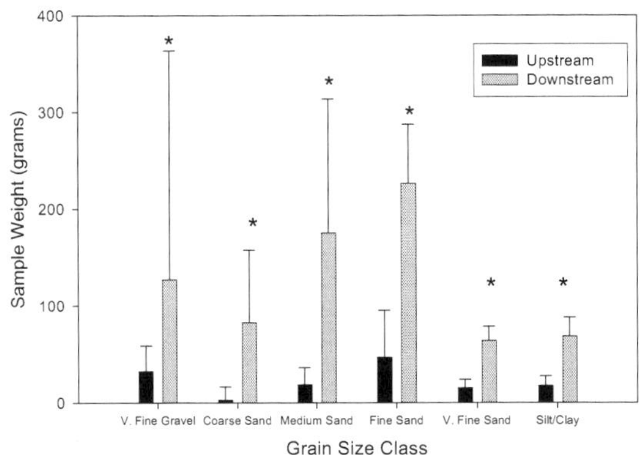

Fig. 9. Gravel bucket mean weights and their upper 95% confidence limits from samples collected at Cluculz Creek during the construction period. An asterisk highlights those grain sizes where there is a significant difference between sites.

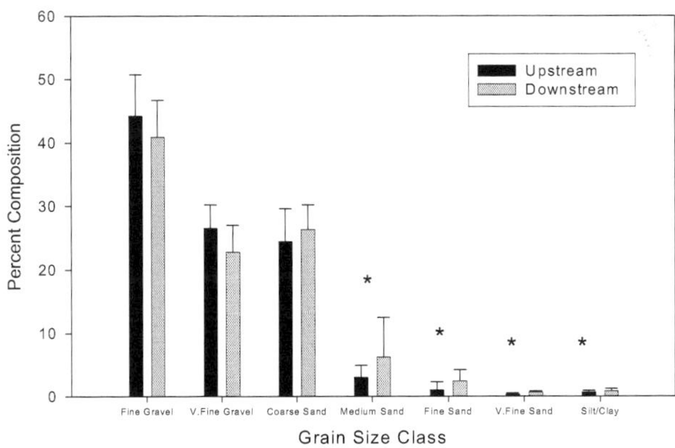

Fig. 10. McNeil core sample means and their upper 95% confidence limits five weeks after culvert replacement at Cluculz Creek. Asterisks highlight significant differences between sites.

Some of the gravel bucket and infiltration bag samplers were lost at the downstream site during October as a result of high flows. Neither the buckets nor bags show a significant difference between sites, possibly due to the low number of samples at the downstream site. Six gravel buckets were used for sampling, whereas eight are recommended for this stream width (Table 2). Eleven weeks following construction, buckets show that the crossing was still acting as a sediment source for the sand and silt/clay fractions (Fig. 12). Although the infiltration bag samples for November 16 captured higher sands and silt/clay at the downstream location, the difference was not statistically significant but it does point to an increase in infiltrating fines at that site (Fig. 13).

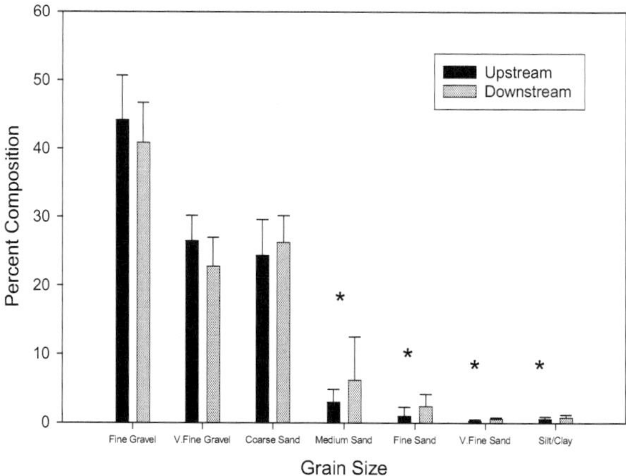

Fig. 11. McNeil core sample means and their upper 95% confidence limits 11 weeks after culvert replacement in Cluculz Creek. Asterisks highlight significant differences between sites.

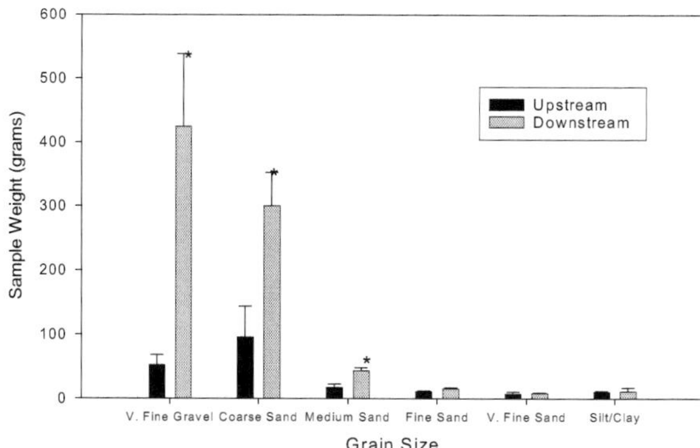

Fig. 12. Gravel bucket sample means and their upper 95% confidence limits for November 6 (11 weeks after culvert replacement in Cluculz Creek). Asterisks highlight significant differences between sites.

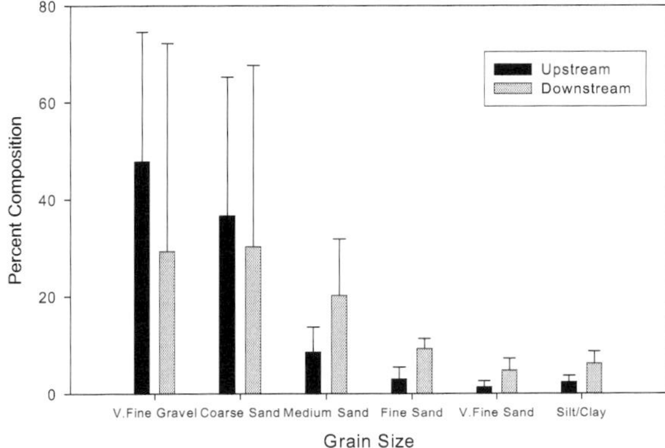

Fig. 13. Infiltration sample bag means and their upper 95% confidence limits for November 6. (11 weeks after culvert replacement on Cluculz Creek)

4. Discussion

The objective of this project was to design and evaluate a sampling protocol to quantify increases in the storage of fine sediments downstream of forest road construction and maintenance activities. Forest roads were highlighted in this study because they are the predominant type of unpaved road within the central interior of British Columbia. The study results confirm that the protocol described here is capable of detecting these increases. One or all of the techniques employed were able to detect significant increases in the fine sediment concentration for seven of the eight case studies presented.

It is suggested that this protocol would be useful for investigating unpaved roads used for other activities including agricultural development, oil and gas exploration, or mining. The latter, may be of particular interest because of high vehicular weight and traffic over wet and dry seasons during the mine's operating life. Mine haul roads support traffic with vehicular weights upwards of 290 tons and deterioration of the road can be accelerated by high traffic volumes and wear-resistance of road materials (Thompson and Visser, 2006).

Observations gathered during this study complement the literature, which often highlights unpaved roads as a major contributor of sediment to streams. Sediment delivery pathways include road and ditch runoff during road construction or following significant road bed deterioration (Cafferata and Spittler, 1998). Beschta (1978) demonstrated that road construction activities increased sediment load to a magnitude similar that of mass wasting in coastal Oregon streams. Bilby *et al.* (1989) determined in their study of a southwestern watershed that 34% of surveyed road drainage points entered streams directly. Further, in most cases fine sand (< 0.25 mm) was delivered to the streams by these roads but as gradient increased there was a shift to the larger grain sizes of sand. Sediment contribution from roads may vary based upon particle size composition of the road, with roads composed of finer sediment contributing more sediment from surface erosion. Turton *et al.* (2009) calculated that 35% of the sediment yield from Stillwater Creek, Oklahoma was generated from sandy loam to clay loam roads that comprise only 1.3% of the watershed

area. Although glacial till roads have higher erosion rates than coarser roads constructed from metamorphic rock, erosion rates are also influenced by road slope and time since the road was last graded (Sugden and Woods, 2007).
During the data analysis process and subsequent presentation of results, two other issues were identified that require more detailed discussion. These include technique sensitivity and error analysis.

4.1 Technique sensitivity

Technique sensitivity as defined here refers to the ability of a sampling technique to detect a difference in sediment storage between sites and for it to provide repeatable results. Prior to assessing sensitivity it is necessary to review the sampling protocol for each technique to highlight external influences on sample collection and to clarify each technique's strength and weakness.

4.1.1 McNeil corer

The McNeil corer is a sediment corer that penetrates the streambed and provides a bulk sample of the bed to the depth that the core is driven. It is the most environmentally representative sampling technique presented here because the natural streambed is sampled directly. Further, Young *et al.* (1991) determined in laboratory trials with known sediment mixtures that the McNeil core provided more accurate and precise samples than the single or tri-probe freeze corer and shovels.
A potential problem with the McNeil core is the under-sampling of fine sediments due to the disturbance of interstitial fines during the coring process (Young *et al.*, 1991). Similar to other coring techniques, this disturbance of interstitial fines may bias the sampler to larger grain sizes. Further, the coarser fine sediments (e.g. sands) that have settled out at the bottom of the core may not be adequately suspended or they can settle out again just prior to collection of the 1 L core water sample. However, by ensuring the consistency of sampling personnel and procedure it is defensible to compare sediment concentrations between sites using this technique.
The McNeil core has distinct advantages over other corers and the trap techniques including its ease of use, portability, and adaptability. The core tube can be exchanged for narrower or broader tubes to best suit the range of particle sizes the researcher wants to collect. It collects natural streambed material making it a better measure of streambed conditions than sediment traps.

4.1.2 Gravel buckets

Gravel buckets are impermeable walled containers that trap depositing and/or saltating sediment that settles on the reference gravel's surface. They provide a standardized measure of sedimentation within a known grain size matrix as related to a specific monitored activity. The gravel bucket is a trap and so it may not be representative of the natural environment because the bucket walls prevent exchange with the surrounding streambed. Further, if the reference gravel has a different grain size composition than the natural substrate, their trapping efficiencies will differ so bucket results may not be indicative of the retained portion of settled solids in the sample area. However, it is important to recognize that the gravel bucket is not meant to simulate the streambed. Instead, its purpose is to measure the contribution of a specific activity to the depositing sediment load within a stream. It quantifies the addition of sediment to the streambed and not streambed alteration. To assess streambed alteration the bucket should be deployed with another sampler that samples the streambed directly such as the McNeil corer.

4.1.3 Infiltration bags

Infiltration bags are also traps but unlike the buckets they are collapsed and buried at the bottom of a column of reference gravel within the streambed. They have an advantage over gravel buckets because the column of reference gravel is open to exchange with the surrounding streambed and therefore collects sediment depositing from above as well as moving vertically or horizontally through the streambed. The reference gravel is clean and its size and shape ensures that pore spaces are numerous and capable of retaining settled fines by reducing interstitial flow, all of which make it an effective sampler. Assuming that the sample collected when the bag is removed from the streambed represents the fine sediment burden at that location and time, this technique may be best applied over short periods before and after an event. When it is left for longer periods the reference gravel may come to equilibrium with the fine sediment composition of the bed but this time will be unknown.

4.1.4 Summary

Each technique focuses on sampling a different portion of the depositing sediment load and as such it can be expected that they may provide different results for the same site as shown in table 5. As previously stated, the McNeil core may be biased toward sampling of the larger grain sizes (> 1 mm) and so may not show increases in finer sediments while the traps do detect a difference in the smaller size range. This was observed at Big Bend, Cluculz, and Spruce Creek sample locations. Infiltration bags incorporate subsurface and surface sediment movement and can therefore differ from gravel buckets as shown in Young's Creek. Further, McNeil core samples and gravel buckets can show a significant increase in fine sediments due to surface loading whereas the infiltration bags can receive surface inputs but lose stored material via intergravel flows because they are open to horizontal and vertical exchange. Fine sediment can move further downstream with intergravel flow so bags may return lower fine sediment mass than buckets. This was observed at Young's Creek where both buckets and cores showed higher fine sediment loads than the bags.
Insufficient sample numbers may also explain the apparent discrepancy between the data collected by different samplers. Many of the case studies did not have sufficient sample numbers collected as was later determined by the sample number estimate program initiated in Cluculz, Spruce, and Youngs Creek. For example, four McNeil cores and three buckets were originally collected at Nithi River but according to our sample size estimates, six cores and eight buckets would have been more representative for that stream width (Tables 2 and 5). Although neither the McNeil or gravel bucket data set consistently has the appropriate number of replicates, the McNeil core sample number was often closer than the buckets, which may explain why cores determined there to be higher sand concentrations downstream and the gravel buckets did not (Table 5).
The sample size requirement information indicates that the infiltration bag will return the least variable data with the fewest number of samples for the 5 m wide stream while the McNeil core will do so for the 9 and 11 m wide streams (Table 2). Generally however, the sample numbers are comparable across techniques indicating that each is capable of returning repeatable and complementary data with relatively few replicates. Where it is not

possible to collect the suggested number of replicates due to site restrictions the sample size should be no less than three samples, which was shown in the Nithi River, Big Bend, and Government Creek site to be sufficient enough to determine a difference between sites.
To summarize, technique sensitivity is subjective because each technique was found to return repeatable data with similar sample numbers per given stream width. Further, there was general agreement between sample results for deployed techniques in each of the eight case studies. Sensitivity then is a consideration best decided upon by the sampler and the monitoring requirements of the project. For example, it is likely that the trapping techniques will be more sensitive to subtle increases in the depositing sediment load when the source is constant because their reference gravels have been cleaned and are of a size that optimizes trapping of fine sediments (Meehan and Swanston, 1977). However, while they may be more sensitive to increased fine sediments the data collected by them may not represent a similar change to the natural streambed and so the information gathered may be more relevant if partnered with the McNeil corer. Finally, the McNeil core and infiltration bag techniques may be a more sensitive measure of compositional changes with depth. The core will detect changes over the depth of the tube and the infiltration bag is open to horizontal and vertical exchange to the depth it is buried.

4.2 Potential errors

Two sources of error have the potential to affect project results, namely sampling and measurement error. Sampling error refers to errors in the sampling method, which is the selection of sites and techniques. Measurement error refers to error in sampling extraction and analysis.
Sampling error was addressed in two manners, 1) site establishment data ensured could be consistency in sample site conditions between sampling locations within the study area and 2) techniques were deployed in accordance with available standards. The site establishment data assured maximum sample site similarity given available field conditions. Although it may have been possible to find more comparable sites further up or downstream of the selected study area, there was a spatial sampling constraint. Specifically, if similar sites were chosen that were more than a couple of reach lengths from each other data interpretation is made more complex because we would have to account for the influence of tributaries, springs, or any sites of increased streambank erosion and sedimentation between the stations. Finally, where the number of sample locations within a site were limited and a sample had to be collected in a location having depth and velocity levels well outside the mean for that site, its data might have to be could be excluded from future analysis should it be shown to be an outlier.
There was limited information to draw from when developing this sampling protocol. Specifically, there was no sampling guidebook that discussed the theoretical and practical considerations necessary to design an effective sampling program. As such, the design and sampling process provided educational opportunities to improve the protocol. Starting with basic information available from the literature on how to use these sampling techniques, we were able to modify the technique to suit our specific needs and the assumptions behind these modifications were verified in the field. For example, the McNeil core has steel teeth that must be driven into the streambed by exerting pressure from above. It appears that the best approach is to torque the handles forcing the teeth to cut the streambed. When sampling in the field it is immediately obvious that when you torque the handles the core can rock, particularly on coarse substrate. This rocking results in the formation of fine

sediment plumes behind the corer due to bed disturbance. To counteract this condition two adjustments were needed. First, sampling was conducted in an upstream manner to minimize contamination of sample sites downstream with excess fines caused by streambed disturbance during the sampling process. Secondly, to maximize capture of fines during the coring process the operator had to be tall enough to rest their body on top of the core and also had to be sufficiently strong so as to force a good seal between the core tube and the streambed.

Measurement error refers specifically to sample analysis procedures and field instruments. A commercial laboratory analyzed sediment samples in accordance to ASTM standards for gravimetric sieving. In addition to the adherence of this sampling protocol 5-10% of the samples sent were re-sieved and the values compared. If the original and re-sieved values were more than 5% different from each other the sample was again re-sieved and if the difference held true, the samples from that batch were excluded. Fortunately, no re-sieved samples lay outside the acceptable level of difference.

Field instruments included a measuring tape, velocity meter, clinometer, and ruler. The same field equipment was used throughout the study to ensure consistency between sample stations within and between streams. The velocity meter was calibrated prior to the field season and its maintenance procedures were adhered to throughout the study period. The combination of these activities ensured that collected data were of good quality.

5. Conclusion and recommendations

The results of this program show that forest road construction and maintenance activities can increase downstream streambed concentrations of sediment less than 6 mm. Further, it was confirmed that these increases can be quantified with the use of the McNeil core, gravel buckets, and infiltration bags. Although each technique was found to have environmental limitations and sampling situations to which they are best applied, it is reasonable to conclude that each of them can detect forestry affected increases in depositing sediment but that they are more robust when used in combination.

Incorporating a geochemical analysis of captured sediments will enhance findings generated from the fine sediment monitoring protocol. Geochemical fingerprinting can increase the reliability of sampling results and may broaden sampling possibilities (Taylor and Owens, 2009). Fletcher and Christie (1999) used inductively coupled plasma mass-spectrophotometry (ICP-MS) to identify tracer elements for several newly formed sediment sources in six small streams (<5 m wide) in the Baptiste Watershed near Fort St. James, BC. They identified an element as useful when there are substantial differences between the sediment source being traced and the streambed sediment. Element composition differences between stream sediment and tracked sources were evaluated using the geochemical contrast (ratio) between concentration of the element within a sediment source and stream sediment above the source as well as by testing mean values of the sediment and source for significant differences.

Several elements were found to be acceptable tracers for the studied basins including calcium, chromium, iron, manganese, nickel, phosphorous, strontium, and titanium. Although not originally identified as a tracer, zinc was also found to be useful because it gave very high concentrations downstream of new stream crossings. They concluded this to be a result of sediment abrading the new galvanized culverts. Once identified, tracer elements were used to determine mixing or dilution of the new sediment into the streambed downstream. They found that within 200 m of the new sediment source the added sediment concentrations had fallen to less than 10% on these small streams. This technique would benefit assessment sampling around specified forest harvesting activities because it would be possible to designate the source material of the increased sediment. Further, by sampling at distances downstream from the investigated activity it would be possible to determine the total streambed area affected and the period of effect.

Once increases in fine sediment are documented it is possible to hypothesize biological effects with reference to water quality criteria and the available literature (Culp *et al.*, 1986; Shaw and Richardson, 2001; Cover *et al.*, 2008). Although possible to infer biotic response, it is recommended that these sediment infiltration studies be conducted in conjunction with biological monitoring programs to determine the susceptibility of monitored populations to any observed increases in fine sediment deposition. Candidate populations include periphyton and invertebrates but if fish are the resource concern it is suggested that redds, eggs, or survival-to emergence of fry be used because unlike the more transient adults these forms reside in the streambed and will be more greatly effected by temporally constrained sediment pulses. Periphyton and invertebrate populations can be collected relatively quickly and interpreted with reference to accepted techniques including the rapid bio-assessment protocols of the U.S. EPA or other biologic indices such as the index of biological integrity (Karr and Chu, 1998). When used in combination, the results of the sedimentation and biological community assessment will be more conclusive.

Finally, this protocol should not be limited to assessing forest harvesting effects alone. Instead, it should be applied to assess the effect of all unpaved roads and any land use activity that can increase fine sediment deposition in streambeds. The protocol may also prove useful where the study goal is to capture sediment for the analysis of sediment bound contaminants. This includes those programs focusing on the quantification of pesticides, hydrocarbons, heavy metals, organo-chlorines and other substances that may be released from industrial activities such as agricultural activities, mining, and pulp mills.

6. Acknowledgment

The authors wish to acknowledge Joselito Arocena, Doug Baker and Richard Woodsmith, for reviewing the thesis (Rex) from which this chapter was generated. Thanks to Dave Mouland, Jean Beckerton, and Greg Warren for help in the field and commiserating about sore backs. This chapter stems from work of completed for the British Columbia Ministry of Environment, Lands and Parks. Accordingly we would like to thank Bruce Carmichael, Dave Sutherland, and Rich Girard for their guidance and support.

7. References

Anderson, P.G., Taylor, B. R., and & Balch, G. C. (1996). Quantifying the effects of sediment release on fish and their habitats. *Canadian Manuscript Report of Fisheries and Aquatic Sciences* No.2346. 110 pp.

American Public Health Association (APHA). (1995). Standard Methods of Water and Wastewater. 19th ed. American Public Health Association, American Water Works Association, Water Environment Federation publication. APHA, Washington D.C.

Bábek, O., Hilscherová, K., Nehyba, S., Zeman, J., Famera, M., Francu, J., Holoubek, I., Machát, J., & Klánová. J. (2008). Contamination history of suspended river sediments accumulated in oxbow lakes over the last 25 years Morava River (Danube catchment area), Czech Republic. *Journal of Soils and Sediments,* Vol 8, No.3. pp. 165-178.

Beschta, R. L. (1978). Long-term patterns of sediment production following road construction and logging in the Oregon Coast Range. *Water Resources Research* Vol. 14, No. 6, pp. 1011-1016.

Bevenger, Gregory, S., & King, R. M. (1995). A pebble count procedure for assessing watershed cumulative effects. Res. Pap. RM-RP-319. Fort Collins, CO: U.S. Department of Agriculture, Forest Service, Rocky Mountain Forest and Range Experiment Station. 17 p..

Bilby, R.E., Sullivan, K., & Duncan, S. H. (1989). Generation and fate of road surface sediment in forested watersheds in western Washington. *Forest Science* Vol. 35, No.2, pp. 453-468.

Brookes, A. (1988). Channelized Rivers: Perspectives for Environmental Management. John Wiley, Chichester.

Bunte, K. & MacDonald, L.H. (1999). Scale considerations and the detectability of cumulative watershed effects: National Council for Air and Stream Improvement; Technical Bulletin No. 776.

Bunte, K., Abt, S. R., Potyondy, J. P., & Swingle, K. W. (2009) Comparison of Three Pebble Count Protocols (EMAP, PIBO, and SFT) in Two Mountain Gravel-Bed Streams. *Journal of the American Water Resources Association* Vol. 45, No. 5, pp.1209-1227.

Burns, J.W. (1970). Spawning bed sedimentation studies in northern California streams. *California Fish and Game* Vol. 56, No. 4, pp. 253-270.

Cafferata, P.H. and T.E. Spittler. (1998). Logging impacts of the 1970's vs. the 1990's in the Caspar Creek Watershed. *Proceseding of the Conference on Coastal Watersheds: The Caspar Creek Story.* pp.103-116 Ukiah California, USA. May 6 1998, USDA Forest Service Gen. Tech. Rep. PSW-GTR-168.

Cederholm, C.M., Reid, L.M. & Salo, E. O. (1981). Cumulative effects of logging road sediment on salmonid populations in the Clearwater River, Jefferson County, Washington. *Proceedings from the Conference Salmon-Spawning Gravel: A Renewable Resource in the Pacific Northwest?* P. 38-74. Wat. Res. Cen., Washington State University, Pullman, WA.

Chapman, D. W. (1988). Critical review of variables used to define effects of fines in redds of large salmonids. *Transactions of the American Fisheries Society* Vol. 117, No. 1, pp. 1-21.

Church, M. A., McLean, D. G. & Wolcott, J. F. (1987). River Bed Gravels: Sampling and analysis. In *Sediment Transport in Gravel Bed Rivers.* C. R. Thorne, J. C. Bathurst, R. D. Hey (Eds.) 43-87. John Wiley & Sons.

Corner, R. A., Bassman, J. H., & Moore, B. C. (1996). Monitoring timber harvest impacts on stream sedimentation: Instream vs. upslope methods. *Western Journal of Applied Forestry* Vol. 11, No. 1, pp. 25-32.

Cover, M. R., May, C. L., Dietrich, W. R., & Resh, V. H. (2008) Quantitative linkages among sediment supply, streambed fine sediment, and benthic macroinvertebrates in northern California streams. *Journal of the North American Benthological Society* Vol. 27, No. 1, pp. 135-149.

Culp, J.M., S.J. Wade, and R.W. Davies. 1983. Relative importance of substrate particle size and detritus to stream benthic macroinvertebrate microdistribution. *Canadian Journal of Fisheries and Aquatic Science* Vol. 40, pp. 1568-1574.

Culp, J.M., Wrona,F. J, & Davies, R. W. (1986). Response of stream benthos and drift to fine sediment deposition versus transport. *Canadian Journal of Zoology* Vol. 64, pp. 1345-1351.

Davies-Colley, R. J., C.W. Hickey, J.M. Quinn,& P.A. Ryan. (1992). Effects of clay discharges on streams: 1 Optical properties and epilithon. *Hydrobiologia* Vol. 248, pp. 215-234.

Fletcher, K. & Christie, T. (1999). Application of multi-element geochemical methods to identifying sediment sources and tracing trransport of sediment in small (S4) streams before and after watershed disturbance. Department of Earth and Ocean Sciences, University of British Columbia. 24 pp.

Fransen, P. J. B., Phillips, C. J., & Fahey, B. D. (2001). Forest road erosion in New Zealand: Overview. *Earth, Surface Processes and Landforms* Vol. 26, pp. 165-174.

Gadgil, A., 1998. Drinking water in developing countries. *Annual Review of Energy and the Environment* Vol. 23, pp. 253–286.

Gibbins, C., Damià, V., Batalla, R.,& Gomez, C. M. (2007). Shaking and moving: low rates of sediment transport trigger mass drift of stream invertebrates. *Canadian Journal of Fisheries and Aquatic Sciences* Vol. 64, No. 1. pp. 1-5.

Gomi, T., Kobayashi, S., Negishi, J., & Maizumi, F. (2010). Short-term responses of macroinvertebrate drift following experimental flushing in a Japanese headwater channel. *Landscape and Ecological Engineering* Vol. 6, No. 2, pp. 257-270.

Harrelson, C. C., C. L. Rawlins, and J. P. Potyondy. 1994. Stream channel reference sites: An illustrated guide to field technique. USDA For. Serv. Gen. Tech. Rep. RM-245.

Hurlbert, S.H. 1984. Pseudoreplication and the design of ecological field experiments. *Ecological Monographs* Vol. 54, No. 2, pp. 187-211.

Hedrick, L. B., Welsh, S. A. & Anderson, J. T. (2007). Effects of Highway Construction on Sediment and Benthic Macroinvertebrates in Two Tributaries of the Lost River, West Virginia. *Journal of Freshwater Ecology, 22*(4), pp.561-569.

Heywood, M. J. T. & Walling, D. E. (2007). The sedimentation of salmonid spawning gravels in the Hampshire Avon catchment, UK: implications for the dissolved oxtgen content of intragravel water and embryo survival. *Hydrological Processes* Vol. 21, No. 6, pp. 770-788.

Karr, J. R. and E. W. Chu. (1998). Restoring Life in Running Waters: Better Biological Monitoring. Island Press.

King, D.L. and R.C. Ball. (1967). Comparative energetics of a polluted stream. *Limnology and Oceanography* Vol.12, pp. 27-33.

Knighton, D. (1998). Fluvial Forms & Processes. John Wiley & Sons, New York.

Kreutzweiser, D.P & Capell, S.S., 2001. Fine sediment deposition in streams after selective forest harvesting without riparian buffers. *Canadian Journal of Forest Research* Vol. 31, No. 12, pp. 2134–2142.

Lane, P.N.J., Sheridan, G.J., (2002). Impact of an unsealed forest road stream crossing: water quality and sediment sources. *Hydrological Processes* Vol. 16, No. 13, pp. 2599–2612.

Larkin, G. A., P.A. Slaney, P. Warburton and A. S. Wilson. (1998). Suspended sediment and fish habitat sedimentation in central interior watersheds of British Columbia. Province of British Columbia, Ministry of Envrionment, Lands and Parks, and Ministry of Forests. Watershed Restoration Management Report No.7.

Larsen, S. & Ormerod, S. J. (2010). Low-level effects of inert sediments on temperate stream invertebrates. *Freshwater Biology* Vol. 55, 476-486.

Leopold, L.B., M.G. Wolman and J.P. Miller. (1992). Fluvial Processes in Geomorphology. Dover Publications, New York.

Leopold, L.B. 1994. A View of the River. Harvard University Press.

Leopold, L.B. 1997. Water, Rivers, and Creeks. University Science Books.

Lisle, T.E. 1989. Sediment transport and resulting deposition in spawning gravel, north coastal California. *Water Resources Research* Vol. 25, No. 6, pp. 303-1319.

Lisle. T.E. and R. E. Eads. (1991). Methods to measure sedimentation of spawning gravels. USDA Forest Service. Research Note PSW-411.

Luce, C.H., & T. Black. (1999). Sediment production from forest roads in western Oregon. *Water Resources Research* Vol. 35, No. 8 pp. 2561-2570.

MacDonald, L.H., Smart, W. R., & Wissmar, R. C. (1991). Moniotring guidelines to evaluate effects of forestry activities on streams in the Pacific Northwest and Alaska. U.S. EPA/910/9-91-001.

MacDonald, D.D. &. McDonald, L. E. (1987). The influence of surface coal mining on potential spawning habitat in the Fording River, British Columbia. *Water Pollution Research Journal of Canada* Vol. 22, No. 4 pp. 584-596.

Manly, B.F. J. (2001). Statistics for Science and Management. Chapman & Hall/CRC Press

McCaffrey, M., Switalski, T. A., & Eby, L. (2007). Effects of road decommissioning on stream habitat characteristics in the South Fork Flathead River Montana. *Transactionsof the American Fisheries Society* Vol. 136 pp. 533-561.

McNeil, W.J. & Ahnell, W. H. (1964). Success of pink salmon spawning relative to size of spawning bed materials. US Fish and Wildlife Service, Special Scientific Report - Fisheries No. 469.

Meehan, W.R. & Swanston, D.N. (1977). Effects of gravel morphology on fine sediment accumulation and survival of incubating eggs. USDA For. Serv. Res. PNW-220.

Megahan, W. F., Wilson, M., & Monsen, S. R. (2001). Sediment prodution from grantic cutslopes on forest roads in Idaho, USA. *Earth Surface Processes and Landforms* Vol. 26 pp. 153-163.

Minshall, G.W., T.V. Royer, & C.T. Robinson (2001). Response of the Cache Creek macroinvertebrates during the first 10 years following disturbance by the 1998 Yellowstone wildfires. *Canadian Journal of Fisheries and Aquatic Science* Vol. 58 pp. 1077-1088.

Montgomery, D.R., Buffington, J. M., Peterson, N. P., Schuett-Hames, D. & Quinn, T. P. (1996). Stream-bed scour, egg burial depths, and the influence of salmonid spawning on bed surface mobility and embryo survival. *Canadian Journal of Fisheries and Aquatic Science* Vol. 53 pp.1061-1070.

Mosley, M. P. & McKerchar, A. I.(1993). Streamflow. I Handbook of Hydrology D. R. Maidment (Ed.) 8.1-8.39. McGraw Hill Publishers, New York.

Nakamoto, R.J. (1998). Effects of timber harvest on aquatic vertebrates and habitat in the North Fork Caspar Creek, p. 87-95. *Proceeding of the Conference on Coastal Watersheds: The Caspar Creek Story* USDA Forest Service Gen. Tech. Rep. PSW-GTR-168, Ukiah California, May 6, 1998.

Newcombe, C.P. & MacDonald, D. D. (1991). Effects of suspended sediment on aquatic ecosystems. *North American Journal of Fisheries Management* Vol. 11 pp. 72-82.

Parkhill, K. L. & Gulliver, J. S. (2002). Effect of inorganic sediment on whole-stream productivity. *Hydrobiologia* Vo. 471 No.1-3, pp. 5-17.

Peckarsky, B. L., Fraissinet, P. R., Penton, M.A., & Conklin, D. J. (1990). Freshwater Macroinvertebrates of Norteastern North America. Comstock Publishing Associates, Cornell University Press.

Petticrew, E.L., Krein, A. & Walling, D. E. (2007). Evaluating fine sediment mobilization and storage ina gravel-bed river using cotrolled reservoir releases. *Hydrological Processes* Vol. 21 No. 2, pp. 198-210.

Platts, W.S., Megahan, W.F., & Minshall, G.W. (1983). Methods for evaluating stream, riparian, and biotic conditions. US Forest Service General Technical Report INT-138.

Platts, W. S., Torguemada, R. J., McHenry, M. L., & Graham, C. K. (1989). Changes in salmon spawning and rearing habitat from increased delivery of fine sediment to the South Fork River, Idaho. *Transactions of the American Fisheries Society* Vol. 118, pp. 274-283.

Quinn, J. M., Davies-Colley, R. J, Hickey, C.W., Vickers, M. L., & Ryan, P. A. (1992). Effects of clay discharges on streams: 2 Benthic Invertebrates. *Hydrobiologia* Vol. 248, pp. 235-247.

Reid, L.M. & Dunne, T. (1984). Sediment production from forest road surfaces. *Water Resources Research*. Vol. 20, No. 11 pp. 1753-1761.

Reiser, D. W. & White, R. G. (1988). Effects of two sediment size classes on survival of steelhead and chinook salmon eggs. *North American Journal of Fisheries Management* Vol. 8 pp. 432-437.

Rosenberg, D. M. & Wiens, A. P. (1978). Effects of sediment addition on macrobenthic invertebrates in a northern Canadian River. *Water Resources Research* Vol. 12 pp.753-763.

Rutherford, J.E. & Mackay, R. J. (1986). Patterns of pupal mortality in field populations of *Hydropsyche* and *Cheumatopsyche* (Trichoptera, Hydropsychidae). *Freshwater Biology* Vol.16 pp.337-350.

Schuett-Hames, D., Pleus, A., Bullchild, L., Hall, S. (1994). Ambient monitoring program manual. Northwest Indian Fisheries Commission. TFW-AAM9-94-001.

Sear, D. A. (1993). Fine sediment infiltration into gravel spawning beds within a regulated river experiencing floods: ecological implication for salmonids. *Regulated River Research and Management* Vol. 8 pp. 373-390.

Shaw, A. E. & Richardson, J. S. (2001). Direct and indirect effects of sediment pulse duration on stream invertebrate assemblages and rainbow trout (*Oncorhynchus mykiss*) growth and survival. *Canadian Journal of Fisheries and Aquatic Science* Vol. 58, pp. 2213-2221.

Sidle, R. C., Sasaki, S., Otsuki, M., Noguchi, & Nik, A. R. (2004). Sediment pathways in a tropical forest: effects of logging roads and kid trails. *Hydrological Processes* Vol. 18, pp. 703-720.

Slaney, P.A. (1975). Impacts of forest harvesting on streams in the Slim Creek watershed in the central interior of British Columbia. British Columbia Fish and Wildlife Branch.

Sokal, R.R. & Rohlf, F. J. (1995). Biometry: The Principles and Practice of Statistics in Biological Research. W.H. Freeman & Company, San Francisco.

South, G. R. & Whittick, A. (1987). Introduction to Phycology. Blackwell Scientific Publications. Oxford, UK.

Sugden, B. D. & Woods, S. W. (2007). Sediment producton from forest roads in western Montana. *Journal of the American Water Resources Association* Vol 43. No. 1 pp. 193-206.

Suren,A. M. (2005). Effect of deposited sediment on patch selection by two grazing stream invertebrates. *Hydrobiologia* Vol. 549. pp. 205-218.

Taylor, K. G., & Owens, P. N. (2009). Sediments in urban river basins: a review of sediment-contaminant dynamics in an environmental system conditioned by human activities. Journal of Soils and Sediments Vol. 9, No. 4, pp. 281-303.

Tripp, D. 1998. Problems, prescriptions and compliance with the coastal fisheries-forestry guidelines in a random sample of cutblocks in coastal British Columbia. In *Carnation Creek and Queen Charlotte Islands Fish/Forestry Workshop: Applying 20 Years of Coastal Research to Management Solutions*. D. L. Hogan, P. J. Tshcaplinski, & S. Chatwin (Eds.) British Columbia Ministry of Forests Research Program: Land Management Handbook 41.

Thompson, R. J. and Visser, A. T. (2009). Selection and maintenance of mine haul road wearing course materials. *Mining Technology* Vol. 115 No. 4 pp. 140-153.Turton, D. J., Smolen, M. D., and Stebler, E. (2009). Effectiveness of BMPs in reducing sediment from unpaved roads in the Stillwater Creek, Oklahoma Watershed. *Journal of the American Water Resources Association* Vol. 45 No. 6 pp. 1343-1351.

Waters, T. F. 1995. Sediment in Streams. American Fisheries Society Monograph 7.

Watschke, D. A. & McMahon, T. E. (2005). A lightweight modifcation of the McNeil Core Substrate Sampler. *Journal of Freshwater Ecology* Vol. 20, No. 4 pp. 795-797.

Wemple, B. C., Swanson, F. J, & Jones, J. A. (2001). Forest roads and geomorphic process interactions, Cascade Range, Oregon. *Earth Surface Processes and Landforms* Vol. 26, pp. 191-204.

Wood, P.L. & Armitage, P. D. (1997). Biological effects of fine sediment on the lotic environment. *Environmental Management* Vol. 21, No. 2, pp. 203-217.

Young, M. K., Hubert, W. A., & Wesche, T. A. (1991). Biases associated with four stream substrate samplers. *Canadian Journal of Fisheries and Aquatic Science* Vol. 48 pp. 1882-1886.

Zimmerman, A. E, Pontibriand, M. C., & LaPointe, M. (2005). Biases of submerged bulk and freeze-core samples. *Earth Surface Processes and Landforms* Vol. 30, No. 11, pp. 1405-117.

Part 2

Numerical Modelling of Sediment Transport

6

The Filling Dynamics of an Estuary: From the Process to the Modelling

Sylvain Guillou[1], Jérôme Thiebot[2], Julien Chauchat[3], Romuald Verjus[4], Anthony Besq[4], Duc Hau Nguyen[4] and Keang Sé Pouv[4]
[1]*University of Caen Lower-Normandy*
[2]*BRGM*
[3]*University of Grenoble*
[4]*University of Caen Lower-Normandy*
France

1. Introduction

Estuaries are submitted to a natural filling caused by the settling of cohesive sediments (ϕ < 63 μm). Those sediments, coming mostly from the sea, are transported in estuaries by the tidal currents during ebb and flood flows. During slack water, fluid velocities vanish and particles are no more suspended by turbulent dispersion. Sediment particles settle towards the bottom, and then deposit on the bed and consolidate. Those particles could be resuspended by erosion when velocities are maximal. Flocculation processes (aggregation), erosion, deposition and compaction are major phenomena that must be considered for an accurate prediction of long term behaviour of estuarine sedimentary dynamics.

Fig. 1. A mudflat located in the up-stream part of the Rance estuary (France).

Several modelling approaches exist. A common way is based on the calculation of diluted particles-water mixture motion with a transport equation for the sediment concentration considered as passive (classical approach). A supplementary model is then used to model the phenomena in the sedimentary bed (i.e. the layer of mud with a high sediment concentration). A second way consists in modelling the behaviour of the particles and the water as two interacting effective fluids from the consolidated rigid bed to the water free surface: this is the two-phase approach.
The aim of this chapter is to summarise the main hydrosedimentary processes that take place in estuaries and to present different approaches for their modelling. Section 2 is dedicated to the description of the physical processes involved in estuaries. Section 3 presents the classical (single phase) approach focussing on different ways to model the bed evolutions; the rheological and erosion properties of mud are discussed. Section 4 deals with the two-phase model for sediment transport. Several applications are presented. Finally, the shortcomings of each modelling strategy are discussed and the perspectives are given in section 5.

2. Physical processes in estuaries

Cohesive sediments are constituted of granular organic and mineral solids. They are qualified of sticky or muddy materials (Winterwerp & Van Kesteren, 2004, and references herein). Cohesive sediments, or mud, are a mixture of clay, silt, fine sand, water, organic materials and colloidal particles. Sometimes the colloidal fraction is treated as a part of the clay. The percentages of all these fractions determine the specific properties of the mud. Due mainly to the size and shape of the particles as well as the electrical charge distribution, the clay minerals are largely responsible for cohesion. The clay particles are bound together because of the neutralisation of negative electrical charges on the particles by the sodium ions in seawater: Van Der Vaals attractive forces become predominant then. In presence of water, cohesive sediment particles aggregate and form flocs. A floc is constitued by thousands of clay particles and has high water content.
Particles agglomerate to produce flocs of greater size inducing a modification of the settling velocity. Flocculation process results from the mutual collisions and adherences during Brownian, settling or turbulent motions. The latter has the highest effect (Verney et al., 2006). High shears produced by turbulence lead to break up of flocs. The settling velocity of cohesive sediment is difficult to estimate because it is highly dependent on the time evolution of the flocs' sizes, their spatial and size distributions.
Fluvial water and seawater have different characteristics especially in terms of density. The mixing of those two types of waters creates a particular circulation that favours the appearance of very turbid zones called « turbidity maximums ». When the cohesive sediment contained in these zones settles rapidly on the bottom, mainly during slack periods, a layer called « fluid mud » is formed. It is a highly concentrated suspension that sometimes moves on the cohesive bed according to its slope or by entrainment by the current.
When the flocs reach the bottom, they are crushed by the flocs that deposit above them and the water contained in is putting out. A structuration appears which changes the properties of the material. This is the consolidation. The mud becomes a material close to a saturated

soil. It is more resistant to erosion. The main hydrosedimentary processes of cohesive sediments are schematized in Figure 2. According to their concentration, the different types of water-sediment mixtures can be classified into 3 types: mobile suspension, fluid mud and cohesive bed (Ross & Metha, as cited in Winterwerp & Van Kesteren, 2004). Mud suspensions with a concentration smaller than 10 g/l are called "mobile suspension". This type of mixture corresponds to the blue zone in Figure 2. The upper limit (10 g/l) corresponds to a value that can be reached in turbidity maximum. Mobile suspensions can be considered as Newtonian fluids. Values of the order of the gel-point concentration ($C_{gel} \approx$ 100 g/l) are generally encountered near the bottom. Typically, mass concentration between 20 and 200g/l are reported for the fluid mud. Fluid mud appears frequently in navigation channel or in harbour basins. It can be horizontally mobile or stationary. Fluid mud is a plastic and shear thinning material (Coussot, 1997).

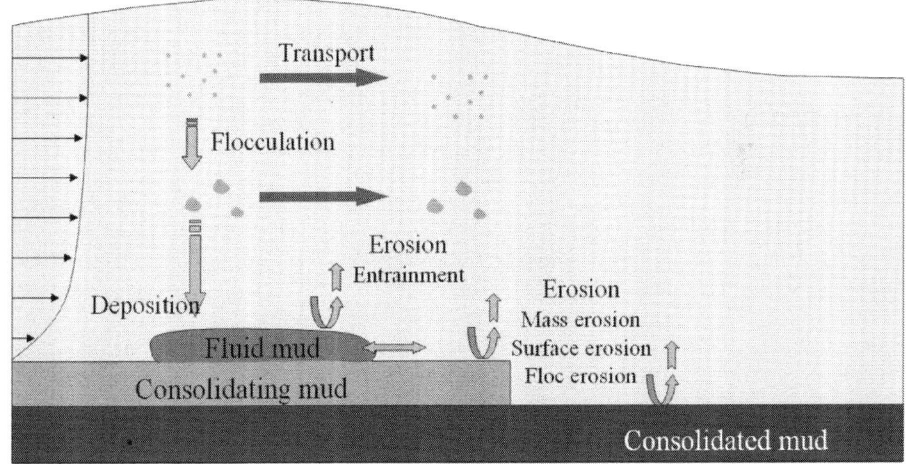

Fig. 2. Macroscopic description of the cohesive sediment transport processes.

For concentration higher than the gelling concentration, the water-sediment mixture has a solid (weak) structure. Its compaction under gravity effect is driven by consolidation process. Such a mixture is referred as consolidating bed or cohesive bed. Pore water is expulsed from the cohesive bed while the solid structure strengthens: this is the self-weight consolidation leading to a slow compaction/densification of the bed. The consolidated bed is a pseudo plastic or a viscoelastic material.

Depending on the nature of the bed, four major types of erosion are distinguished. Erosion of a soft mud layer by turbulent water flow is called "entrainment". It concerns the fluid mud. "Floc erosion" appears when flocs or part of flocs from the bed are individually disrupted or broken-up. It appears when there are peaks of shear stresses (Burst or sweep). "Surface erosion" is a consequence of a drained failure process (swelling of the surface) and of the liquefaction of the top of the bed. With the "mass erosion", lumps of material are removed.

3. Modelling estuarine morphodynamics: The classical approach

The classic approach used to model estuarine morphodynamics consists in dividing the calculation domain into subdomains corresponding to different types of water-sediment mixtures. Each subdomain is associated to particular processes. The level of complexity and the realism of the model depend on the number of subdomains and on the way the interactions between them are accounted for. Following Ross and Metha, a description in three subdomains can be used. The first one is the mobile suspension (C < 10 g/l) in which the cohesive particles are transported by the water flow. The second one is the fluid mud. It can be considered as a buffer that plays an important role in the exchanges between the suspension and the cohesive bed (erosion, deposition). The horizontal displacement of the fluid mud can be neglected in certain estuarine configurations. The lower subdomain is the cohesive bed. The dynamics of the bed is governed by the consolidation process. It is important to model consolidation because it modifies the resistance to erosion (the higher is the bed concentration, the higher is its erosion threshold) and because it determines the thickness of the bed (compaction). Under the cohesive bed, the non-erodible bed (or rigid bed) defines the lower limit of the calculation domain. The cohesive bed and the fluid mud layer constitute the stock of sedimentary materials which could be resuspended by erosion.

3.1 Modelling the mobile suspension dynamics

In the mobile suspension, the actions of the sediments on the fluid are neglected. Fluid motion is therefore calculated using Navier-Stokes Equations or Shallow Water Equations. Then the suspended sediment transport is modelled by the equation (1) where C_s represents the sediment concentration, $\vec{u}$ is the water velocity, $\vec{w}_s$, is the vertical settling velocity, $\vec{\nabla}$ is the gradient operator, D_{cs} is the diffusivity coefficient (horizontal and vertical diffusivities can be used), S is the source term, t is the time. The boundary conditions at the surface and the bottom are given by equations (2). The flux at the surface is nil whereas the flux between mobile suspension and fluid mud is equal to the sedimentation flux given by the difference between the deposition and the erosion fluxes F_d and F_e. Classically, those fluxes are modelled using the Krone's (1962) and the Partheniades' (1965) formulas respectively (4) and (5) where τ_b is the bed shear stress, τ_{cd} and τ_{ce} are the critical shear stress for deposition and erosion respectively. The latter depends on the properties of the material to erode (density, structuration, rheology ...). The flocculation and the hindered settling processes are basically accounted for by linking the settling velocity to the concentration. The formulation of Thorn (1981) is an example of such a formulation (3). When the concentration is low, the fall velocity increases with the concentration as a consequence of flocculation whereas for higher concentrations, the fall velocity decreases as a consequence of hindered settling. Linking the settling velocity to the concentration only is a rough approximation because flocculation is known to depend on other parameters especially the salinity or the turbulence effects (Verney et al., 2006). Further investigations are required concerning the parameterisation of the settling velocity of cohesive sediments.

$$\frac{\partial C_s}{\partial t} + \left(\vec{u} - \vec{w}_s\right) \cdot \vec{\nabla} C_s = \vec{\nabla} \cdot \left(D_{cs} \vec{\nabla} C_s\right) + S \tag{1}$$

$$\left(K \frac{\partial C_s}{\partial z} + W_s C_s\right)\Bigg|_{surface} = 0 \qquad \left(-K \frac{\partial C_s}{\partial z} - W_s C_s\right)\Bigg|_{bottom} = F_e - F_d \tag{2}$$

$$\begin{cases} W_s = 0,513 C_s^{1,3} & for\ C_s \leq 3g.l^{-1} \\ W_s = 2,6.10^{-3}(1-0,008C_s)^{4,65} & for\ 3g.l^{-1} < C_s < 100g.l^{-1} \end{cases} \tag{3}$$

$$F_d = W_s.C_s\left(1 - \frac{\tau_b}{\tau_{cd}}\right) \quad for \quad \tau_b \leq \tau_{cd} \tag{4}$$

$$F_e = M\left(\frac{\tau_b - \tau_{ce}}{\tau_{ce}}\right) \quad for \quad \tau_b \geq \tau_{ce} \tag{5}$$

3.2 Modelling the fluid mud dynamics

In some estuaries (e.g. the Loire estuary in France), the fluid mud is known to move horizontally. In that case the motion of the fluid mud on the cohesive bed should be considered in the modelling. The fluid mud can be seen as a dense mixture layer with a moving free surface and specific rheological properties. A model based on the Saint-Venant Equations (depth averaged equations) was proposed by Le Normant (2000). It uses the system of equations (6), in which h_{fm} is the thickness of the fluid mud layer, ρ_w and ρ_{fm}, the water density and fluid mud density, z_{cb} and z_s, the positions of the cohesive bed and of the free surface flow, ν_{fm}, the viscosity coefficient of the fluid mud, $\vec{f}_c$, the Coriolis force, $\vec{\tau}_{cb}$ and $\vec{\tau}_s$, the shear stresses at the interface cohesive bed - fluid mud and at the interface mobile suspension - fluid mud. $dMfm/dt$ represents the rate of mass exchange with the mobile suspension and with the cohesive mud.

$$\begin{cases} \dfrac{\partial h_{fm}}{\partial t} + \vec{\nabla}_h \cdot \left(h_{fm}\vec{U}_{fm}\right) - \dfrac{1}{C_{fm}}\dfrac{dMfm}{dt} = 0 \\ \dfrac{\partial \vec{U}_{fm}}{\partial t} + \left(\vec{U}_{fm}\cdot\vec{\nabla}_h\right)\vec{U}_{fm} = -\dfrac{\rho_{fm}-\rho_w}{\rho_{fm}} g\vec{\nabla}_h\left(z_{cb}+h_{fm}\right) - \dfrac{\rho_w}{\rho_{fm}} g\vec{\nabla}_h\left(z_s\right) + \upsilon_{fm}\Delta_h\vec{U}_{fm} + \dfrac{\vec{\tau}_{cb}-\vec{\tau}_{sx}}{\rho_{fm}h_{fm}} + \vec{f}_c \end{cases} \tag{6}$$

The concentration of the layer C_{fm} is assumed to be constant vertically. In estuaries where the fluid mud does not move horizontally, there is no need to use such a model. Then only vertical exchanges with the mobile suspension and the cohesive bed must be taken into account.

3.3 Modelling the cohesive bed

Various strategies can be used to model the bed evolution. The simplest strategy consists in neglecting the consolidation. In this case, the cohesive bed is a sedimentary stock with a homogeneous concentration along the vertical. Basically, the thickness of the bed evolves according to erosion and deposition only. Using such a simple model, the compaction of the bed by consolidation is not accounted for. Moreover, the erosion threshold is described very roughly because it is assumed to be constant. A more precise approach is used by Teisson (1991), Teisson et al. (1993) and Lumborg & Windelin (2003). It consists in discretising the bed using a stack of layers having a fixed concentration and a variable thickness. The bed is divided into several layers characterized by their concentration C_c their residence time T_s (time after which the sediment passes to the layer of higher concentration) and their critical shear stress for erosion. These parameters are fixed. When deposition occurs, the mud

remains in the upper layer m until the time $T_s(m)$ is reached. Then the mud is transferred to the layer located beyond $(m-1)$ which has a higher concentration (more consolidated).
The thickness of the layer m increases as: $E(m-1) = E(m-1) + E(m)\ C_c(m) / C_c(m-1)$ and the thickness of the layer $m-1$ is then modified accordingly $E(m) = 0$. The residence time and the layers' concentrations are obtained from the consolidation curve (mean concentration versus the time; this curve is obtained from settling curve experiments). The critical shear stress for erosion associated with each layer follows a law deduced from experiment. One layer can be eroded only if all the upper layers are empty. The concentration of the deposition layer is set to *100 g/l* corresponding to the fluid mud layer. This model was applied by Denot & Lang (2000) for the Rance estuary and by Le Normant (2000) for the Loire estuary.
Gibson's (1967) theory (7) constitutes the reference for soft soil consolidation modelling (Toorman, 1996) where k is the permeability, σ' is the effective stress, C_c is the sediment mass concentration, t is the time, ρ_w and ρ_s are the water and solid particle densities, g is the gravity acceleration, Eulerian coordinate z is taken as positive in the upward direction. Gibson's theory consists in describing the upward water flow using Darcy - Gersevanov's law and in characterizing the strength of the mud skeleton using Terzaghi's principle. Hypothesis is made that both the permeability k and the effective stress σ' depend on the sediment concentration only. Gibson's models require the constitutive relationships $k(C_c)$ and $\sigma'(C_c)$ to be determined. Equations (8) are commonly used. The A_i and B_i coefficients are dependent on the mud characteristics and should be determined from experiments following the technique proposed by Been & Sills (1981). The optimisation method of Thiébot et al. (2010) can be also used. Gibson's theory has been used in many consolidation models (Townsend & Mc Vay, 1990; Toorman, 1999; Bürger, 2000; Lee et al., 2000; Le Normant, 2000; Bürger & Hvistendahl, 2001; Bartholomeeusen, 2003; De Boer et al., 2007, Jeeravipoolvarn et al., 2009; Thiébot et al., 2010).

$$\frac{\partial C_c}{\partial t} - \frac{\rho_s - \rho_w}{\rho_s \rho_w} \frac{\partial}{\partial z}\left(kC_c^2\right) - \frac{1}{g\rho_w} \frac{\partial}{\partial z}\left(kC_c \frac{\partial \sigma'}{\partial z}\right) = 0 \tag{7}$$

$$\begin{cases} k(C_c) = k_0 \exp\left(\dfrac{\rho_s/C_c - A_1}{A_2}\right) \\ k_0 = 1 \text{ m/s} \end{cases} \qquad \sigma'(C_c) = B_1\left(B_2 - \rho_s/C_c\right)^{B3} \tag{8}$$

The MDM (Mud Deposit Model) has been designed to simulate the evolution of homogeneous mud deposits in hydro-sedimentary simulations (Thiébot & Guillou, 2006). The MDM uses the same physics as Gibson's theory but the equations have been rewritten to account for the multilayer discretisation. In the MDM the mud deposit is represented by a stack of layers of specified concentrations Cc_i (low concentrated layers on top of the deposit) and variable thicknesses $Ep_i(t)$. At each time step, $Epi(t)$ varies according to the solid mass fluxes at the boundaries of the layer i: $F_i(t)$ and $F_{i+1}(t)$. F_i is negative as it is downward-oriented. The evolution of the thicknesses is given by equation (9). The calculation of the mass flux between two layers is made with relation (10). Interested readers should refer to Thiébot et al. (2010).

$$Ep_i(t + \Delta t) = Ep_i(t) + \frac{(F_i(t) - F_{i+1}(t))\Delta t}{Cc_i} \tag{9}$$

$$F_i(t) = -\frac{(V_{s,i-1}(t) - V_{s,i}(t))\, Cc_i\, Cc_{i-1}}{Cc_{i-1} - Cc_i} \tag{10}$$

$$V_{s,i}(t) = -k(Cc_i)\, Cc_i \left(\frac{1}{\rho_s} - \frac{1}{\rho_w}\right) \tag{11}$$

$$V_{s,i}(t) = -k(Cc_i)\, Cc_i \left(\frac{1}{\rho_s} - \frac{1}{\rho_w}\right) + \frac{k(Cc_i)}{g\rho_w} \frac{\sigma'(Cc_i) - \sigma'(Cc_{i-1})}{(Ep_i(t) + Ep_{i-1}(t))/2} \tag{12}$$

The MDM was originally developed to simulate the vertical dynamics of the cohesive bed under consolidation effect. However, it can also be used to simulate the sedimentation of particles in the fluid mud. Been & Sills (1981) pointed out that Gibson's equation becomes a sedimentation equation when the effective stress is neglected. This is used in the MDM. A sediment concentration value C_t is used to distinguish sedimentation from consolidation. In the layers where Cc is smaller than C_t, the effective stress is neglected and the MDM resolves a sedimentation equation. Otherwise, the MDM resolve a consolidation equation. From a physical point of view, C_t is the representative value of the transition between a fluid-supported suspension and a (soft) soil which has a solid structure (i.e. the concentration is greater than the gelling concentration). So, in the MDM, the flux is linked to the solid particle velocities in layers *i* and *i-1* which are estimated with relation (11) during the sedimentation process (when $Cc < C_t$) and with the relation (12) during the consolidation process (when $Cc > C_t$).

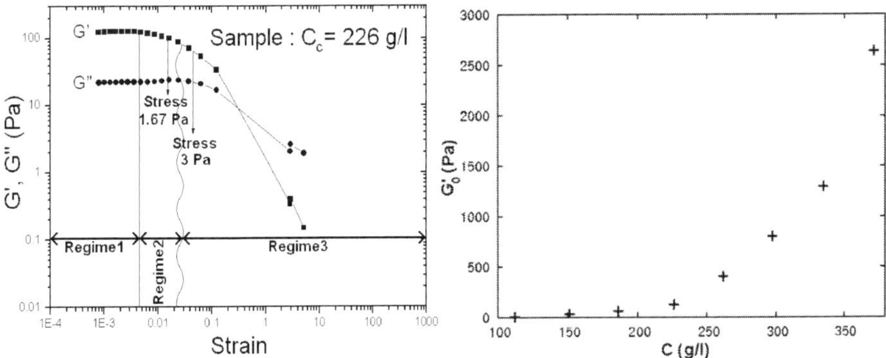

Fig. 3. Dynamic response of a mud sample of the Rance estuary: Evolution of the storage module G' and of the loss module G'' relating to the imposed deformation for one concentration (left); Evolution of the storage modulus plateau G'_0 at low deformation for different concentrations (right).

For the mud of Rance, the existence of C_t has been justified using rheometric tests performed with samples of mud collected from the Rance estuary (Thiébot et al., 2006).

Dynamic oscillation tests have been performed. This type of experiments aims at characterising the viscoelastic behaviour of materials. When the material is submitted to oscillatory solicitations, a part of the energy is transmitted to the structure of the material (it is characterised with the elastic modulus G'), the other part is dissipated (it is characterised with the loss module G"). Basically, a dynamic oscillation test involving cohesive sediments consists in studying the response of the materials for increasing strain where the sediment changes from having predominantly solid-like properties immediately after a rest period, to having predominantly liquid-like properties once set. The test represented in Figure 3 is an example. For lower strains, the response of the material is mostly elastic which gives rise to a plateau on the storage modulus (regime 1 in Fig. 3a). For intermediate strains, the storage module decreases revealing a weakening of the structure (regime 2 in Fig. 3a). One deals with rearrangements but the strength of 3D network is not broken. Finally, for great strain, the structure of the material is destroyed: it is a liquefaction process (regime 3 in Fig. 3a).

For the present purpose, we intend to find a transition behaviour for a given mass concentration value. The aim is to justify the use of a concentration which characterises the appearance of effective stress. Dynamical oscillation tests have been performed with samples of increasing concentrations. For each sample, the storage module during the regime 1 (i.e. when the solid structure is intact) has been estimated: it is G'_0. The results are presented in Figure 3b. G'_0 becomes significant and increases steeply when the concentration exceeds approximately *200 g/l*. The change in the trend of $G'_0(C)$ indicates an evolution of the material structure when a given concentration value is exceeded. From a physical point of view, the *200 g/l* value is interpreted as the appearance of a "sufficiently stiff" solid structure. This justifies the use of a concentration transition C_t in our sedimentation-consolidation model.

The MDM has been validated using the experimental data of Bartholomeeusen et al. (2002). An example of comparison between numerical and experimental results is represented in Figure 4.

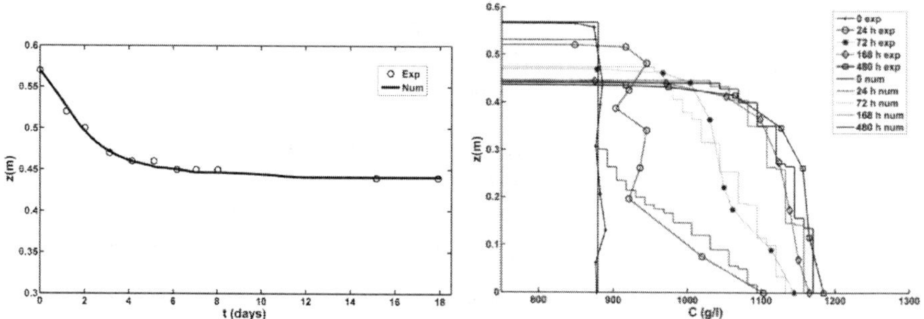

Fig. 4. Experimental ("Exp") and simulated ("Num") settling curves (left) and mass concentration profiles (right) along the time. The initial height of the settling test is $H_0 = 0.57$ *m*; the initial concentration is $C_0 = 879\ kg/m^3$. The relations (8) where used with following parameters: $A_1 = 7.28$; $A_2 = 0.298$; $B_1 = 2590$ kg/m/s2; $B_2 = 3.31$ and $B_3 = 9.25$.

3.4 Modelling the filling the Rance estuary: An example

In this section, a way of modelling the filling of estuary is presented. The Rance estuary model is used to illustrate. The Rance estuary (20 km long) is located in the north of Brittany (France). A tidal power station was built in the sixties. It modifies the hydrodynamic regime and the sedimentary processes (Bonnot-Courtois et al., 2002). As the silting up is intense in the most upstream part of the Rance estuary, dredging works are regularly carried out. Originally, the hydro-sedimentary model of the Rance aims at optimizing these operations (Denot & Lang, 2000). Simulations are performed with the modelling system Telemac (Hervouet, 2007). The mesh contains 6250 nodes. The mesh contains 6250 nodes. Telemac2d and Subief2d are used. The hydrodynamics is calculated from the flows given at the Châtelier lock (upstream boundary of the calculation domain) and at the tidal power plant (downstream boundary of the calculation domain).

$$\begin{cases} \tau_{ce}(C_c) = 2,238.10^{-6}.C_c^{\ 1,953} \ \text{if } C_c < 367 \ g/l \\ \tau_{ce}(C_c) = 4,110.10^{-10}.C_c^{\ 3,41} \ \text{otherwise} \end{cases} \tag{13}$$

In the Rance model, the erosion flux is calculated with the widely used Partheniades (1962) formula (5). The use of this formula implies to define a critical shear stress for erosion. Equation (13) is used. It has been determined from experiments on the mud of Rance following Migniot (1968). The Partheniades formula (5) also involves a parameter M which has to be tuned according the mud characteristics. Several values have been tested until the concentration of suspended sediment calculated by the model fits well to the suspended sediment concentration measured in-situ. Using M = 1.875 10^{-3} kg m^{-2} s^{-1}, the mean concentration reaches 200 mg/l in the turbidity maximum which is consistent with measurements (Thiébot, 2008). In our model, the settling velocity of suspended sediment W_s is supposed to be constant and equals *0.1 mm/s*. The sediment concentrations at the boundaries of the domain are set to 0 because turbidity measurements indicate that the sediment inputs are quasi-negligible except for particular meteorological conditions.
The suspended sediment concentrations in the Rance estuary (typically smaller than *200 mg/l*) are small in comparison to many estuaries. As a consequence, the fluid mud layer is generally thin and appears only during slack periods. During the ebb and flood flows, the fluid mud rapidly disappears by erosion. In such configuration, the horizontal movements of the fluid mud can be neglected and therefore the fluid mud can be considered as a part of the bed. In the Rance model, the bed evolution is simulated with the MDM: ten layers are used. The upper layer corresponds to the fluid mud; its concentration is set to *100 g/l* which is consistent with the earlier works of Ross & Metha. The concentrations in the other layers vary uniformly from *200 g/l* to *1000 g/l*. The constitutive relationships $k(C_c)$ and $\sigma'(C_c)$, given by (14), have been determined from experimental results obtained with mud of Rance following the method presented in Thiébot et al. (2010). The effective stress σ' is set to zero in the fluid mud layer because this material does not have any solid structure (according to the rheological tests, the solid structure appears for a concentration value of approximately *200 g/l*). The maximum concentration of the multilayer bed is *1000 g/l*. This value is the concentration beyond which the mud does not evolve much neither by consolidation nor by erosion.

$$k(C_c) = \left(\frac{1}{181,5} \left(\frac{\rho_s}{C_c} - 1 \right) \right)^{4,62} \qquad \begin{cases} \sigma'(C_c) = 0 \ \ \text{if } C_c < 200 \ g/l \\ \sigma'(C_c) = 5,81.10^{-3}.C_c^{\ 1,81} \ \ \text{otherwise} \end{cases} \tag{14}$$

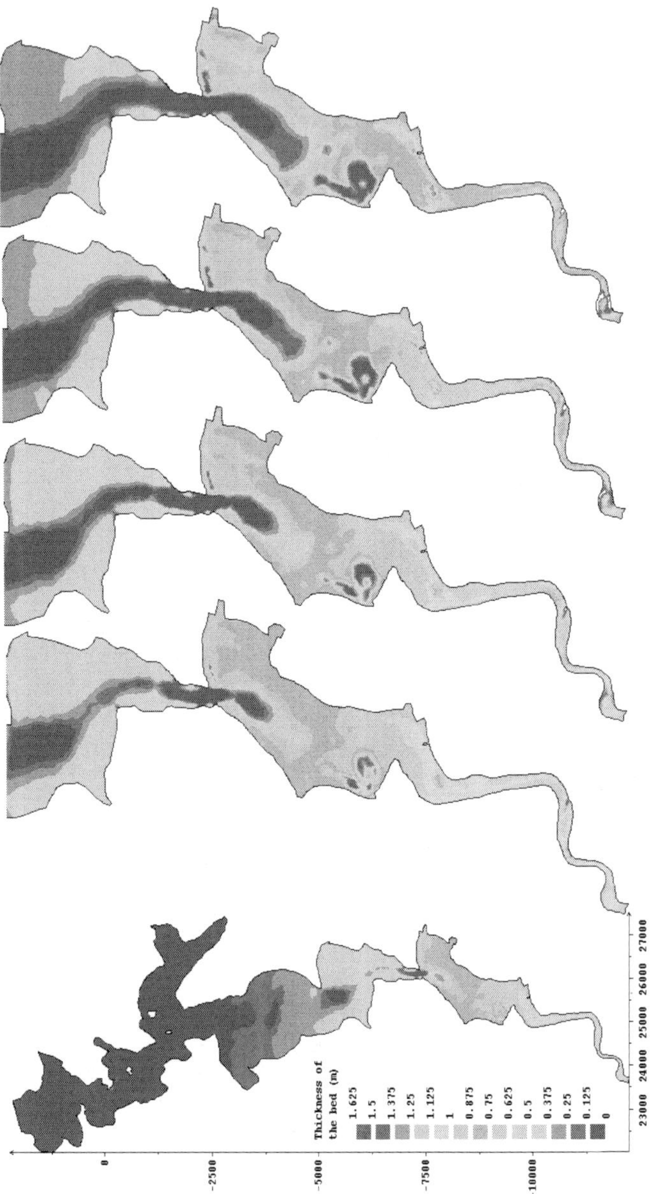

Fig. 5. Thicknesses of the solid bed in the Rance estuary: left) initial deposit in the complete estuary; right) evolution of the bed thickness in the up-stream part after 1, 2, 5 and 10 lunar cycles.

For the morphodynamics modelling of the estuary, a preliminary simulation is used in order to introduce the sediment in the system and to obtain a configuration representative of the actual state of the estuary. This state is characterised by a greater stock of cohesive sediment in the upstream part of the estuary and by a turbidity maximum with sediment concentration of the order of *200 mg/l*. At the beginning of the preliminary simulation, the sediment is introduced in the *500 g/l* -layer. A variable thickness has been used to account for the fact that the sediment stock is greater in the upstream part of the estuary. At the beginning of the preliminary simulation, the bed thickness varies from *0* to *1 m* from the downstream to the upstream boundaries of the estuary. The preliminary simulation lasts a lunar cycle (≈ 29.5 days). At the end of this simulation, we consider that the influence of the initial procedure vanishes and start to simulate a 10 lunar cycles period.
Figure 5 shows the evolution of the bed thickness during the simulation. The model shows that the sediment is progressively re-distributed within the estuary. In the middle of the estuary, the channel progressively widens while in the most upstream part, the channel and the mudflat accrete. This evolution is consistent with the in-situ observations. Additional investigations are required in order to validate the model results quantitatively.

4. Two-phase model for sediment transport

In the two-phase approach, the flow is composed of two phases: the fluid phase and the solid phase (sediment). Continuity and momentum equations are solved for each phase with the introduction of interaction terms between the two phases (fluid-solid particles, particle-particle interactions, particle-wall collision). Wallis (1969) was the first, who applied two-phase equations to the sedimentation problem in the late 60s. More recently, Toorman (1996) has presented a unifying theory of sedimentation-consolidation derived from the two-phase equations that allows to recover Kynch's sedimentation theory at low sediment concentration and Gibson's consolidation theory at high sediment concentration.
The two-phase approach was then applied to modelling sediment transports in the 90s (Teisson et al., 1992, Vilaret & Davis, 1995, Greimann et al., 1999, Barbry et al., 2000). The key idea is that the computing domain extends from the "true" non-erodible bottom to the free water surface which is a great advantage compared with the classical three-layer approach. More recently, the interest in this approach has increased and led to numerous publications mainly for non-cohesive sediment transport (*i.e.* sand) (Greimann & Holly, 2001, Dong & Zhang, 2002, Hsu et al., 2003, Hsu & Liu, 2004, Jiang et al., 2004, Amoudry et al., 2005, 2008, Longo, 2005, Chauchat & Guillou, 2008, Nguyen et al., 2009). These studies show encouraging results concerning the suspended-load transport mainly by integrating the influence of the sediment particles on the fluid turbulence and the collisions between particles (two-way and four-way coupling). Hsu et al. (2007) and Torres –Freyermuth & Hsu (2010) have applied a simplified two-phase flow model to boundary layer and gravity-driven fine sediment transport under tidal and wave forces. At the same time, first applications of 2-D X/Z two-phase model for fine-sediment transport in estuaries (Nguyen et al., 2009) have been published.

4.1 Mathematical modelling

In an Eulerian-Eulerian or two-fluid formulation, a set of equations (continuity and momentum equations) are written for each phase. If k is the index of the phase (k is "f" for the fluid phase and "s" for the solid phase), they are:

$$\begin{cases} \dfrac{\partial(\alpha_k)}{\partial t} + \vec{\nabla}\cdot\left(\alpha_k \vec{u}_k\right) = 0 \\ \dfrac{\partial\left(\alpha_k \vec{u}_k\right)}{\partial t} + \vec{\nabla}\cdot\left(\alpha_k \rho_k \vec{u}_k \otimes \vec{u}_k\right) = \dfrac{1}{\rho_k}\vec{\nabla}\cdot\left(\alpha_k\left(-p_k \overline{\overline{I}} + \overline{\overline{\tau_k}} + \overline{\overline{\tau_k^{Re}}}\right)\right) + \alpha_k \vec{g} + \dfrac{1}{\rho_k}\overrightarrow{M_k} \end{cases} \tag{15}$$

where α_k represents the volume fraction of the k-phase, with $\alpha_s+\alpha_f=1$, ρ_k is the averaged density, and $\vec{u}_k$ is the averaged velocity vector of k-phase. $\vec{g}$ is the gravity acceleration and $\vec{M}_k$ is the momentum exchanged between these two phases. p_k is the pressure, $\overline{\overline{\tau_k}}$ is the shear stress tensor and $\overline{\overline{\tau_k^{Re}}}$ is the turbulent Reynolds stress tensor for k-phase. The viscous shear stresses depend on the strains of the fluid and of the solid ($\overline{\overline{D_f}}$ and $\overline{\overline{D_s}}$) by relation (18). μ_{ff}, μ_{fs}, μ_{sf} and μ_{ss} designate effective viscosity coefficients (17), which are proportional to the fluid's viscosity μ_f . β is the amplification factor of viscous stresses and depends (18) on ξ - the distance between solid particles. Where $\alpha_{s,max}$ is equivalent to the maximum value of the solid-particle concentration. The interactions between the phases are given by the transfer laws (20). p_{ki} and $\overline{\overline{\tau_{ki}}}$ are the pressure and the shear-stress tensor of k-phase at the interface (21). $\vec{M}'_s = -\vec{M}'_f$ represents all the forces exerted by the fluid on the solid particles. For application to sediment particles in water the Drag force is dominant and given by (22), in which $\vec{u}_r$ is the relative velocity of fluid-particles, and τ_{fs} is the particle relaxation time, which represents the time needed by a particle initially at rest submitted to a constant fluid velocity to reach its steady state. The relative velocity of fluid-particles is defined as $\vec{u}_r = \vec{u}_s - \vec{u}_f - \vec{u}_d$ where $\vec{u}_d = \left\langle \vec{u}'_f \right\rangle_s$, the drift velocity, represents the correlation between the fluctuating velocity of the fluid phase and the instantaneous spatial distribution of the solid phase (Deutch & Simonin, 1991). A complete description of the model was made in Nguyen et al. (2009). A study of several turbulence models was made by Chauchat & Guillou (2008).

$$\alpha_f \overline{\overline{\tau_f}} = \mu_{fs}\overline{\overline{D_s}} + \mu_{ff}\overline{\overline{D_f}} \qquad \alpha_s \overline{\overline{\tau_s}} = \mu_{ss}\overline{\overline{D_s}} + \mu_{sf}\overline{\overline{D_f}} \tag{16}$$

$$\mu_{ff} = \alpha_f \mu_f \quad \mu_{fs} = \alpha_s \mu_f \quad \mu_{ss} = \alpha_s \beta \mu_{fs} \quad \mu_{sf} = \alpha_s \beta \mu_{ff} \tag{17}$$

$$\beta = \frac{5}{2} + \frac{9}{4}\left(\frac{1}{1+\xi/d}\right)\left[\frac{1}{2\xi/d} - \frac{1}{1+2\xi/d} - \frac{1}{(1+2\xi/d)^2}\right]\frac{1}{\alpha_s} \tag{18}$$

$$\xi/d = \frac{1-\left(\alpha_s/\alpha_{s,max}\right)^{1/3}}{\left(\alpha_s/\alpha_{s,max}\right)^{1/3}} \tag{19}$$

$$\vec{M}_f = p_{fi}\vec{\nabla}\alpha_f - \overline{\overline{\tau}}_{fi}\vec{\nabla}\alpha_f + \vec{M}'_f \qquad \vec{M}_s = p_{si}\vec{\nabla}\alpha_s - \overline{\overline{\tau}}_{si}\vec{\nabla}\alpha_s + \vec{M}'_s \tag{20}$$

$$\overline{\overline{\tau}}_{si} = \overline{\overline{\tau}}_{fi} = \beta\overline{\overline{\tau}}_f \tag{21}$$

$$\vec{M}_s \approx \vec{F}_D = \frac{\alpha_s\rho_s}{\tau_{fs}}\vec{u}_r \qquad \vec{u}_r = \vec{u}_s - \vec{u}_f - \vec{u}_d \qquad \tau_{fs} = \frac{4d\rho_s}{3\rho_f C_D \|\vec{u}_r\|} \tag{22}$$

4.2 Applications to non-cohesive sediment transport

Dredging operations of navigation channels and harbours are regularly planned to maintain the nautical depth and ensure therefore the navigation safety. The dumping operation of dredged sediment could affect the environment by increasing the turbidity of water or burying the biological habitats. After the release, the sediments settle under a cloud of very high concentration (more than 350 g/l at the beginning). This settling step is followed after impact on the bed by the formation of turbidity current. In Guillou et al. (2011), we used our two-phase model to study this phenomenon by performing simulations of the experiment of Villaret et al. (1998). In this paper only sand release with no horizontal current is considered (Fig. 6). The initial concentration was *350 g/l* with an injection velocity of 0.6m/s. The different steps of the process are qualitatively simulated by the model such as the formation and the displacement of the turbidity current.

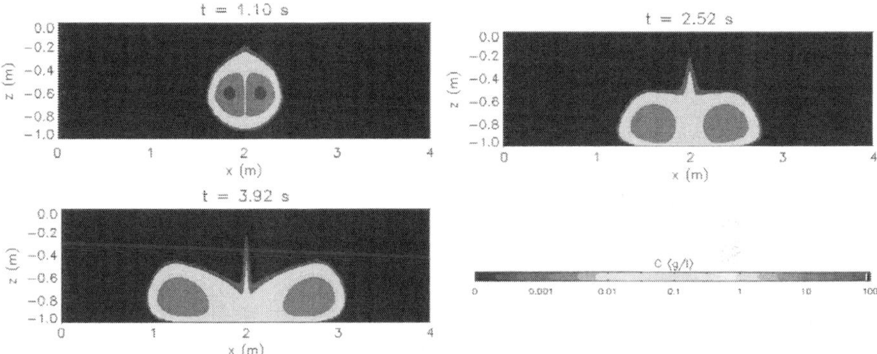

Fig. 6. Isoconcentration of sand for several instants (test e6).

One application was tried in the Seine estuary (Chauchat et al., 2009) to simulate motion of the turbidity maximum dynamics. The computational domain extends from the extremity of the semi-submersible dykes to the dam of Poses (*160 km*). A mixing length model is used to model the turbulence. The particle's diameter is equal to *16 μm* with a density of *1700 kg/m³*. The simulations have been performed over a semi-lunar cycle with a river discharge of *300 m³/s*. A one meter thickness layer with a concentration of *25 g/l* between *20* and *60 km* from the river mouth is imposed initially. The hydrodynamics is well reproduced. A Turbidity Maximum is simulated in the estuary. Its location is in coherence with observations. One main interest of this simulation is the appearance of a very high concentration layer close to the bottom with concentrations in relation with the one of a fluid mud and with a horizontal motion (Fig. 7). One major default of this model is its computational cost.

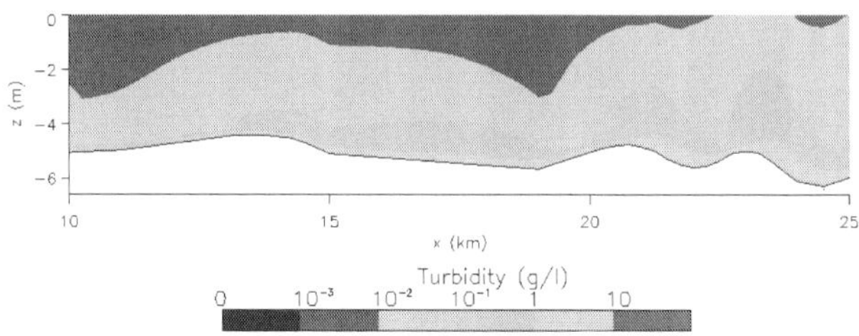

Fig. 7. Zoom near the bottom during spring tide (High water).

The concentration in the TM is not as high as in the reality. This may be in relation with the turbulent model used. Thus Chauchat & Guillou (2008) have shown that the way the turbulence is modelled had a very important impact on the capacity of the flow to keep in suspension particles. A better modelling of the turbulence in estuarine application could certainly overcome this problem. Other phenomena, not considered here, such as flocculation or consolidation processes should provide more realistic results.

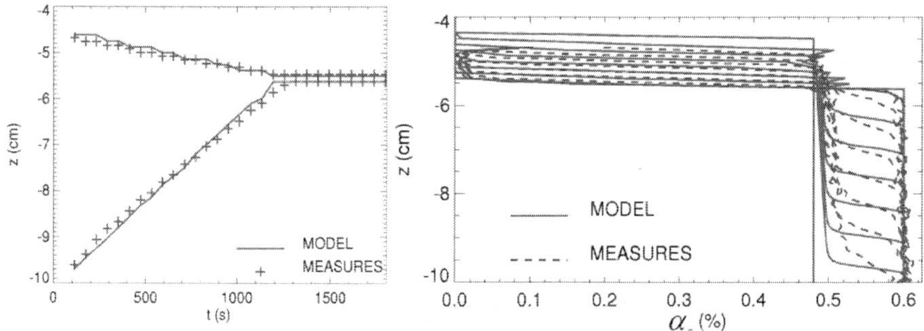

Fig. 8. Sedimentation of spherical particles: Left) Time evolution of the interface clear water and mixture and of the lutocline; right) instantaneous profiles of the solid volume fraction at different times. Comparison with the experimental data of Pham Van Bang et al. (2006).

A way to consider the latter is to study sedimentation in settling column. Nguyen et al. (2009) have published some results about the sedimentation of suspended polystyrene particles (diameters of 290 ± 30 µm, density of 1.05 kg.m-3), falling through a tank of silicon oil (viscosity of 20 mPa.s-1, density of 0.95 kg.m-3). It initial solid volume fraction was 0.48. The process in that case is well reproduced (Fig. 8).

4.3 Study of cohesive sediment transport

In the case of cohesive particles, the previous formulation of the two-phase flow model fails to simulate the hindered settling process as well as the consolidation process. Only the hindered settling regime is usually considered and parameterized using a hindrance

function (Richardson & Zaki, 1954). Improvements in modelling sedimentation and consolidation processes are needed for progressing two-phase modelling of sediment transport in estuaries. In particular, closure laws for the two-phase equations are required and need to be checked by comparison with experiments. The consistency of the excess pore pressure calculated from the two-phase equations should be checked as well.
In a recent work, Chauchat et al. (2011) proposed closure law for the two-phase flow model that allows to capture the essential features of sedimentation-consolidation processes.

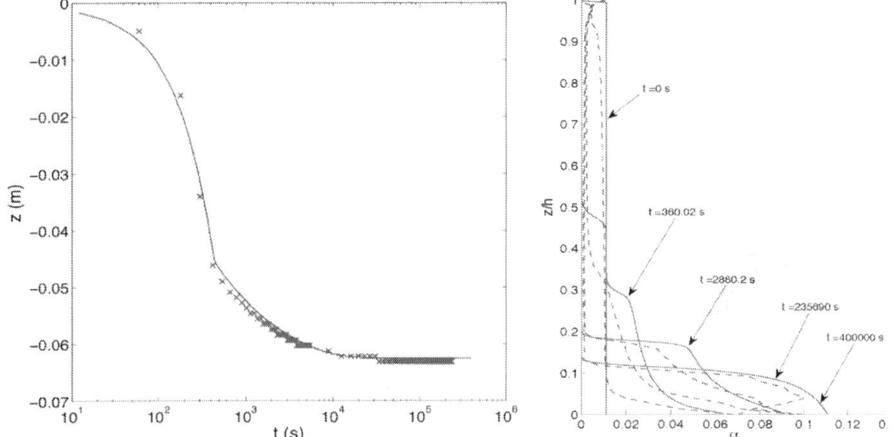

Fig. 9. Comparison of two-phase model results (____) with experiments (xxx) (Pham Van Bang, 2007) for initial concentrations $\alpha_s^0=1.2$ %: a) Settling curves: time evolution of the mud-clear water interface position and b) solid volume fraction profiles.

In a dense mixture the pressure of the solid phase p_s is the sum of the fluid pressure, the so-called pore pressure p_f, and of a particle pressure or effective stress σ' induced by inter-particle contacts in the solid network (23) (Concha et al., 1996, Burger, 2000). The particle pressure exists when the solid volume fraction exceeds a percolation value: it is the maximum packing $\alpha_{s,max}$ for non-cohesive particles, and is the gelling point for cohesive ones. The particle pressure should vanish rapidly when α_s decreases below this percolation threshold. For a high value of the volume fraction of solid-particles, the fluid flow can be viewed as the flow in a porous media constituted by the particles network. In this case, the total pressure of the mixture p is partially supported by the solid skeleton and partially by the fluid filling the pores, and leads to the relation (24) corresponding to the principle of Terzahi.

$$p_s = p_f + \sigma' / \alpha_s \tag{23}$$

$$p = \alpha_f p_f + \alpha_s p_s = p_f + \sigma' \tag{24}$$

$$\vec{M}_s \approx \frac{\rho_f g}{K}\left(\vec{u}_s - \vec{u}_f\right) \tag{25}$$

$$\sigma' = \begin{cases} 0 & if \quad \alpha_s < \alpha_s^{gel} \\ \sigma_0 \left[\left(1 - \dfrac{\alpha_s - \alpha_s^{gel}}{\alpha_{s,\max}} \right)^{-\frac{2}{3-n}} - 1 \right] & if \quad \alpha_s \geq \alpha_s^{gel} \end{cases} \tag{26}$$

Following Toorman (1996) the Darcy-Gersevanov's semi-empirical expression for the drag force (25) is used in the two-phase model, where K (in *m/s*) represents the permeability. Therefore the closure issue consists in finding closure laws for the permeability and the effective stress. The effective stress represents both permanent contacts between particles in concentrated suspension and inter-particle collisions during sedimentation. The relation (26) is proposed. The permeability is fixed regarding the settling velocity which can be estimated following the method proposed by Camenen & Pham Van Bang (2011). The results presented in Figure 9 indicate that the two-phase model is able to reproduce almost quantitatively the experimental. This work is still in progress and will be implemented in a 2D vertical two-phase model to study mud-flow interactions in estuaries.

5. Shortcomings of the modelling and perspectives

With the classic approach the different kind of estuarial water - sediment mixtures are represented by three subdomains associated with a given range of sediment concentration (mobile suspension, fluid mud and cohesive bed). When the horizontal displacements of the fluid mud can be neglected, the fluid mud can be treated in the same manner as the cohesive bed. The simulation of the consolidation processes in the cohesive bed is of importance. A convenient way of modelling the vertical evolution of the fluid mud and the cohesive bed consists in considering the bed as a stack of layers having fixed concentrations and having thicknesses that vary according to erosion, deposition, sedimentation and consolidation. The MDM is a model of that type. It has recently been introduced in the *Sisyphe* model (Villaret et al., 2010).
The major shortcoming of the classic approach is that a lot of parameters have to be determined for the formulation of the settling velocity, the permeability, the effective stress, the erosion flux ... Those parameters are specific for a given type of mud that is why, their determination requires us to collect experimental data from either from laboratory experiments (rheology, settling experiments) or field measurements (turbidity measurements). Furthermore, the results of the model are often much sensitive to the values of the parameters. Modelling the estuarine morphodynamics in an operational manner is still a challenge. Additional investigations are therefore required especially regarding the link between the erosive properties of the mud and the structuration of the material, *i.e.,* and the rheological behaviour of the mud (Thiébot et al., 2006). Current experimental works are focussed on that link (Pouv et al., 2010).
Two-phase models provide *a priori* a more general framework that allows the representation of the physical processes involved from the suspension to the consolidating bed such as interactions between fluid-solid particles, fluid-bottom as well as particle-particle interactions. No erosion/deposition fluxes are needed to be empirically prescribed.
First development about two-phase sediment transport model was one dimensional vertical. Recent improvements tend to develop 2D and 3D models. These models are based on the theory of granular flows. Then, the great majority of them suppose the sediment as non-cohesive. Due to this physics, the motions of very high concentrated suspension can be

simulated realistically. It is interesting to note that fluid mud could be simulated by this modelling. New achievement must concern the simulation of cohesive sediments. Our recent works (Chauchat et al., 2011) show that it is necessary to introduce appropriate closure laws to simulate correctly the consolidation process in a settling column test (Darcy drag expression and effective stress). In particular, the explicit calculation of the fluid pressure from the suspension to the consolidated bed represents a major advantage of the two-phase approach compared with the classical Kynch or Gibson approach.
The flocculating and deflocculating processes must be also introduced. Attentions must be paid to fractal description of mud (Merckelbach & Kranenburg, 2004). Turbulent modelling in two-phase models is really complex even for mono-disperse non-cohesive particles (Chauchat & Guillou, 2008). It is a challenge for the future in the case of cohesive particles (Hsu et al., 2007; Torres-Freyermuth & Hsu, 2010). Finally, these two-phase models are very time consuming even in a 2D case (Nguyen et al., 2009). At this time it is not possible to use them as operational tools for studying sediment transport in an open estuary. Nevertheless, it is a powerful tool to study the interaction of particles and turbulence (fluid-particles turbulent interactions, flocculation, structuration of sediment bed, mud-flow interactions).

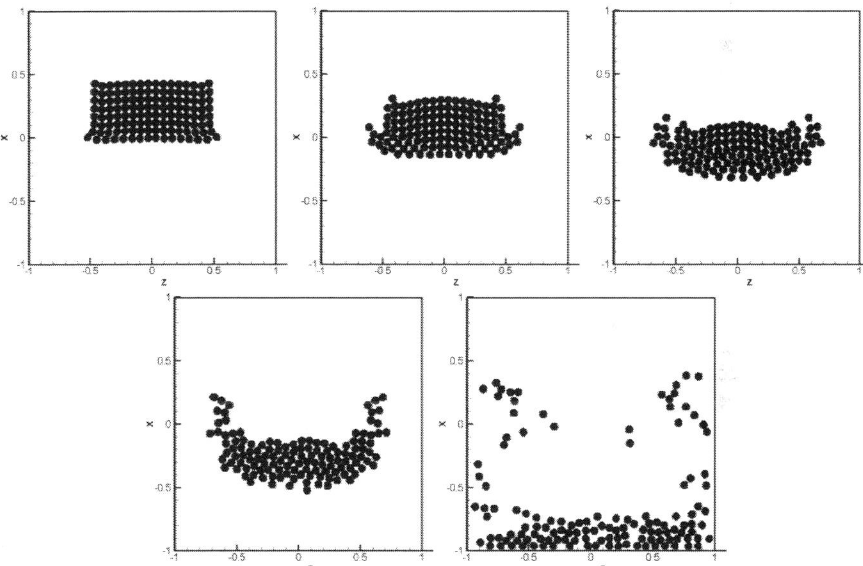

Fig. 10. Snap shots at different time steps of the sedimentation of 128 solid particles in water with Direct Numerical Simulation. The density of particles is 1.5.

Both single and two-phase models need some closure relationships for each process. Those relationships are based either on theoretical basis (*e.g.* kinetic theory of granular flows) or empirical considerations (*e.g.* permeability, effective stress). Thanks to the constant improvements in computer sciences, it is now possible to simulate directly the motion of the particles in a fluid. "Direct" numerical simulations for particulate flows have received a great attention for fifteen years in various domains like chemical engineering, petroleum industry. In these methods the flow field around each particle is resolved and hydrodynamical forces between particles and fluid are results of simulation (Glowinsky et al., 1999, Peskin, 2002, Yu & Shao, 2007). Of course, it will not be possible to apply this method to an estuary, but it can be useful to better understand the physic at the scale of the particles (collision, flocculation ...) and then to improve Eulerian models: classical models or two-phase flow models. Figure 10 provides an example of sedimentation of a cloud of particles with our model (Verjus et al., 2011).

6. Acknowledgment

The authors thank EDF GEH Ouest, IIBSN, AgroParisTech - ENGREF and COEUR who funded J. Thiébot's Ph.D at Laboratory LUSAC, the French research ministry for their funding for the Ph.D. of R. Verjus and K.S. Pouv, The Vietnamese research ministry for their funding for the Ph.D of D.H. Nguyen, and the CETMEF for its financial support of the Ph.D. of J. Chauchat. We thank the Syndicat Mixte du Cotentin for their financial support.

7. References

Amoudry, L.; Hsu, T.J. & Liu, L.F. (2005). Schmidt number and near-bed boundary condition effects on a two-phase dilute sediment transport model. *Journal of Geophysical Research*, Vol. 110, C09003, 12 pp. DOI:10.1029/2004JC002798

Amoudry, L.; Hsu, T.J. & Liu, P. L.F. (2008). Two-phase model for sand transport in sheet flow regime. *Journal of Geophysical Research*, Vol. 113, C03011, 15 pp. DOI: 10.1029/2007JC004179

Barbry, N.; Guillou, S. & Nguyen, K. D. (2000). Une approche diphasique pour le calcul du transport sédimentaire en milieux estuariens. *Comptes Rendus Mécanique*, Série II.b, Vol. 328, pp. 793-799.

Bartholomeeusen, G. (2003). *Compound shock waves and behaviour in sediment beds*. PhD thesis, St. Catherine College, Oxford, 229 p.

Bartholomeeusen, G. ; Sills, G. C. ; Znidarcic, D. ; Van Kesteren, W. ; Merckelbach, L. M. ; Pykes, R. ; Carrier, W. D. ; Lin, H. ; Penumadu, D. ; Winterwerp, H. ; Masala, S.; Chan, D. (2002). Sidere: numerical prediction of large strain consolidation. *Géotechnique*, Vol. 52, No. 9, pp. 639-648.

Been, K. & Sills, G. C. (1981). Self-weight consolidation of soft soils: an experimental and theoretical study. *Géotechnique*, Vol. 31, No. 4, pp. 519-535.

Bonnot-Courtois, C. ; Caline, B. ; L'Homer, A. & Le Vot, M. (2002). *La Baie du Mont-Saint-Michel et l'Estuaire de la Rance – Environnements sédimentaires, aménagements et évolution récente*. Bull. Centre Rech. Elf Explor. Prod., Mém. 26, 256 p.

Bürger, R. & Hvistendahl, K. (2001). On some upwind difference schemes for the phenomenological sedimentation-consolidation model. *Journal of Engineering Mathematics*, Vol. 41, No. 2-3, pp. 145-166.

Bürger, R. (2000). Phenomenological foundation and mathematical theory of sedimentation - consolidation processes. *Chemical Engineering Journal*, Vol. 80, pp. 177-188.

Camenen, B. & Pham van Bang, D. (201). Modelling the settling of suspended sediments for concentrations close to the gelling concentration. *Continental Shelf Research*, doi:10.1016/j.csr.2010.07.003

Chauchat, J. & Guillou, S. (2008). On turbulence closures for two-phase sediment-laden flows models. *Journal of Geophysical Research,* Vol. 113, C11017, doi:10.1029/2007JC004708.

Chauchat, J., Guillou, S., Barbry, N., & Nguyen K. D. (2009). Simulation of the turbidity maximum in the Seine estuary with a two-phase flow model. *Comptes Rendus Geosciences,* Série II.a., Vol. 341, pp. 505-512. doi: 10.1016/j.crte.2009.04.002.

Chauchat, J.; Guillou, S.; Pham Van Bang, D. & Nguyen, K.D. (2011), A one-dimensional two-phase flow approach for sedimentation-consolidation modelling, *Journal Hydraulic Engineering, ASCE,* (submitted)

Concha, F.; Bustos, M.C. & Barrientos, A. (1996). Phenomenological theory of sedimentation. In: *Sedimentation of Small Particles in a Viscous Fluid,* Tory, E. (Ed.), pp. 51-96, Computational Mechanics Publications.

Coussot, Ph (1997). *Mudflow rheology and dynamics,* IAHR monograph series, Balkema, Rotterdam.

De Boer, G. J.; Merckelbach, L. M.; Winterwerp, J. C. (2007). A parameterised consolidation model for cohesive sediments. *Proceedings in Marine Science,* Vol. 8, pp. 243-262.

Denot, T. & Lang, P. (2000). Apports de la modélisation numérique à l'étude de la dynamique hydro-sédimentaire de l'estuaire de la Rance, *Journées Nationales Génie Civil Génie Côtier,* pp. 147-154. DOI:10.5150/jngcgc.2000.013-D

Deutsch, E. & Simonin, O. (1991). Large eddy simulation applied to the motion of particles in stationary homogeneous fluid turbulence. *Turbulence Modification in Multiphase Flows ASME-FED,* Vol. 110, pp. 35-42.

Dong, P. & Zhang, K. (2002). Intense near-bed sediment motions in waves and currents. *Coastal Engineering,* Vol. 45, No. 2, pp. 75 - 87.

Gibson, R. E. (1967). The theory of one-dimensional consolidation of saturated clays. *Géotechnique,* Vol. 17, pp. 261-273.

Glowinski, R.; Pan, T.-W.; Hesla, T.I. & Joseph, D.D. (1999). A Distributed Lagrange multiplier/fictitious domain for particulate flows. *International Journal of Multiphase Flow,* Vol. 25, pp. 755-794

Greimann, B. P. & Holly, F.M. (2001). Two-phase flow analysis of concentration profiles. *ASCE Journal of hydraulic Engineering,* Vol. 127, No. 9, 753-762.

Greimann, B.P., Muste, M., Jr., F. M.H. (1999). Two-phase formulation of suspended sediment transport. *J. Hydraul. Res.,* 37, 479 - 500.

Guillou, S.; Chauchat, J.; Pham Van Bang, D.; Nguyen, D.H. & Nguyen, K.D. (2011). Simulation of the dredged sediment's release with a two-phase flow model, *Bulletin of the Permanent International Association of Navigation Congresses,* Vol. 142, 25-33, ISSN 0374-1001.

Hervouet, J.M. (2007). *Hydrodynamics of free surface flows: modelling with the finite element method.* Ed. John Wiley & Sons, New-York.

Hsu, T.; Jenkins, J.T. & Liu, L.F. (2003). On two-phase sediment transport: Dilute flow. *Journal of Geophysical Research,* Vol. 108, 3057, 14 pp., DOI:10.1029/2001JC001276

Hsu, T.-J. & Liu, P. L.F. (2004). Toward modelling turbulent suspension of sand in the nearshore. *Journal of Geophysical Research,* Vol. 109, C06018, 14pp., DOI: 10.1029/2003JC002240

Hsu, T.J.; Traykovski, P.A. & Kineke, G.C. (2007). On modeling boundary layer and gravity-driven fluid mud transport. *Journal of Geophysical Research*, Vol. 112, C04011. doi:10.1029/2006JC003719

Jeeravipoolvarn, S.; Chalaturnyk, R.J. & Scott, J.D. (2009). Sedimentation–consolidation modeling with an interaction coefficient. *Computers and Geotechnics*, Vol. 36, No. 5, pp. 751-761.

Jiang, J.; Law, A. W.-K. & Cheng, N.-S. (2004). Two-phase modeling of suspended sediment distribution in open channel flows. *J. Hydraul. Res.*, Vol. 42, No. 273 - 281.

Krone, R. B. (1962). *Flume studies of the transport of sediment in estuarine shoaling processes.* Technical report, Hydraulic Engineering Laboratory, University of California: Berkeley, CA, 110p.

Le Normant, C. (2000). Three-dimensional modeling of cohesive sediment transport in the Loire estuary. *Hydrological processes*, Vol. 14, pp. 2231-2243.

Lee, S. R.; Kim, Y. S. & Kim, Y. S. (2000). Analysis of sedimentation - consolidation by finite element method. *Computers and Geotechnics*, Vol. 27, pp. 141-160.

Li, M.Z. & Gust, G. (2000). Boundary layer dynamics and drag reduction in flows of high cohesive sediment suspensions. *Sedimentology*, Vol. 47, pp. 71-86.

Longo, S. (2005). Two-phase flow modeling of sediment motion in sheet-flows above plane beds. *Journal of Hydraulic Engineering*, Vol. 131, No. 5, pp. 366-379.

Lumborg, U. & Windelin, A. (2003). Hydrography and cohesive sediment modelling: application to the Rømø Dyb tidal area. *Journal of Marine Systems*, Vol. 38, No. 3-4, pp. 287-303.

Merckelbach, L.M. & Kranenburg, C. (2004). Determining effective stress and permeability equations for soft mud from simple laboratory experiments. *Géotechnique*, Vol. 54, No. 9, pp. 581-591.

Migniot, C. (1968). Etude des propriétés physiques de différents sédiments très fins et de leur comportement sous des actions hydrodynamiques. *La Houille Blanche*, Vol. 7, pp. 591-620.

Nguyen, K. D.; Guillou, S.; Chauchat, J. & Barbry, N. (2009). A two-phase numerical model for suspended-sediment transport in estuaries. *Advances in Water Resourses*, Vol. 32, pp. 1187-1196, DOI:10.1016/j.advwatres.2009.04.001

Pane, V. & Schiffman, R.L. (1985). A note on sedimentation and consolidation, *Géotechnique*, Vol. 35, No. 1, pp. 69-72

Partheniades, E. (1965). Erosion and deposition of cohesive soils. *Proceedings of the American Society of Civil Engineers, Journal of the Hydraulics Division*, Vol. 91, pp. 105-139.

Peskin, C.S. (2002) – The immersed boundary method. *Acta Numerica*, pp. 479-517.

Pham Van Bang, D. (2007). Rhéophysique des vases : rhéologie et sédimentation. Tech. rep., LCPC-CETMEF.

Pham Van Bang, D.; Lefrançois, E.; Ovarlez, G. & Bertrand F. (2006). Mri experimental and 1d fe-fct numerical investigation of the sedimentation and consolidation, in *7th International Conference on Hydroinformatics.*

Pouv, K. S ; Besq, A. & Guillou, S. (2010). Caractérisation rhéométrique des conditions de transition solide/liquide de sédiments cohésifs : vers un lien avec le comportement en érosion ?, *XIèmes journées nationales Génie côtier et Génie Civil*, Sable d'Olonnes, Juin 2010, p 529-538. DOI:10.5150/jngcgc.2010.062-P

Richardson, J.F. & Zaki, W.N. (1954). Sedimentation and fluidization: Part I. Trans. *Instn. Chem. Engrs*, Vol. 32.

Teisson, C. (1991). Cohesive suspended sediment transport: feasibility and limitations of numerical modelling. *Journal of Hydraulic Research*, Vol. 29, No. 6, pp. 755–769.

Teisson, C.; Ockenden, M. C.; Le Hir, P.; Kranenburg, C. & Hamm, L. (1993). Cohesive sediment transport processes. *Coastal Engineering*, Vol. 21, pp. 129-162.

Teisson, C.; Simonin, O.; Galland, J.C. & Laurence, D. (1992). Turbulence and mud sedimentation: A Reynolds stress model and a two-phase flow model. *Proceedings of 23rd International Conference on Coastal Engineering, ASCE*, pp. 2853-2866.

Thiébot, J. & Guillou, S. (2006). Simulation of processes acting on water-sediment mixtures in estuaries. *Flow simulation in hydraulic engineering* / Dresdner Wasserbaukolloquium, edited by H.-B. Horlacher and K.-U.Graw, Dresden, mars 2006, pp. 141-148.

Thiébot, J. (2008). *Modélisation numérique des processus gouvernant la formation et la dégradation des massifs vaseux – Application à l'estuaire de la Rance et aux berges de la Sèvre Niortaise*. PhD Thesis, AgroParisTech-ENGREF, in French, 139p.

Thiébot, J., Besq, A., Qi, X., Guillou, S. & Brun-Cottan, J.-C. (2006). Sédimentation et consolidation des sédiments cohésifs estuariens : influence des propriétés rhéologiques. *41ème Colloque Annuel de Groupe Français de Rhéologie*, Cherbourg, octobre 2006, pp. 19-22.

Thiébot, J.; Guillou, S. & Brun-Cottan, J.C. (2010). An optimisation method for determining permeability and effective stress relationships of consolidating cohesive sediment deposits, *Continental Shelf Research*. doi:10.1016/j.csr.2010.12.004

Thorn, M.F.C. (1981). Physical processes of siltation in tidal channels, *Proceedings of Hydraulic Modelling applied to Maritime Engineering Problems*, ICE, London, 47-55.

Toorman, E. A. (1999). Sedimentation and self-weight consolidation: constitutive equations and numerical modeling. *Géotechnique*, Vol. 49, No. 6, pp. 709-726.

Toorman, E.A. (1996). Sedimentation and self-weight consolidation: general, unifying theory. *Géotechnique*, Vol. 46, No. 1, pp. 103-113.

Torres-Freyermuth, A. & Hsu, T.-J. (2010). On the dynamics of wave-mud interaction: A numerical study. *Journal of Geophysical Research*, Vol. 115, C07014, 18 p., DOI: 10.1029/2009JC005552}

Townsend, F. C.; McVay, M. C. (1990). SOA: large-strain consolidation predictions. Journal of Geotechnical Engineering, 116(2), 166-176.

Verjus, R.; Guillou, S. & Ahamadi, M. (2011). Simulation of sedimentation of rigid particle suspensions: Towards a multi scale analysis. *International conference 'Two-pHase ModElling for Sediment DynamIcS'*, Paris, 26-28 April, 4 p.

Verney, R. ; Brun Cottan J. C. ; Lafite R. ; Deloffre J. (2006). Tidally-induced shear stress variability above intertidal mudflats. Case of the macrotidal Seine estuary. *Estuaries and Coasts*, Vol. 29, No. 4, pp. 653-664.

Villaret, C. & A. G. Davies (1995). Modelling of sediment-turbulent flow interactions. *Applied Mechanics Review*, Vol. 48, pp. 601-609.

Villaret, C.; Anh Van, L.; Huybrechts, N.; Pham Van Bang, D. & Boucher O. (2010). Consolidation effects on morphodynamics modelling: application to the Gironde estuary. *La Houille Blanche*, Vol. 6, pp. 15-24.

Villaret, C.; Claude B. & Du Rivau J.D. (1998). Etude expérimentale de la dispersion des rejets par clapage. He-42/98/065/a, LNHE, EDF.

Wallis, G., (1969). *One-dimensional Two-Phase Flow*. McGraw-Hill.

Winterwerp, J. C. & Van Kesteren, W. G. (2004). *Introduction to the Physics of Cohesive Sediment in the Marine Environment.* Developments in Sedimentology Series, Elsevier, Amsterdam.

Yu, Z. & Shao, X. (2007). A direct-forcing fictitious domain method for particulate flows, *Journal of computational physics*, Vol. 227, pp. 292-314.

7

Transport of Sediments in Water Bodies of the Gulf of California

Noel Carbajal[1] and Yovani Montaño-Ley[2]
[1]Instituto Potosino de Investigación Científica y Tecnológica Apartado Postal 3-74, Tangamanga, San Luis Potosí,
[2]Instituto de Ciencias del Mar y Limnología Universidad Nacional Autónoma de MéxicoUnidad Mazatlán, Mazatlán Sinaloa, Mexico

1. Introduction

The knowledge of transport of sediments in coastal ecosystems is relevant to understand forms of marine life. Many marine species find refuge and food in benthic substrates like rocky, sandy and muddy intertidal coastal areas. On the other side, and from an economical point of view, sands dynamics has a large influence on the design of harbors and on the development of industry zones along the coastline like installations for oil storage, refineries, vessel traffic and beach management. One of the principal problems is to find a balance among the conservation of the marine ecosystems and the needs for an additional development of coastal engineering structures like design of dikes and breakwaters, navigation channels and beach protection (Herbich, 2000). Strong mobility of sediments is observed in satellite imagery of many coastal areas of the world. It is an indication of the sediment dynamics and of the complexity of this kind of processes. Fundamentally, the transport of sediments has to do with erosion and accretion phenomena and with the transport of particles in suspension as a consequence of hydrodynamic forces acting on the single particles (Julien, 1998). In fact, these particles are fragmental material accumulated in the geological time as result of physical and chemical disintegration of rocks (Van Rijn, 1993). The continuous erosion, transport and deposition of these particles have modified their geometrical form and through hydrodynamic processes there is a tendency to be separated in particle sizes (very coarse, coarse, medium, fine, very fine). The analysis of source material has revealed that it is dominantly conformed of clay, quartz, silt and sand, among others.

The circulation in the ocean that causes the transport of sediment is the result of several forcing mechanisms; wind induced currents, baroclinic circulation and tidal currents. The dynamics generated by wind stress is an intermittent process but very effective in modifying sediment patterns through the induced currents and surface waves. Baroclinic or density induced currents vary seasonally and may contribute, together with changing wind systems, to reverse the transport of sediment along the coast, the so called littoral transport. Tidal currents are caused by the gravitational force of sun and moon. In the oceans and seas, the intensity of tidal currents varies from a water body to another, but a characteristic of tides is that they are always acting on the marine ecosystems, principally in diurnal, semidiurnal and fortnightly cycles. Since the transport of sediments depends on the intensity of currents (or near bottom currents), regions, where tidal amplitudes are large and there is a sandy sea bottom, are of particular interest. In the North Sea, the Gulf of California, the English Channel, the Bohai Sea, the Patagonian Shelf, etc., tidal currents are very intense. The interaction of tidal currents and a sandy sea bottom may lead to the development of wave-like regular patterns of different spatial and temporal scales (Hulscher, 1995). A series of analytical studies on the formation of sandbanks, sand waves and other rhythmic patterns have been carried out (Hulscher, 1996a; Kamorova & Hulscher, 2000). They apply equations for the transport of sediments which have been derived experimentally (Van Rijn, 1993). They consider the generation of morphological features as a process of instability when currents and the sandy sea bottom interact. The range of scales of these rhythmic sea bottom patterns varies from ripples (0.1 – 1) m, beach cusps (1-100) m, nearshore bars (50-500) m, sand waves (300-700) m, shoreface-connected sand ridges (5-8) km, tidal sandbanks (2-10) km (Dodd et al, 2003). There are manifestation of sandbanks, sand waves and other rhythmic seabed features in important seas of the world. Sandbanks and sand waves formation in the North Sea has been investigated intensively (Huthnance, 1982; Hulscher, 1996a; Hulscher, 1996b; Komarova & Hulscher, 2000). The influence of geometry on the generation of groups of sandbanks in the North Sea was also investigated (Carbajal et. al, 2005). The scale of sandbanks and the presence of sand waves in the Gulf of California has also been investigated (Meckel, 1975, Carbajal & Montaño, 1999). In spite of the findings on the formation process of sandbanks, like the understanding of instability mechanisms, the prediction of wavelengths and other very interesting characteristics of seabed features, many of these studies suppose an infinite sea, i.e. they do not consider boundaries in the calculations. Bottom and geographical boundaries of a water body should play a fundamental role in the sediments dynamics, since satellite imagery of coastal areas of many parts of the world reveals sand features with a wide range of scales. Along the coasts of the world there are a lot of water bodies like semi-enclosed seas, bays, estuaries, inlets and coastal lagoons where a strong mobility of sediments occurs. It suggests that the geometry (geography and bathymetry) is an important factor in the generation of regular and irregular areas of erosion and accretion of sediments. The analytical study of rhythmic seabed features with consideration of realistic boundaries is, of course, extremely difficult. Therefore, numerical models are an important alternative tool to investigate the transport of sediments in complex coastal areas. We believe that a complementary work of experimental, theoretical and numerical research on transport of sediments is necessary.

2. Study area

The Gulf of California is a marginal sea with a length of about 1100 km and an averaged width of approximately 200 km (figure 1). In the central part of the gulf there is an archipelago formed by the islands of Angel de la Guarda, Tiburón, San Esteban, San Lorenzo, Salsipuedes and Partida. In the northernmost part is located the Colorado River Delta, a triangular shallow platform where the water discharge took place in the past. The huge quantities of discharged sediment and the combined tidal and water discharge dynamics led to the formation of the islands of Montague and Gore. The northern part can be considered as a continental shelf with maximum depths of 200 m. In the southern part of the gulf, depths of more than 3000 m can be found. In the Gulf of California, satellite imagery reveals an intense mobility of sediments in the area of the Colorado River Delta and in water bodies along the eastern coast like the bays of Adair, San Jorge and Yavaros and the coastal lagoons of Topolobampo, Santa María la Reforma, among others (figure 1). Tides are very important in the dynamics of the Gulf of California. In the central part, tides have a mixed character but with dominant diurnal signals and maximum tidal ranges of about 1.6 m. In the southern part, mixed tides govern the sea level change, but with a dominance of semidiurnal signals and tidal ranges of approximately 2 m. In the northern part, tides have a semidiurnal character and the tidal ranges reach the largest values of the gulf. We are going to document the bedload sediment transport in the Gulf of California with the numerical study of two representative water bodies: the Colorado River Delta and the Yavaros Bay (see figure 1). These calculations will give an idea of the enormous work which has to be done to study the transport of sediment in all water bodies of the Gulf of California.

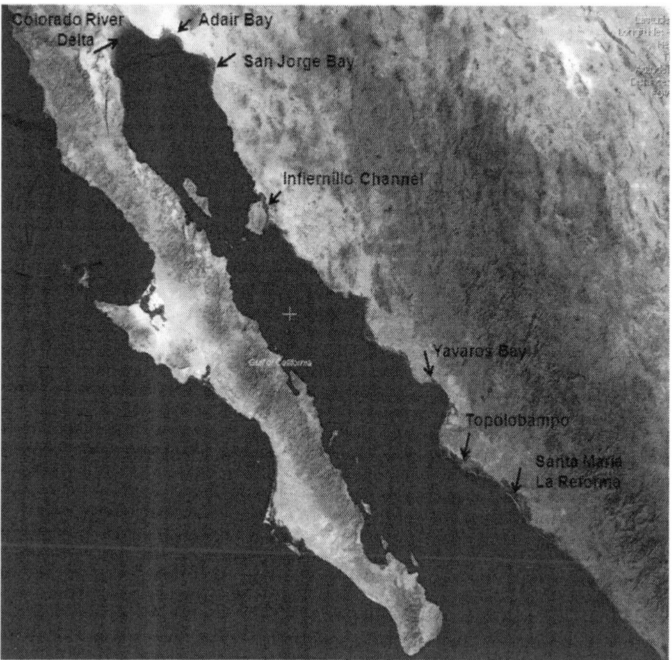

Fig. 1. Some water bodies in the Gulf of California where transport of sediments occurs (Landsat Image).

3. The model

We estimated the bedload sediment transport in several water bodies of the Gulf of California applying a vertically integrated two-dimensional, non-linear, semi-implicit numerical model. The model has been used previously for the study of tidal dynamics and transport of sediments in the Colorado River Delta (Montaño & Carbajal, 2008) and the dynamics of coastal lagoons like Topolobampo (Montaño-Ley et.al, 2007). The model is coupled with a semi-empirical bedload sediment transport equation (Van Rijn 1993). The applied vertically integrated equations of motion are

$$\frac{\partial U}{\partial t}+\frac{U}{(H+\zeta)}\frac{\partial U}{\partial x}+\frac{V}{(H+\zeta)}\frac{\partial U}{\partial y}-fV=$$
$$-g(H+\zeta)\frac{\partial \zeta}{\partial x}+\frac{\partial}{\partial x}\left(A_H\frac{\partial U}{\partial x}\right)+\frac{\partial}{\partial y}\left(A_H\frac{\partial U}{\partial y}\right)-r\frac{U}{(H+\zeta)^2}\sqrt{U^2+V^2} \tag{1}$$

$$\frac{\partial V}{\partial t}+\frac{U}{(H+\zeta)}\frac{\partial V}{\partial x}+\frac{V}{(H+\zeta)}\frac{\partial V}{\partial y}+fU=$$
$$-g(H+\zeta)\frac{\partial \zeta}{\partial y}+\frac{\partial}{\partial x}\left(A_H\frac{\partial V}{\partial x}\right)+\frac{\partial}{\partial y}\left(A_H\frac{\partial V}{\partial y}\right)-r\frac{V}{(H+\zeta)^2}\sqrt{U^2+V^2} \tag{2}$$

The used equation of continuity is

$$\frac{\partial \zeta}{\partial t}+\frac{\partial U}{\partial x}+\frac{\partial V}{\partial y}=0 \tag{3}$$

In this system of equations, U and V are the transports in the x and y directions respectively $(\mathrm{m^2\ s^{-1}})$, $\mathbf{v}$ is the velocity vector $(\mathrm{m\ s^{-1}})$, t is time (s), g is the gravitational acceleration $(\mathrm{m\ s^{-2}})$, H is the water depth (m), ζ is the sea surface elevation (m), $f=2\Omega\sin\varphi$ is the Coriolis parameter $(\mathrm{s^{-1}})$, Ω is the angular velocity of the Earth $(\mathrm{s^{-1}})$, φ is the latitude and A_H is the horizontal coefficient of Eddy viscosity $(\mathrm{m^2\ s^{-1}})$.
At the open boundary, the sea surface elevation is given by

$$\zeta(x,y,t)=A_0\cos(\omega t-\Phi) \tag{4}$$

where ω is the angular frequency of the M_2 tidal component. A_0 and Φ are the observed amplitudes and phases. The boundary condition for the velocity vector $\mathbf{v}$ at the closed sides is $\mathbf{v}\cdot\mathbf{n}=0$, where $\mathbf{n}$ is a unit vector normal to the coast. At the open side the condition is

$$\frac{\partial v_n}{\partial x_n}=0$$

where the velocity v_n and the coordinate x_n are perpendicular to the boundary.
The transport of sediment by a flow of water takes place in the form of bedload and suspended load. It depends on the particles size and on the intensity of the flows. The motion of the bed material particles occurs basically in three ways; rolling and sliding motion, saltation motion and suspended particle motion (Van Rijn, 1993). There are different mathematical formulae for the transport of bedload and suspended load. One of them is when the volumetric sediment flux, $\mathbf{S}=\mathbf{S}(S_x,S_y)$, is proportional to the velocities, i.e. $\mathbf{S}\propto(\overline{\mathbf{v}}-\overline{\mathbf{v}}_{cr})^m$, where $\overline{\mathbf{v}}$ is the depth-averaged velocity, $\overline{\mathbf{v}}_{cr}$ is the depth-averaged critical velocity and $3\le m\le 5$. Another form is $\mathbf{S}\propto(\boldsymbol{\tau}-\boldsymbol{\tau}_{cr})^n$, i.e. it is a function of the bed shear stress $\boldsymbol{\tau}$, with $n\approx 1.5$. We used the following conservation equation for bedload sediment transport,

$$\frac{\partial H}{\partial t}-\nabla\cdot\mathbf{S}=0 \tag{5}$$

And we applied an equation where the volumetric sediment flux $\mathbf{S}$ is proportional to the depth-averaged velocities (van Rijn, 1993; Hulscher, 1996a).

$$\mathbf{S} = \alpha \frac{|\mathbf{v}|^b}{u_c^b} \left(\frac{\mathbf{v}}{|\mathbf{v}|} - k_* \nabla H \right) \tag{6}$$

The volumetric sediment flux is given in $(\mathrm{m}^2\ \mathrm{s}^{-1})$ (cubic meters of sediment per meter of sea bottom width per second) and S_x and S_y are the flux components in the x and y directions. $\alpha = 10^{-6}$ is a function of the sediment properties $(\mathrm{m}^2\ \mathrm{s}^{-1})$, $b = 3.0$ is the potency of the transport, $k_* = 2.0$ is a coefficient for bed slope correction and u_c $(\mathrm{m\ s}^{-1})$ is the critical velocity for bedload sediment transport, i.e. if $|\mathbf{v}| \le u_c$ then no bedload sediment transport occurs.

4. Results

4.1 Colorado River Delta

In the Colorado River Delta, tidal ranges reach values of about 10 m at spring tides, with a dominant semidiurnal signal (Carbajal & Backhaus, 1998). Tidal currents of the order of 3 m/s have been estimated in areas where the water is channeled by the presence of sandbanks or islands. In figure 2, the distribution of sediments in the Colorado River Delta is shown at the time between neap and spring tides. The sediments mobility is very strong and occurs in form of filaments that extend more than 60 km to the south. This sediment was deposited principally by the discharge of the Colorado River in the geological time. It is important to mention that since the construction of the Hoover Dam in the thirties of the last century, the water and sediments discharge through the Colorado River came to an end (Carbajal et al., 1997). The large concentration of suspended sediments, visible in satellite imagery, has been previously discussed by the same authors. Since satellite imagery (LANDSAT TM5, March 2011) clearly shows evidences of high concentrations of suspended sediments (figure 2), it suggests a long-term changing sea-bottom morphology caused by a heavy suspended and bedload sediment transport.

Considerable efforts have been carried out to understand the sedimentation process in the Colorado River Delta. Baba et al. (1991) suggests that sediments in the wide and shallow platforms of the Northern Gulf of California are being intensely reworked, re-suspended and also transported southwards. Based mostly on north-migrating bed-forms and coastal sand bars, Meckel (1975) described a dominant sediment transport along the Sonora coast. Filloux (1960) described a dominant north to south sediment transport along the Baja California coast. Baumgartner et al. (1991) suggest that other sources of sediment have become important such as aeolian input from the northern Mexican desert. Applying principles of analytical geometry and vector analysis of textural data, Carriquiry and Sanchez (1999) estimated the direction of the mean transport vectors. Their results indicate the existence of two opposing littoral transport components along the Sonoran and Baja California coasts. With a few local exceptions, sediment transport occurs from SE to NW along the coast of the State of Sonora and from NW to SW along the Baja California. According to them, sediment supplied to the area comes from three different sources: (1) sediment derived from the actual delta structure (2) sediment provided by the adjacent La Mesa deposits that becomes exposed along the Sonora coastline, and (3) sediment supplied by the coast of Baja California. Other studies concerning sedimentation on the Colorado River Delta were carried out by Baba et al (1991), Carriquiry (1993), Cupul (1994) and Zamora (1993). Most of the above studies have been based on geologic considerations, satellite imagery and textural data as well as measurements of suspended sediment concentration.

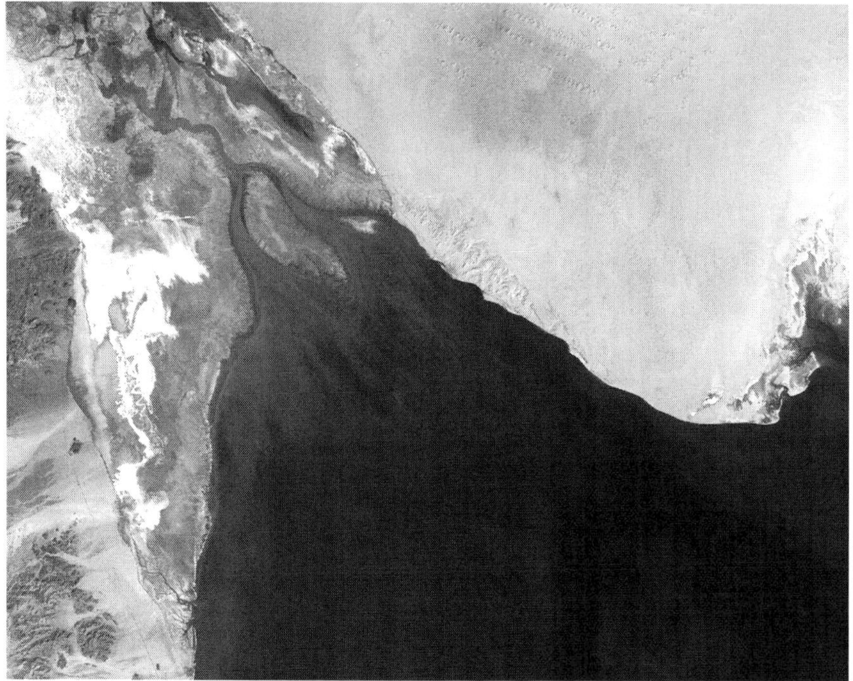

Fig. 2. Sediments distribution in the Colorado River Delta (LANDSAT TM5 March 2011), RGB321.

In figure 3, the bathymetry and the deltaic configuration of the studied area is displayed. The most important features are the mouth of the Colorado River, the islands of Montague and Gore and the Wagner basin in the south. The triangular form and the bathymetric convergence contribute notably trough the continuity equation to increase the tidal amplitudes. To investigate the dynamics and the transport of sediments in the Colorado River Delta and in other areas of the Gulf of California, we applied the vertically integrated numerical model described above.

In the northern part of the Gulf of California, the dominant signal is the M_2 tide (Principal lunar constituent). In figure 4, the calculated sea surface elevation at four different times of a M_2 tidal cycle is shown. Around the island Montague, amplitudes of more than 2 m are found. These large tidal amplitudes are associated to strong tidal currents that cause a bedload transport of sediments through the equations (5) and (6). In our simulations, the bathymetry evolves at each time step, maintaining completely the non linearity of the calculations. Tidal currents in the Gulf of California and particularly in the Colorado River Delta have been investigated in several research works (Carbajal et. al, 1997; Carbajal & Backhaus, 1998); Salas et.al, 2003). We focus here our interest in the bedload transport of sediments.

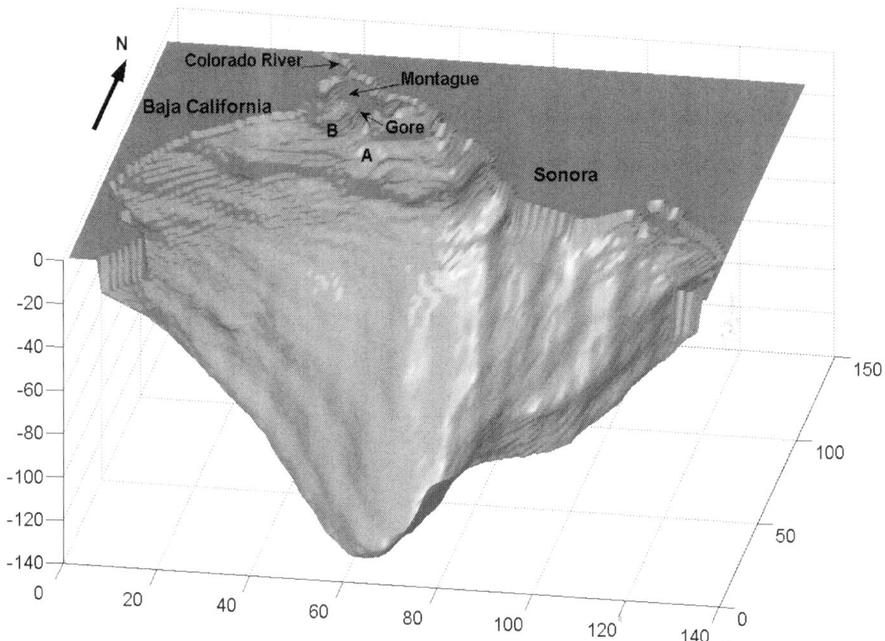

Fig. 3. Bathymetry and the deltaic structure of the northernmost part of the Gulf of California.

To give an idea how the process of erosion and accretion occurs in the Colorado River Delta, we show time series of the bottom evolution at the points A and B whose positions are indicated in figure 3. We simulated the bedload transport of sediments for one year. The variation during two M_2 tidal periods of the accumulated sediment, instantaneous bedload sediment transport, the instantaneous tidal velocities and the tidal elevation are displayed in figures 5a and 5b. All these variables are separated in the x-component (left side) and y-component (right side). The accumulated sediment indicate, at every time step, how much sediment has been built up separately by the x and y components of the volumetric sediment flux vector $\mathbf{S} = \mathbf{S}(S_x, S_y)$. We note that at point A, the x-component and the y-component are out of phase and their contribution cancels each other to some extent. At the point B, the accumulated sediment indicates a decreasing tendency in the x-component and a growing trend in the y-component. The behavior of the accumulated sediment at these two points gives us an idea about the complexity of the transport of sediment process. The instantaneous rate of bedload sediment transport, the components of the velocity vector and the sea surface elevation reflects the nonlinearity of this kind of phenomena. A tidal distortion is observed in the bedload sediment transport, in the accumulated bedload sediment transport and in the two components of the velocity vector. Observe that the curves of these variables deviate clearly from a sinusoidal shape. When all net effects at all grid points are quantified, a general morphological change is then visualized. The largest accumulative sediment transport rate was found in the neighborhood of the island Montague, with a value of about 0.0005 m^2/s. But we have to remember that this parameter is oscillating continuously, sometimes it has a growing character and sometimes it has a decreasing trend.

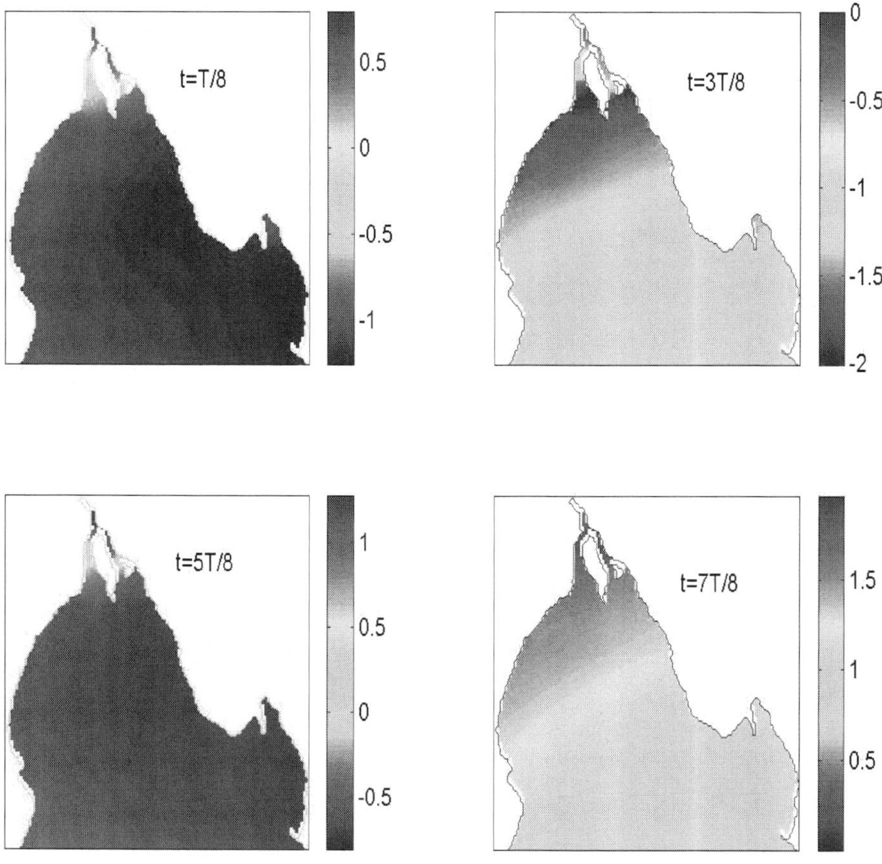

Fig. 4. The calculated sea surface elevation at four different times of a M_2 tidal cycle is shown. The sea surface elevation is given in meters.

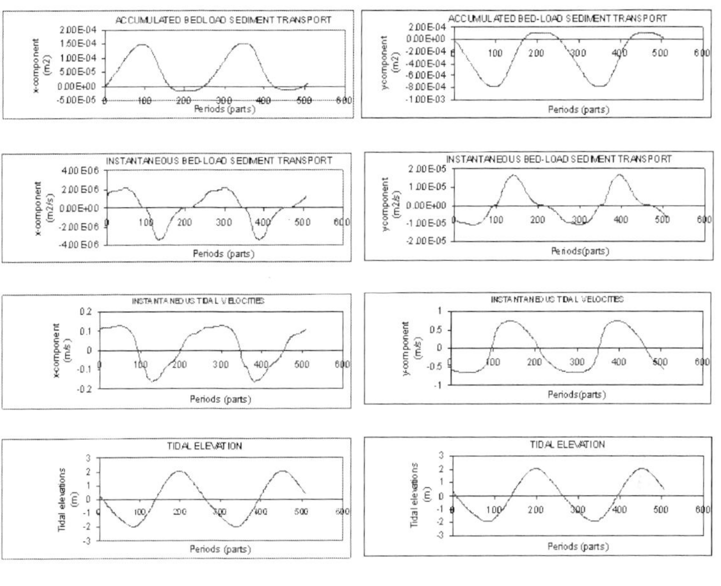

Fig. 5a. Sediment transport and hydrodynamic parameters obtained for two tidal periods at control point A.

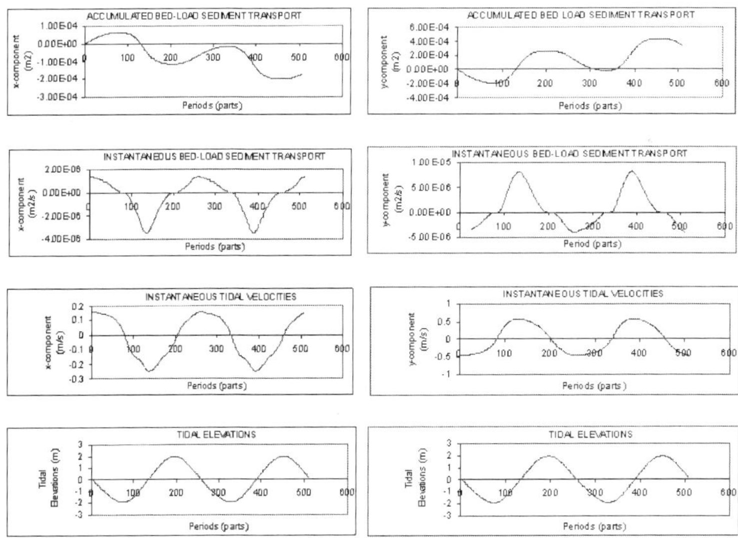

Fig. 5b. Sediment transport and hydrodynamic parameters obtained for two tidal periods at control point B.

In figure 6, the net morphological change caused by the M_2 tide after one year of numerical simulation is displayed. The major morphological evolution takes place around the islands of Montague and Gore, in the northernmost part of the Colorado River Delta. Changes due to erosion and accretion processes larger than 0.1 m were observed. The presence of sandbanks in the Colorado River Delta has been detected by bathymetric measurements (Meckel, 1975; Alvarez et al., 2009). It is interesting to observe that in our calculation there is a tendency to the formation of sandbanks to the south of Montague Island. Although we carried out calculations for the bedload sediment transport alone, there are some dynamic similarities among the distribution of suspended sediment shown in figure 2 and the pattern of the morphological change by bedload sediment transport after one year of simulation of the M_2 tide. The modeled areas of erosion and accretion show a similar tendency like the direction of the sediment filaments depicted in figure 2. The general morphological pattern calculated in this research work agrees with that showed by Montaño and Carbajal (2008). However, we make here emphasis in the morphological changes described in the time series of the accumulated sediment and of the instantaneous sediment transport (Figure 5).

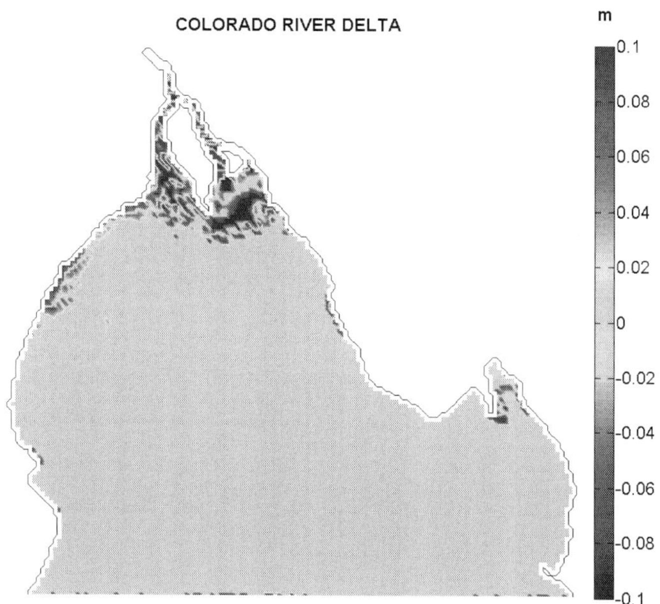

Fig. 6. Areas of erosion and accretion caused by the M_2 tide after one year of simulation.

4.2 Yavaros Bay

The Yavaros Bay is located in the central-eastern coast of the Gulf of California, between 26° 40′ N and 26° 45′ 33″ N and 109° 25′ 21″ W and 109° 34′ 31″ W. The region is characterized by a semi-arid climate with winds predominantly from the southwest in the summer months and from the northeast in winter. As many coastal lagoons, this water body arose by the developing of sand bars, keeping on an opening of about 1.7 km. The Yavaros Bay can be considered a well connected coastal lagoon with the adjacent Gulf of California. The formation of coastal lagoons in the Gulf of California was investigated by Lankford (1976). He stated that most of these water bodies are shallow embayments associated with deltaic systems. Yavaros Bay is located in a region where the rivers Yaqui, Mayo and El Fuerte have discharged huge quantities of sediment to this part the Gulf of California. Located not so far away from the tropic of cancer, the direct solar radiation is intense with large evaporation rates. In this region, evaporation exceeds precipitation by far, since average annual evaporation values from 1500 to 2000 mm have been estimated whereas the average annual rainfall varies among 300 and 500 mm (Dworak & Gómez-Valdés, 2003). With exception of the months from July to September, rainfall is in general scarce. The bottom topography of the Yavaros Bay is very complex with a dominant navigation channel from the mouth to the northwest direction (figure 7). Maximum depths of about 8 meters are found and very shallow areas of one meter or less are situated on the northeastern side.

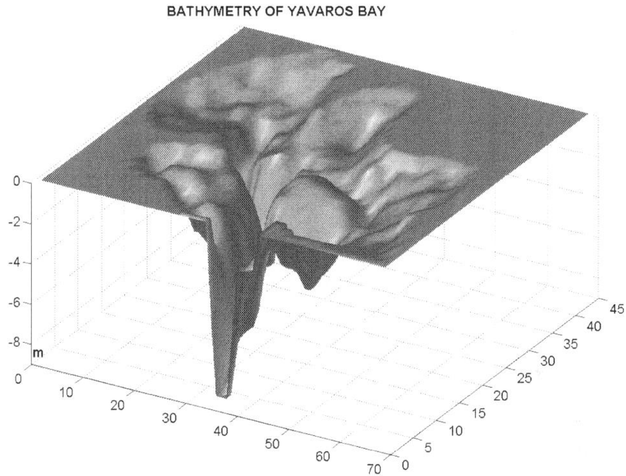

Fig. 7. Bathymetry of the Yavaros Bay.

The central part of the Gulf of California is characterized by mixed tides with predominance of diurnal signals. Tides in the Yavaros Bay have been studied in detail (Dworak & Gómez-Valdés, 2003; Dworak & Gómez-Valdés, 2005). Through non-linear processes, the propagation of tides in very shallow areas leads to generation of high harmonics that are controlled by astronomical configurations. The generation and behavior of tides in shallow waters like this coastal lagoon is affected, for example, by the lunar and solar declination effects. Since tides in this coastal lagoon have a mixed character, semidiurnal, diurnal and fortnightly oscillations can be distinguished in measurements. Semidiurnal tides are dominantly originated by the principal solar (S_2) and lunar (M_2) constituents, by solar and lunar declination effects (K_2) and by longer lunar elliptic effect (N_2). Diurnal tides are due to solar and lunar declination effects (K_1), to main lunar (O_1) and solar (P_1) contributions and to lunar elliptic effect (Q_1). Fortnightly oscillations are due to the phases of the Moon (synodic), declinational variations (tropical), and the time taken for the Moon to move from perigee to perigee (anomalistic) (Dworak & Gómez-Valdés, 2005). The numerical modeling of tides and, of course, measurements indicate velocities of the order of 1 m/s. In figure 8, tidal currents produced by the soli-lunar declinational K_1 are displayed. Velocities larger than 0.3 m/s are found in the navigation channel and adjacent areas.

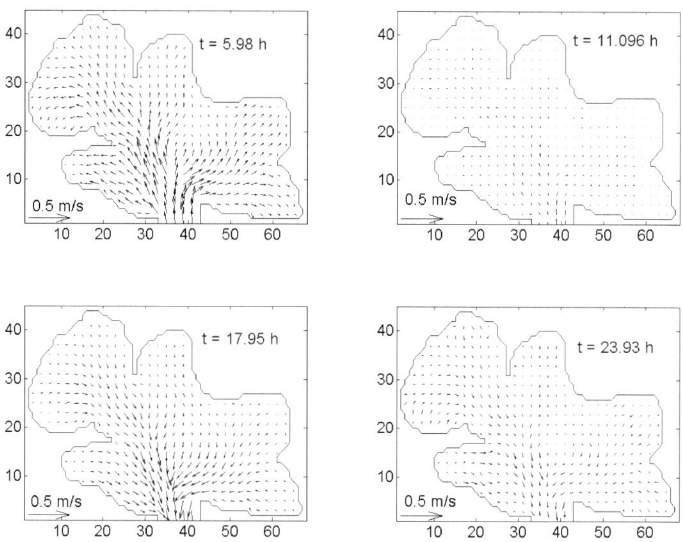

Fig. 8. Numerical modeling of currents produced by the soli-lunar declinational K_1 tide in the Yavaros Bay.

As an example, we document the bedload transport of sediment with a calculation where the K_1 tidal constituent is forcing the dynamics in the Yavaros Bay. The amplitude of the K_1 tide at the entrance of the lagoon system is 0.249 m. Although recent research work on tides in the Yavaros Bay reveal that 13 tidal constituents (M_{sf}, O_1, K_1, M_2, S_2, N_2, MK_3, SK_3, M_4, MS_4, $2SK_5$, M_6 and $2SM_6$) (Dworak and Gómez-Valdés, 2003), are of relative importance, the amplitude of the K_1 is the largest and to some extent, the dominant signal. In figure 9, the morphological changes after one year of simulation are shown. Areas of erosion (blue) and accretion (red) reach for one year heights of a few centimeters. If one considers that the time scale of sandbanks formation is of centuries, then a linear extrapolation to one century of the results shown in figure 9 would reach heights of a few meters. However, this linear extrapolation is, of course, not correct due to the intrinsic non-linearity of the hydrodynamic and transport of sediment processes. An accretion or erosion area, formed by the hydrodynamics, modifies the flow and the flow modifies again the transport of sediment and so on. The area of intense mobility of sediment is at the entrance of this coastal lagoon, therefore it is very important to understand the sedimentation process and to develop mathematical models to predict correctly the evolution of the morphology in this kind of coastal lagoons.

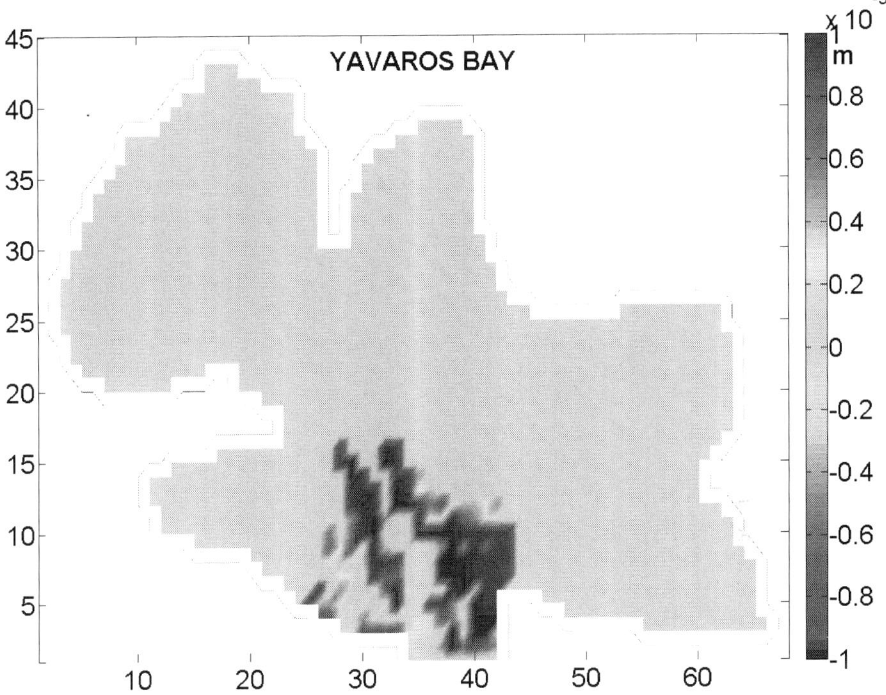

Fig. 9. Morphological changes caused by 13 tidal constituents in the yavaros Bay after one year of simulation.

5. Conclusions

In general terms, we have described the problematic associated with the transport of sediments in coastal areas. We commented several methods (theoretical, experimental and numerical) to investigate the mechanisms involved in transport of sediments. We mentioned several seas of the world where tidal amplitudes and the transport of sediment are large. We have shown that satellite imagery reveals large sand patterns in the northernmost part and along the eastern coast of the Gulf of California, where an important number of water bodies (Bays of Adair, San Jorge, Yavaros) and coastal lagoons (Topolobampo and Santa María la Reforma) are located. We presented a methodology to investigate the transport of sediments based on the application of a vertically-integrated two-dimensional numerical model together with the use of a volumetric sediment flux vector and an equation of conservation of sediment. We documented the application of this methodology with the numerical modeling of the bedload sediment transport in two water bodies of the Gulf of California, namely the Colorado River Delta and the Yavaros Bay. For the Colorado River Delta, we calculated a morphological change (figure 6) which is similar to the distribution of suspended sediment depicted in figure 2. In time series of accumulated bedload sediment transport and of instantaneous bedload sediment transport we explained the complexity associated with this kind of non-linear processes. Erosion and accretion areas arise as a consequence of the interaction of the tidal induced flow and a sandy bottom. These areas may grow or decrease with the time, or even the rate of grow. We presented results of the bedload sediment transport in the Yavaros Bay caused by the diurnal tide K_1. We calculated the morphological change after one year of numerical simulation, finding changes of the order of a few centimeters. These grow rates are in concordance with observed and theoretically calculated grow rates. Finally, there is a huge quantity of research work to be done to understand and to quantify the transport of sediments in the Gulf of California. There are many water bodies, particularly along the eastern coast, interacting with the littoral transport of sediments, or simply coastal lagoons interacting or exchanging sediment with Gulf of California.

6. Acknowledgements

Thanks are due to Dr. Cristina Noyola Medrano, Department of Applied Geoscience, Instituto Potosino de Investigación Científica y tecnológica, for his valued and productive assistence. We thank Dr. J.A. Dworak for the bathymetric data.

7. References

Álvarez, L.G.; Suarez-Vidal, F.; Mendoza-Borunda, R. & González-Escobar, M. (2009). Bathymetry and activ geological structures in the Upper Gulf of California. *Boletín de la Sociedad Geológica Mexicana,* Volumen 61, No. 1, p. 129-141.

Baba, J.; Peterson, C.D. & Schrader, H.J. (1991). Fine- grained terrigenous sediment supply and dispersal in the Gulf of California during the last century. In: J.P. Dauphin and B.R.T. Simoneit (Editors), The Gulf and Peninsular Province of the California. *Am. Assoc. Petrol. Geol. Mem.,* 47, 589-602.

Baumgartner, T. R., Ferreira, V., Schrader, H. J. (1991). Varve formation in the Central Gulf of California : a reconsideration of the origin dark laminae from the 20th century varve record. In: Dauphin, J. P., Simineit. B. R. T. (Eds.), The Gulf and Peninsular Province of the California . *Am. Assoc. Petrol. Geol. Memo.* 47, 617-635.

Carbajal, N.; Souza, A. & Durazo, R. (1997). A numerical study of the exROFI of the Colorado River, *Journal of Marine Systems* 12 1/4.

Carbajal, N. & Backhaus, J.O. (1998). Simulation of tides, residual flow and energy budget in the Gulf of California. *Oceanologica Acta,* 21, pp. 429-446.

Carbajal, N. & Montaño, Y. (1999). Tasas de Crecimiento y Escalas de Bancos de Arena en el Delta del Río Colorado. *Ciencias Marinas.* 25.

Carbajal, N.; Piney, S. & Gómez-Rivera, J. (2005). A numerical study on the influence of geometry on the formation of sandbanks. *Ocean Dynamics,* 55, pp. 559-568.

Carriquiry, J. D. (1993). Dynamics of sedimentation in the deltaic system of the Colorado River. *Proc. 2nd Int. Mtg. Geology of the Baja California Peninsula* . Ensenada Baja California.

Carriquiry, J.D. & Sánchez, A. (1999). Sedimentation in the Colorado River delta and Upper Gulf of California after nearly a century of discharge loss. *Marine Geology* 158, 125-145.

Cupul, A. L. (1994). Flujos de sedimentos en suspensión y nutrientes en la cuenca estuarina del Rio Colorado. M. Sc. Thesis. Facultad de Ciencias Marinas, Universidad Autónoma de Baja California.

Dworak, J.A. & Gómez-Valdés, J. (2003). Tide-induced residual current in a coastal lagoon of the Gulf of California. *Estuarine Coastal and Shelf Science*, 57, pp. 99-109.

Dworak, J.A. & Gómez-Valdés, J. (2005). Modulation of shallow water tides in an inlet-basin system with a mixed tidal regime. *Journal of geophysical Research*, vol. 110, C01007, doi:10.1029/2003JC001865.

Dodd, N.; Blondeaux, P.; Calvete, D.; De Zwart H.E.; Falqués, A.; Hulscher, S.J.M.H.; Różyński, G. & Vittore, G. (2003). Understanding Coastal Morphodynamics Using Stability Methods. *Journal of Coastal Research*, 19, 4, pp. 849-865.

Filloux, J. H. (1973). Tidal pattern and energy balance in the Gulf of California. *Nature* 243, 217-221.

Herbich, J. B. (2000). *Handbook of Coastal Engineering*. McGraw-Hill. Editor.

Hulscher, S.J.M.H. (1995). Tidal-induced large-scale regular bed forms patterns in a three-dimensional shallow water model. *Report R95-11, Institute for Marine and Atmospheric Research Utrecht.*

Hulscher, S.J.M.H. (1996a). Formation and migration of large-scale rhythmic sea-bed patterns. Ph.D. thesis, Utrecht University.

Hulscher, S.J.M.H. (1996b). Tidal-induced large-scale regular sea bed form patterns in a three-dimensional shallow water model. *Journal of Geophysical Research*, 101, 20, 727-20, 744.

Huthnance, J.M. (1982). On one mechanism forming linear sand banks. *Estuarine Coastal and Shelf Sciences*, 14, 79-99.

Julien, P. Y. (1998). *Erosion and Sedimentation*. Cambridge University Press, 280 pp.

Komarova, N.L. & Hulscher, S.J.M.H. (2000). Linear instability mechanism for sand wave formation. *Journal of Fluid Mechanics*, vol. 413, pp. 219-243.

Lankford, R.R. (1976). Coastal lagoons of Mexico: their origin and classification. In: M. Wiley (Editor), *Estuarine Processes*. Vol. II. Academic Press, New York, pp. 182-215.

Meckel, L.D.(1975). Holocene sand bodies in the Colorado River Delta area , Northern Gulf of California . In: Broussard , M.C. (De.) Deltas, Models for Exploration. Houston Geological Society, TX, pp. 239-265 .

Montaño, Y. & Carbajal, N.(2008). Numerical experiments on the long term morphodynamics of the Colorado River Delta. *Ocean Dynamics*. 58, 19-29.

Montaño-Ley, Y.; Peraza-Vizcarra, R. & Paez-Osuna, F. (2007). The tidal hydrodynamics modeling of the Topolobampo coastal lagoon system and the implications for pollutant dispersion. *Environmental Pollution*, Volume 147, Issue 1, pp. 282-290.

Salas de León, D.A.; Carbajal, N. & Monreal-Gómez, A. (2003). Residual circulation and tidal stress in the Gulf of California. *Journal of Geophysical Research-Oceans*. 108(C10): 3317-3327.

Van Rijn, L. C. (1993). *Principles of sediment transport in rivers, estuaries and coastal seas.* Aqua Publications.

Zamora, C. (1993). Comportamiento del seston en la desembocadura del Río Colorado, Sonora-Baja California. *Tesis, Facultad de Ciencias Marinas, Universidad Autónoma de Baja California.*

8

Sediment Transport Modelling and Morphological Trends at a Tidal Inlet

Sandra Plecha[1], Paulo A. Silva[1],
Anabela Oliveira[2] and João M. Dias[1]
[1]*CESAM &University of Aveiro*
[2]*National Laboratory of Civil Engineering*
Portugal

1. Introduction

Currently it is of great concern the morphological changes observed at distinct coastal systems. Frequently, drastic episodes of coastal erosion, threatening the houses built near the shore, and land loss are reported. At coastal lagoons and estuaries, increasing accretion can interfere with the water renewal and therefore with the local ecosystems. At inland harbors, the accretion of inlet and channels can restrict the navigation and consequently the harbor conditions and its management.

The sediment transport, identified by patterns of erosion or accretion, can be evaluated by single formulae that compute the bedload and suspended load or by sediment transport models.

In the literature are published formulae for bed-load transport of sediments in conditions characteristic of coastal waters, covering current alone, current plus symmetrical waves, current plus asymmetrical waves alone and integrated longshore transport.

Due to the fact that sand transport models are often based on semi-empirical equilibrium transport formulae that relate sediment fluxes to physical properties, such as velocity, depth and grain size, it is crucial to perform sensitivity analysis of the formulae used.

Pinto *et al.* (2006) compared four sediment transport formulations considering only the tidal current only: Ackers and White (1973); Engelund and Hansen (1967); van Rijn (1984a,b,c) and Karim and Kennedy (1990). The authors concluded that the van Rijn formula is the most sensitive to basic physical properties. Hence, it should only be used when physical properties are known with precision.

The sediment transport modules, coupled with hydrodynamic and wave modules compose the morphodynamic models, resulting in very complex systems emergent in the last years. In the past 30 years, morphodynamic models have been developed (Nicholson *et al.*, 1997). Among them the MIKE21, developed by the Danish Hydraulic Institute - DHI (Warren and Bach, 1992) and DELFT3D, by WL | Delft Hydraulics - DH (Roelvink and Banning, 1994), are the most popular. Most of the morphodynamic studies are generally related on engineering methods and techniques for coastal defense.

The DELFT3D model was applied by several authors, such as Grunnet *et al.* (2004), Xie *et al.* (2009) and Tung *et al.* (2009), between others. Grunnet *et al.* (2004) applied DELF3D to hindcast the morphological development of shoreface nourishment along the barrier island of Terschelling, The Netherlands. The morphodynamic results showed a dependency on the spatial scale: on the scale of the bed level evolution with respect to bar migration and growth, model predictions are poor as the nearshore bars are predicted to flatten out. On the other hand, the model satisfactory predicted the overall effects of the nourishment taking into account the mass volumes integrated over larger spatial scales.

Successful results were obtained by Xie *et al.* (2009) in the analysis of the physical processes and mechanisms essential to the formation and evolution of a tidal channel in the macro-tidal Hangzhou Bay, China. This work shows that the model results reproduced accurately the real morphological features. The results showed that spatial gradients of flood dominance, caused by boundary enhancement via current convergences, are responsible for the formation of the channel system, due to, between other effects, the funnel-shaped geometry.

DELFT3D was also applied to model inlets morphodynamics. Tung *et al.* (2009) investigate the migration and closure of an idealized tidal inlet system due to wave driven longshore sediment transport. The authors reproduced a typical example of a migrating tidal inlet due to oblique waves that include features such as ebb channel formation, migration and welding to the downdrift barrier. It was also reproduced the inlet closure due to the prolongation of the inlet channel and infilling with littoral-drift material.

Other morphodynamic modeling systems were developed and applied to coastal lagoons and inlets. For example, Ranasinghe *et al.* (1999) developed a morphodynamic model capable of simulating the seasonal closure of inlets, including both longshore and cross-shore transport processes. This model was successfully applied to two idealized scenarios demonstrating its ability to produce realistic results. Later, Ranasinghe and Pattiaratchi (2003) used field experiments and this numerical model to study a real case, the Wilson Inlet at Western Australia. With this study they identified the morphodynamic processes governing seasonal inlet closure and determined the effect of the seasonal closure of the inlet on the morphodynamic characteristics of the adjacent estuary/lagoon. These authors concluded that the inlet closure is due to the onshore sediment transport induced by persisting swell wave conditions during summer.

Cayocca (2001) developed and applied a two-dimensional horizontal morphodynamic model, combining modules for hydrodynamics, waves, sediment transport and bathymetry updates, to Arcachon lagoon, at the French coast. The results showed that the tide is responsible for the opening of a new channel at the extremity of the sand pit, while waves induce a littoral transport responsible for the longshore drift of sand bodies across the inlet.

Work *et al.* (2001) described and applied a modeling system (Mesoscale Inlet Morphology, MIM) that include coupled modules of hydrodynamic, wave transformation, shoreline change and sediment transport, to the undredged and unstabilized Price Inlet, South California. The predicted bathymetric changes were similar in sign and location to the observations, but the model tends to underpredict the magnitude of changes.

Silvio *et al.* (2010) used a long term model of planimetric and bathymetric evolution of Venice lagoon. These authors applied this model to a schematic system consisting of an inlet area and a lagoon basin and considering the interactions between tidal currents, longshore currents and sea waves. It was found that the planimetric evolution of the channel network was much faster than the bathymetric evolution of the lagoon.

One aspect that contributes to the patterns of the sediment transport in a tidal inlet is the ebb/flood dominance. The processes that contribute to determine this dominance were studied by Robins and Davies (2010) through a 2D model (TELEMAC modeling system) applied to Dyfi estuary in Wales, UK. The authors concluded that shallow water depths lead to flood dominance in the inner estuary whilst tidal flats and deep channels cause ebb dominance in the outer estuary.

Two morphodynamic modeling systems developed in Portugal are described in the literature: MOHID and MORSYS2D.

MOHID model simulates the non-cohesive sediment-dynamics in lagoons driven by tide, waves and river flows (Malhadas *et al.*, 2009). It integrates MOHID hydrodynamic (Aires *et al.*, 2005; Neves *et al.*, 2000), sand transport modules (Silva *et al.*, 2004) and the wave model STWAVE (Smith *et al.*, 2001). This model was applied to Óbidos lagoon in order to evaluate the effect of the bathymetric changes on the hydrodynamic and residence time, suggesting strategies to avoid the lagoon's accretion and to prevent the safety of local houses at the lagoon margins. The conclusions confirmed that tidal propagation depends strongly on the bathymetric configuration.

The MORSYS2D model is composed by the hydrodynamic models ADCIRC or ELCIRC, the wave model SWAN and the sediment transport and bottom update model SAND2D (Fortunato and Oliveira, 2003, 2007; Bertin *et al.*, 2009b). The performance of MORSYS2D was assessed in the morphodynamic simulation forced both by current and waves of the Óbidos lagoon and is described in Oliveira *et al.* (2004). The morphodynamic of this lagoon and its inlet have been studied and published in several works: Fortunato and Oliveira (2006, 2007); Bertin *et al.* (2007, 2009a,b); Fortunato *et al.* (2009). In these studies the authors concluded, that the sediment grain size and the choice of the empirical sediment transport formulae affect the morphological predictions. The development of a meander and the formation of sandbars in the wave dominated inlet of Óbidos lagoon had been successfully simulated also with the recently partially parallelized MORSYS2D (Bruneau *et al.*, 2010). Besides Óbidos, other inlets morphodynamics were also studied with this model, the Ancão inlet in Ria Formosa (Bertin et al., 2009c) and the Ria de Aveiro lagoon inlet (Oliveira et al., 2007, Plecha et al., 2007,2010,2011). Other studies have been performed with this modeling system: scenario test cases of an idealized dredged sandpit (Ramos *et al.*, 2005), a dredged area evolution in southern Portugal (Rosa *et al.*, 2011), the Aljezur coastal stream morphological variability (Guerreiro *et al.*, 2010).

This work presents a sensitivity analysis of several formulations usually applied for estimating the sediment transport rates in coastal lagoons and estuaries. Morphodynamic simulations are then performed at a study case inlet, the northwest coastal lagoon inlet of Ria de Aveiro, considering the selected formulations to compute the sediment transport rates with the MORSYS2D modeling system. The numerical results are analyzed concerning the morphological changes (erosion and accretion areas) induced at that site.

2. Sediment transport

The sediment transport processes are very important in estuaries and inlets being essential to describe their morphologic changes. These processes are generally complex and are function of the hydrodynamic circulation and sediment characteristics of the bed.

When describing the sediment balance at an inlet, it has to be taken into account the longshore currents induced by the waves that approach the coast at an oblique angle and the tidal currents alone or coupled with waves. The following subsections present several formulations to compute the longshore and load sediment transport rates published in literature and used frequently by researchers.

2.1 Longshore sediment transport

The sediment transport induced by longshore currents along the coast is easily identified through coastal erosion or accretion around structures. Along the coast, considerable amounts of sediment are transported, depending on the height, period and direction of the waves. Additionally, the sediment size and the bottom slope are important parameters that determine the longshore transport.

The longshore sediment transport can be calculated by means of several longshore sediment transport formulations. Herein are presented six of them. The first one was presented by Vongvisessomjai *et al.* (1983) (Chonwattana *et al.*, 2005), and was adapted from the Coastal Engineering Manual (2002) formula. In this formulation the longshore sediment transport is proportional to the wave characteristics in deep water. The remaining five formulations are published in literature (e.g. Larangeiro and Oliveira (2003)) and are proportional to the wave characteristics in the breaker line, with different dependences in wave breaker height, wave period and incident wave breaker angle.

2.1.1 Formulation CERC

This formulation is based on the assumption that the longshore transport rate depends on the longshore component of wave energy flux in the surf zone:

$$Q_\ell = 0.064208 H_0^{5/2} F(\alpha_0) f \tag{1}$$

with

$$F(\alpha_0) = \cos\alpha^{1/4} \sin 2\alpha_0 \tag{2}$$

where H_0 is the wave height in deep water, a_0 is the wave angle in deep water and f is the wave frequency.

2.1.2 Formulation C2

Valle *et al.* (1993) proposed a formulation for the longshore transport rate given by:

$$Q_\ell = \frac{k}{16\sqrt{\gamma}(\rho_s - \rho)a'} \rho H_b^{5/2} \sin 2\alpha_b \tag{3}$$

with $k = 1.4e^{-2.5 d_{50}}$, the medium particle diameter is d_{50} in mm and $a' = 1 - n$ is the breaker index, ρ_s is the sand density, ρ is the water density, g is the gravity, a_b is the wave angle at breaker point and n is the sediment porosity. The wave height in this formulation is the root mean square wave height H_{rms} given by:

$$\begin{gathered} H_{rms} = \sqrt{8m_0} \Rightarrow \sqrt{m_0} = \frac{H_{rms}}{\sqrt{8}} \\ H_s = 4\sqrt{m_0} \Rightarrow \sqrt{m_0} = \frac{H_s}{4} \\ \frac{H_{rms}}{\sqrt{8}} = \frac{H_s}{4} \Leftrightarrow H_{rms} = \frac{\sqrt{8}}{4} H_s \Leftrightarrow H_{rms} = \frac{\sqrt{2}}{2} H_s \end{gathered} \tag{4}$$

where H_s is the significant wave height.

2.1.3 Formulation K&I

Komar and Inman (1970) proposed a formulation for the longshore transport rate given by:

$$I_\ell = K'(Ec_g)_b \cos\alpha_b \frac{\overline{V_\ell}}{u_m} \quad (5)$$

with

$$\overline{V_\ell} = 20.7i(gH_b)^{1/2} \sin 2\alpha_b \quad (6)$$

And

$$u_m = \frac{\gamma}{2}(gH_b)^{1/2} \quad (7)$$

where K' is a dimensionless coefficient, E is the wave energy, c_g is the wave group velocity and i is the bottom slope. By the Coastal Engineering Manual (2002):

$$I_\ell = (\rho_s - \rho)g(1-n)Q_\ell \Leftrightarrow Q_\ell = \frac{I_\ell}{(\rho_s - \rho)g(1-n)} \quad (8)$$

Thus:

$$Q_\ell = \frac{K'(Ec_g)_b \cos\alpha_b \frac{\overline{V_\ell}}{u_m}}{(\rho_s - \rho)g(1-n)} \quad (9)$$

2.1.4 Formulation Kr88

Kraus *et al.* (1988) proposed a formulation for the longshore transport rate given by:

$$I_\ell = 2.7(R - R_c) \quad (10)$$

With $R = \overline{V_\ell} W H_b$, being W the total width of the cross section and R_c a constant. By Kraus *et al.* (1988) it is known that I_ℓ has units of N s^{-1} and the coefficient 2.7 is not adimensional. By Larangeiro *et al.* (2005):

$$Q_\ell = \frac{2.7(R - R_c)}{(\rho_s - \rho)g(1-n)} \quad (11)$$

2.1.5 Formulation K86

Kamphuis *et al.* (1986) proposed a formulation for the longshore transport rate given by:

$$Q_\ell = \frac{1.28}{(\rho_s - \rho)(1-n)} H_b^{7/2} i d_{50}^{-1} \sin 2\alpha_b \quad (12)$$

2.1.6 Formulation K91

Kamphuis (1991) proposed a formulation for the longshore transport rate given by:

$$Q_\ell = \frac{2.27}{(\rho_s - \rho)(1-n)} H_b^2 T_p^{1.5} i^{0.75} d_{50}^{-0.25} \sin^{0.6} 2\alpha_b \qquad (13)$$

where the peak period is $T_p = 2.07T_z - 4.02$ and T_z is the zero-upcrossing wave period.

2.2 Sensitivity analysis to longshore sediment transport formulations

To choose which one of the previous formulations is the most adequate to use in a study case, it is mandatory to perform a sensitivity analysis of the sediment transport formulations to several characteristics: sediment size d_{50} and bottom slope i. The analysis of the results should be performed knowing à priori a range of longshore transport values in order to compare the values obtained.

In this case, the longshore sediment transport is calculated by means of the six longshore sediment transport formulations presented in the previous subsection. Each one of this formulation has a different dependency on the wave height and angle at the breaker line and on d_{50} and bottom slope. These dependencies are presented in Table 1.

Formulation	$Q_\ell (m^3 year^{-1})$	$K \propto$
C2	$K_1 H_{rms}^{5/2} \sin 2\alpha_b$	$e^{d_{50}}$
K&I	$K_2 H_b^{5/2} \cos\alpha_b \sin 2\alpha_b$	i
Kr88	$K_3 H_b^{3/2} \sin 2\alpha_b W$	i
K86	$K_4 H_b^{7/2} \sin 2\alpha_b$	$d_{50}^{-1} i$
K91	$K_5 H_b^2 T_p^{1.5} \sin^{0.6} 2\alpha_b$	$d_{50}^{-0.25} i^{0.75}$

Table 1. Longshore sediment transport formulations.

It should be noted that all these formulae make use of significant wave height, except the C2 formulation which accounts for the root mean square wave height, H_{rms}. The K&I, Kr88, K86 and K91 formulations depend on the bottom slope and only the C2, K86 and K91 have dependencies on the sediment size d_{50}. The longshore sediment transport is computed considering a wave regime of 11 years long of the study area and typical values for the sediment size d_{50} and bottom slope, presented in bibliography or measured in situ. However, as these two parameters are not known accurately, a sensitivity analysis of the several formulations to typical values of d_{50} and i was performed. Figure 1 illustrated the results obtained for the formulations presented for typical ranges of d_{50} and bottom slope of the northwest Portuguese nearshore

It is observed that there is a wide spread of results for the longshore sediment transport. The K86 formulation revealed to be highly sensitive to the value of d_{50} considered, due to its inverse dependency on the sediment size. Also, the C2 formulation, with an exponential dependency on d_{50} revealed a strong dependency to this parameter. Analyzing the results of the dependency of longshore sediment transport to the bottom slope, again the K86 formulation shows the higher dependency due to the linear dependency on slope i. The same dependency and sensible behavior is obtained when using the K&I formulation to compute the longshore sediment transport.

The reference value for the longshore sediment transport of the Portuguese Aveiro west coast considered was presented by Larangeiro and Oliveira (2003) and is $Q_\ell = 1\times10^6$ $m^3 year^{-1}$.

Analyzing the values obtained in Figure 1, it is concluded that the longshore sediment transport computed for the complete wave regime by all transport formulae is overestimated when compared with the reference value.

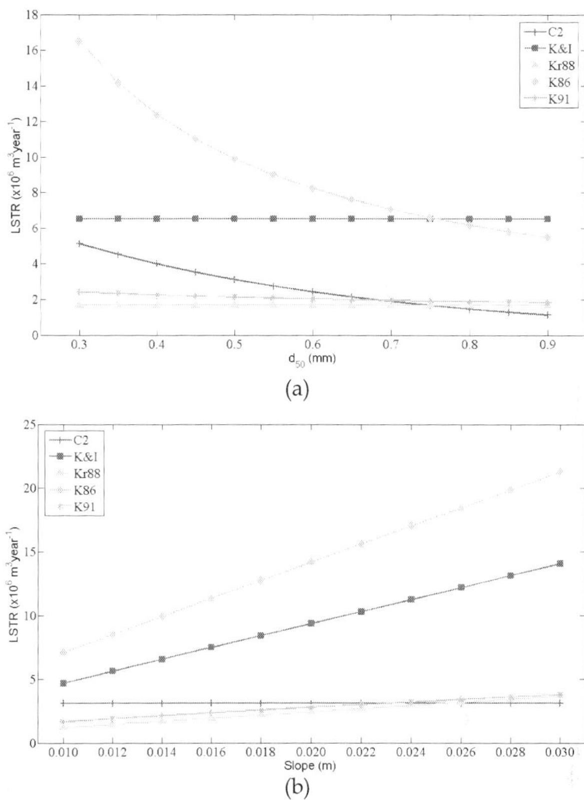

Fig. 1. Sensitivity analysis of the longshore sediment transport formulations to (a) d_{50} (mm) and (b) the bottom slope (m).

2.3 Sediment transport formulations

Usually the sediment load is subdivided into bedload and suspended load. The bedload is defined as the part of the total load that is in more or less continuous contact with the bed during the transport. It primarily includes grains that roll, slide or jump along the bed. The suspended load is the part of the total load that is moving without continuous contact with the bed as a result of the agitation of fluid turbulence (Fredsoe and Deigaard, 1992).

The nonlinear dependency of the sediment transport to the current velocity makes the net transport through inlets very sensitive to ebb/flood tidal asymmetries.

The initiation of bed sediments movement occurs when the bottom shear stress exceeds a critical shear stress defined by the gravitational force, the frictional forces on the bed and the sediment size.

The rates of sediment transport can be computed by several formulations that compute the total load considering the forcing of tidal currents or the coupled effect of tidal currents and waves. Herein are presented 8 formulations. A sensitivity analysis of these formulations is performed in subsection 2.4. These formulations integrate the morphodynamic modelling system used in this study (MORSYS2D), as presented in Section 3.

2.3.1 Formulation Bha

Bhattacharya *et al.* (2007) proposed a formulation to compute the total transport directly, for the tidal current forcing:

$$\frac{q_s}{\sqrt{(s_d-1gd_{50}^3)}}=\begin{cases}0.072078\dfrac{T^{0.893}}{d_*^{0.353}}\left(\dfrac{H}{d_{50}}\right)^{0.486} & T>2.22\\ 0.0000782\dfrac{T^{0.54}}{d_*^{0.00407}}\left(\dfrac{H}{d_{50}}\right)^{1.16} & T\leq 2.22\end{cases} \tag{14}$$

where $T=(\theta'-\theta_{cr})/\theta_{cr}$, is the transport stage parameter, θ_0 is the mobility parameter relative to grain roughness, θ_{cr} is the Shield critical shear stress, $d_*=d_{50}((s_d-1)g/\nu^2)^{1/3}$ is the dimensionless grain size, d_{50} is the medium particle diameter and ν is the cinematic viscosity of the water.

2.3.2 Formulation EH

Engelund and Hansen (1967) proposed a formulation that computes the total load directly. The threshold for initiation of motion is not considered. The sediment flux forced by tidal currents is given by:

$$\vec{q_s}=\frac{0.05}{s^2 d_{50} C^3\sqrt{g}}U^4\vec{u} \tag{15}$$

where C is the Chézy coefficient given by $\sqrt{c_f/g}$, with c_f the friction coefficient and U is the modulus of the depth-averaged velocity.

2.3.3 Formulation kk

Karim and Kennedy (1990) proposed a formulation that computes the total load transport directly considering the effect of tidal currents. The sediment flux is given by:

$$\vec{q_s}=10^{-2.821+3.369\log(v1-0.84\log v11)}\sqrt{(s-1)gd_{50}^3}\,\frac{\vec{u}}{U} \tag{16}$$

Where $v_1=U/\sqrt{(s-1)gd_{50}}$ and $v_1=(u_*-u_{*cr})/\sqrt{(s-1)gd_{50}}$. U_* is the stress velocity and u_{*cr} is the critical stress velocity.

2.3.4 Formulation MPM

Meyer-Peter and Muller (1948) proposed a formulation based on experimental studies only valid for bed-load transport. Carmo (1995) improved this formulation in order to include the influence of bed-slope. The bed load sand flux forced by tidal currents is then given by:

$$q_s = \frac{8}{(s-1)g}\left(\frac{\left|\bar{\theta}\right| - \theta_c}{\rho_s}\right)^{1.5} \tag{17}$$

where s is the dimensionless sand density, θ_c is the critical Shields parameter and ρ_s is the sand density.

2.3.5 Formulation vR

van Rijn (1984a,b,c) proposed a formulation valid for bed load and suspended load transport.
The bottom slope is considered in the evaluation of the threshold for initiation of motion. The sediment flux forced by tidal currents is given by:

$$\vec{q_s} = \left(\frac{0.053}{U}\sqrt{g\frac{\rho_s - \rho}{\rho}\frac{d_{50}^{2.1}}{d_*^{0.3}}}T_{sp}^{2.1} + FHc_a\right)\vec{u} \tag{18}$$

where T_{sp} is a transport stage parameter $(=(u_*^2 - \tau_c/\rho)/(\tau_c/\rho))$, c_a is the reference concentration $(=0.015d_{50}T_{sp}^{1.5}/(ad_*^{0.3}))$ where a is the reference level given by:

$$a = \begin{cases} 3d_{90} + 1.1\Delta\left(1 - \exp\left(-25\Delta/\Lambda\right)\right) & T_{sp} < 25 \\ 3d_{90} & T_{sp} \geq 25 \end{cases} \tag{19}$$

With $\Delta = 0.11H(d_{50}/H)^{0.3}(1 - \exp(-T_{sp}/2))(25 - T_{sp})$ is the dune height and $\Lambda = 7.3H$ is the dune length.
A minimum value of a is set to H/100. In equation 2.21 F is given by:

$$F = \frac{\left(\frac{a}{H}\right)^Z - \left(\frac{a}{H}\right)^{1.2}}{\left(1 - \frac{a}{H}\right)^Z(1.2 - Z)} \tag{20}$$

Where $Z = \omega/(\kappa\beta u_*) + \phi$ is the suspension parameter, where κ is the von Karman constant (=0.4), $\beta = 1 - 2(\omega^2/u_*^2)$ and $\phi = 1$ if $u_* \geq 100\omega$ and $\phi = 2.5(\omega/u_*)^{0.8}(c_a/0.65)^{0.4}$ if $u_* < 100\omega$.

2.3.6 Formulation AW

The formulation presented herein is an adaptation by van de Graaff and van Overeem (1979) of the Ackers and White (1973) formulation for currents to take into account the effect of waves. The total load is given by:

$$q_{st} = U_c\frac{1}{1-\lambda}d_{35}\left[\frac{U_{cw}}{U_{*cw}}\right]^n\frac{C_{dgr}}{A^m}\left[\frac{C_d^nU_{cw}\left(\frac{U_{*cw}}{U_{cw}}\right)^n}{C_dg^{n/2}\sqrt{\left(\frac{\rho_s}{\rho} - 1\right)d_{35}}} - A\right]^m \tag{21}$$

where λ is the sediment porosity, d_{35} the particle diameter exceeded by 65% of the weight and $A = 0.23 / \sqrt{d_{gr}} + 0.14$, $n = 1 - 0.2432\ln(d_{gr})$, $m = 9.66 / d_{gr} + 1.34$, $C_{dgr} = \exp(2.86\ln d_{gr} - 0.4343[\ln d_{gr}]^2 - 8.128)$. In Equation 2.23, U_{cw} and U_{*cw} are the current velocity and shear velocity modified and are given by:

$$U_{cw} = U_c\sqrt{1 + 0.5\left(\xi'\frac{U_\omega}{U_c}\right)^2} \qquad U_{*cw} = U_{*c}\sqrt{1 + 0.5\left(\xi\frac{U_\omega}{U_c}\right)^2} \tag{22}$$

With

$$\xi = 18\log\left(\frac{12H}{r}\right)\sqrt{\left(\frac{f_\omega}{r}\right)} \qquad \xi' = 18\log\left(\frac{10H}{d_{35}}\right)\sqrt{\left(\frac{f_\omega^{'}}{2g}\right)} \tag{23}$$

where r is the bed roughness and f_w and f_w' are the wave friction coefficient using r and d_{35} as bed roughness, respectively.

2.3.7 Formulation Bi

Bijker (1971) derived a formulation for bedload transport where the total load is expressed as the sum of a bedload q_{sb} term and a suspended load q_{ss} term, considering the coupled effect of tidal currents and wave regime:

$$\begin{aligned} q_{sb} &= C_b d_{50}\frac{U_c}{C}\sqrt{g}\exp\left[\frac{-0.27(\rho_s - \rho)gd_{50}}{\mu\tau_{cw}}\right] \\ q_{ss} &= 1.83q_{sb}\left[I_1\ln\left(\frac{33H}{\delta_c}\right) + I_2\right] \end{aligned} \tag{24}$$

where C_b is a wave breaking parameter (1.0 for non breaking waves and 5.0 for breaking waves with a ramp function between the two situations), C is the Chézy coefficient based on d_{50}, μ is a ripple factor (= $(C/C_{90})^{1.5}$, where C_{90} is the Chézy coefficient based on d_{90}), I_1, I_2 are integrals and τ_{cw} is the combined shear stress due to waves and currents given by:

$$\tau_{c\omega} = \tau_c\left[1 + 0.5\left(\zeta\frac{U_c}{U_\omega}\right)^2\right] \tag{25}$$

where τ_c is the bed shear stress due to currents only, U_w the wave orbital velocity, $\zeta = C\sqrt{f_\omega / (2g)}$ is a parameter for wave-current interaction, f_w a wave friction factor and U_c is the current velocity.

2.3.8 Formulation SvR

The formulation of Soulsby and van Rijn (Soulsby, 1997) compute sediment transport under the combined action of wave and currents. The total load is given by:

$$q_{st} = A_s U_c\left[\sqrt{U_c^2 + \frac{0.018}{C_d}U_{\omega rms}} - U_{cr}\right]^{2.4}(1 - 1.6\tan\beta) \tag{26}$$

where U_{wrms} is the root mean square wave orbital velocity and β is the local bottom slope, C_d is a drag coefficient and U_{cr} is the threshold velocity. $A_s = A_{sb} + A_{ss}$ with A_{sb} and A_{ss} the terms for bedload and suspended load, respectively, given by:

$$A_{sb} = \frac{0.005H\left(\frac{d_{50}}{H}\right)^{1.2}}{\left(\left(\frac{\rho_s}{\rho}-1\right)gd_{50}\right)^{1.2}} \qquad A_{ss} = \frac{0.012 d_{50} d_*^{-0.6}}{\left(\left(\frac{\rho_s}{\rho}-1\right)gd_{50}\right)^{1.2}} \tag{27}$$

where C_d is the drag coefficient $(= (0.40/(\ln(H/z_0)-1))^2)$, where z_0 is the bed roughness length. In Equation 2.28, U_{cr} is the threshold current velocity equal to $0.19 d_{50}^{0.1} \log(4H/d_{90})$ if $0.1 \text{ mm} \leq d_{50} < 0.5 \text{ mm}$ and to $8.5 d_{50}^{0.6} \log(4H/d_{90})$ if $0.5 \text{ mm} \leq d50 \leq 2.0 \text{ mm}$.

2.4 Sensitivity analysis of the sediment transport formulations

A sensitivity analysis of the sediment transport rates (q_s), computed from the aforementioned formulae, to the median sediment grain size d_{50}, the water depth h, and depth-averaged velocity U, is performed to better understand the response of the numerical solutions concerning bathymetric changes. These computations are made using a single point formulation that was retrieved from the numerical module of MORSYS2D that computes the sediment transport (Silva *et al.*, 2009) and considering values for d_{50}, h and U characteristics of inlets.

In Figure 2(a) is illustrated the transport rate q_s as a function of the medium sediment grain size, d_{50}, for a constant water depth of 2, 10 and 20 m and considering a steady current with a depth-averaged velocity value of 1.0, 1.5 and 2.0 ms^{-1}. In Figure 2(b) the added effect of a single wave with significant height of 1 m and 7 s of wave period was considered.

The SvR, EH, kk, vR and AW formulations show a decrease of the transport rate with increasing d_{50}, either in presence of tidal currents only (Figure 2(a)) or coupled with a monochromatic wave (Figure2(b)). Therefore, the dependency of these formulae in d_{50} is very similar. It can be noted that this dependency is non-linear; the range of variation of q_s for the fine and medium grain sizes is higher than for the coarser sand.

In Figure 2(a) is visible that the formulation by Bha predicts higher transport rates, except for the higher velocities (U = 2 ms^{-1}) or lower depths, where the results are close to those obtained with the other formulations.

The MPM formulation predicts systematically lower values of q_s, and changes with d_{50} are not very significant. Note that this formulation only takes into account the bed load transport.

The transport rates computed from Bi (in Figure 2(a) and (b)) and vR formulations present some oscillations with d_{50} that are not observed with any other formulae, in particular, for the finer sediments. When compared to formulae SvR, EH, kk and AW, the vR formula over-predicts the transport rates for finer sediments. The results predicted by all the formulations and from the ensemble of tests considered show a larger spread for the lowest values of U and finer sand than for the highest velocities and coarser sand, respectively. These results stand equally valid for the large range of depth-average tidal flow velocities observed within the inlet.

As an example, Figure 3 represents the computed net transport rates for a tidal cycle (for a neap and spring tide periods) for a local mean depth of 10 m and 20 m. The depth-averaged velocity intensity considered is illustrated in the upper panel of Figure 3, characteristic of this inlet. The transport rates were computed considering two values for the d_{50}: 0.3 and 1.5 mm.

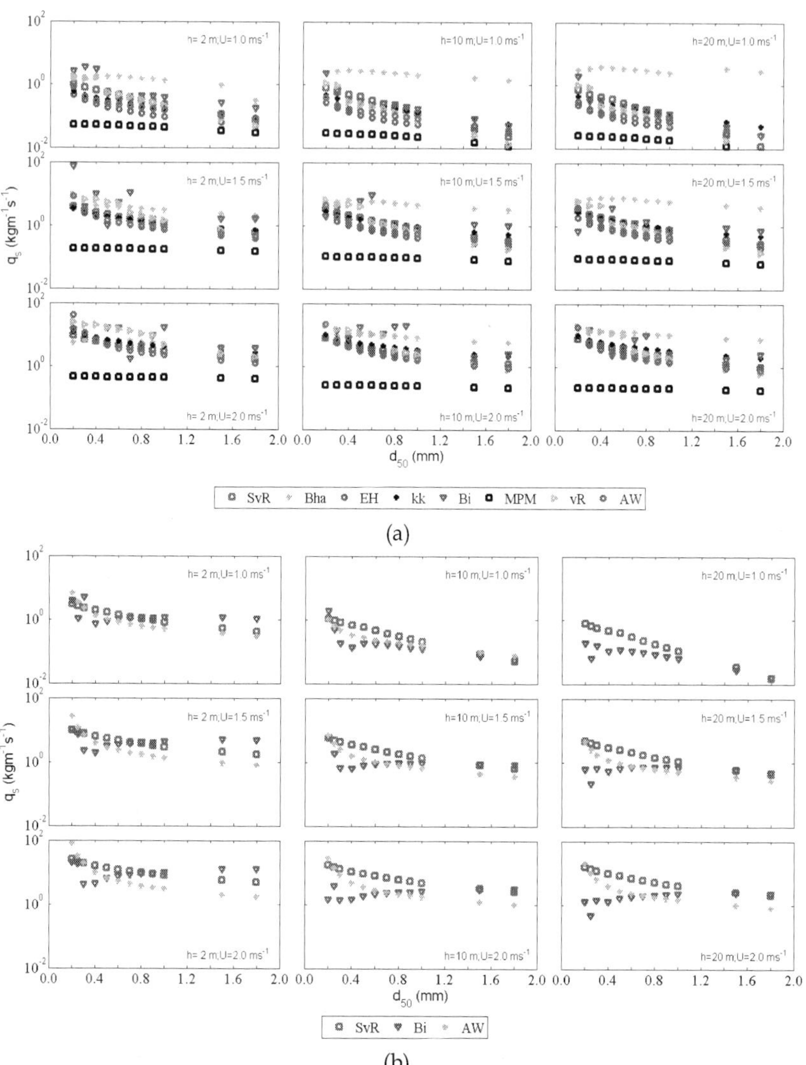

Fig. 2. Computed transport rate q_s (kgm^{-1}s) in function of d_{50} (mm), considering (a) only the tidal currents and (b) tidal currents coupled to a wave.

For any sediment grain size and any sediment transport formulation, the maximum values of q_s increase as the maximum depth-average velocities increase from neap to spring tides, and for small velocities, the transport is null. An exception to this behavior is with EH formulation, because it does not include a threshold velocity for sediment motion. Also, the sediment transport rates computed from the Bi and vR formulations present some perturbations for the highest velocities.

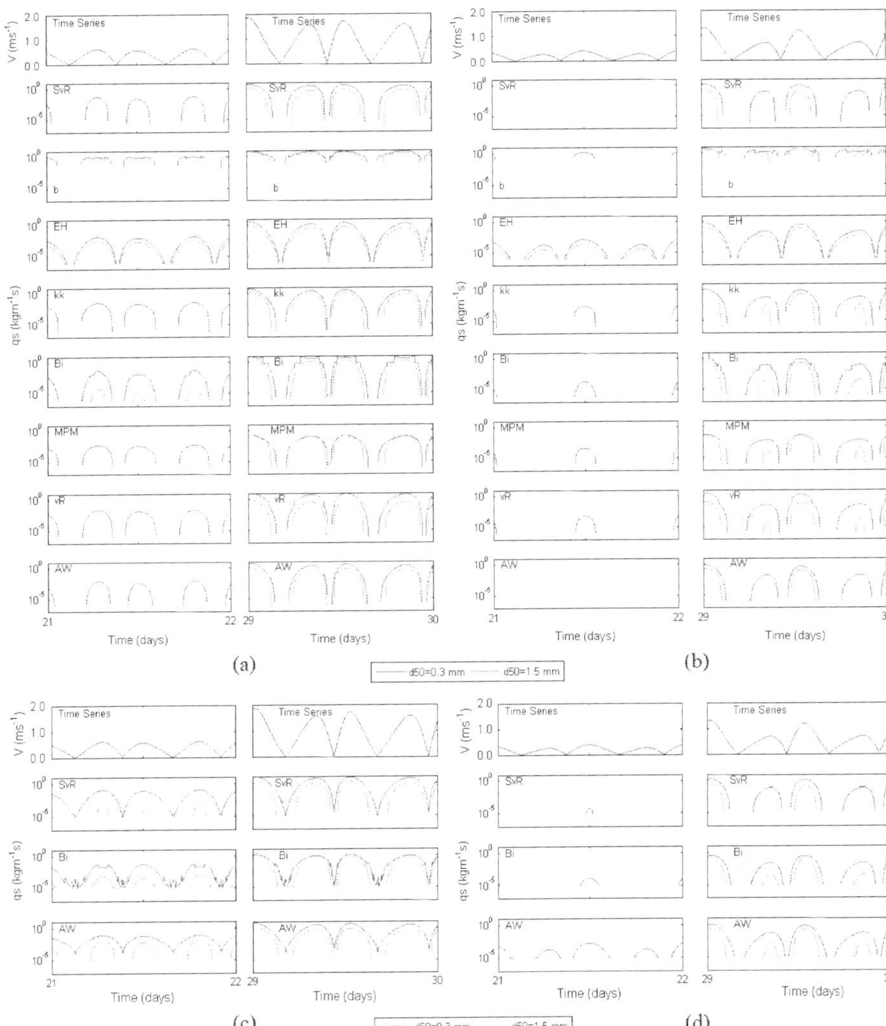

Fig. 3. Computed transport rate q_s (kgm-1s) in function of d_{50} (millimetre) for velocity tidal flow computed by a hydrodynamic model forced by tidal currents only (a) and (b) and forced by tidal currents coupled with a wave regime (c) and (d), for a depth of (a) and (c) 10 m and (b) and (d) 20 m. The left (right) columns in (a), (b), (c) and (d) represent the transport rates at neap (spring) tide condition.

Analyzing Figure 2(b) it is possible to perceive that the presence of a monochromatic wave induce continuous sediment transport for lower depths (10 m in this example). For this water depth, the generated currents have the capacity to move coarser sediments, even in neap tide conditions. For Bi formulation, are observed higher instabilities in the sediment transport rates.

For the deeper water case, the effect of the wave is not strongly noticeable. However, for neap tide condition exists a slightly increasing in the capacity of transporting finer sediments.

3. Morphodynamic simulations

To the study presented herein, the morphodynamic modelling system used is the MORSYS2D (Fortunato and Oliveira, 2004, 2007; Bertin *et al.*, 2009c). This modelling system integrates the hydrodynamic model ELCIRC (Zhang *et al.*, 2004), which calculates tidal elevations and currents, the wave model SWAN (Booij *et al.*, 1999), which computes wave propagation and the model SAND2D (Fortunato and Oliveira, 2004, 2007; Bertin *et al.*, 2009c) that computes sand transports and updates the bottom topography.
This model was applied to an inlet and adjacent nearshore of a coastal lagoon in the northwest of the Iberian Peninsula, the Ria de Aveiro.
Due to the different response of the several sediment transport formulations on the sediment size and water depth as presented in the previous section, it is important to analyze the morphodynamicof the study area itself, with d_{50}, depths and currents variable in space and time.

3.1 Dependency on sediment transport formulation

In order to compare the numerical predictions using all the sediment transport formulations mentioned in Subsection 2.3, numerical simulations for a period of 1 year were performed, considering a variable d_{50} distribution, a tidal current and a wave regime variable in time. In this work a comparison of the residual sediment fluxes obtained by averaging the sand fluxes for two MSf constituent periods (2×14.78 days) was performed.
For the simulations forced only by tidal currents, the SvR, EH, kk and AW formulations predict similar patterns, contrary to the obtained with Bha (Figure 4(a)), vR (Figure 4(b)) and MPM (Figure 4(c)). The residual sediment fluxes obtained with Bha formulation, Figure 4(a), differ substantially from the other numerical solutions, predicting strong residual sediment fluxes that will induce over-prediction of the bathymetric variations and an unrealistic bathymetry. This over-prediction is consistent with previous studies presented by Fortunato *et al.* (2009) and Silva *et al.* (2009) and with that obtained in Figure 2(a).
For the vR and Bi formulations are expected oscillations in the sediment fluxes due to the patterns illustrated in Figure 3 for finer sediments and intermediate tidal velocities. In Figure 4(b) are illustrated the residual sediment fluxes computed with vR formulation, where are visible the oscillations that induce instabilities in the bathymetric predictions. On the other hand, when considering the Bi formulation, the oscillations are not easily noticeable in the residual sediment fluxes (not presented here). However, due to its sensitivity to sediment size d_{50}, this formulation should be used carefully and only when physical properties are known with precision.
In contrast to the previous results, the solutions obtained with MPM formulation (Figure 4(c)) under-predict the sediment fluxes and consequently will under-predict the bathymetric changes. This under-prediction is consistent with that obtained in Figure 3.
The numerical results obtained with the other four formulations are similar between them (SvR, EH, kk and AW (Figure 4(d))) and more realistic. The sediment fluxes at the inlet and offshore area, at the center of the navigation channel and at the beginning of the bifurcation channel, are all outward, denoting ebb dominance. Also, the residual sediment fluxes

pattern observed identifies four regions with higher values: at the inlet and offshore area, at the center of the navigation channel and at the beginning of the channels at the right side of the domain. The two first patterns referred are located in areas of strong bathymetric changes. The first one, which present the more intense flux obtained, is located in the transition from the deepest zone located between the breakwaters to shallow zones offshore. At the centre of the navigation channel, it is also observed an intense residual sediment flux, although with low intensity when compared to the former. Once again, the transition from deeper to shallower bottom originates convergence and consequently an intensification of the residual sediment flux.

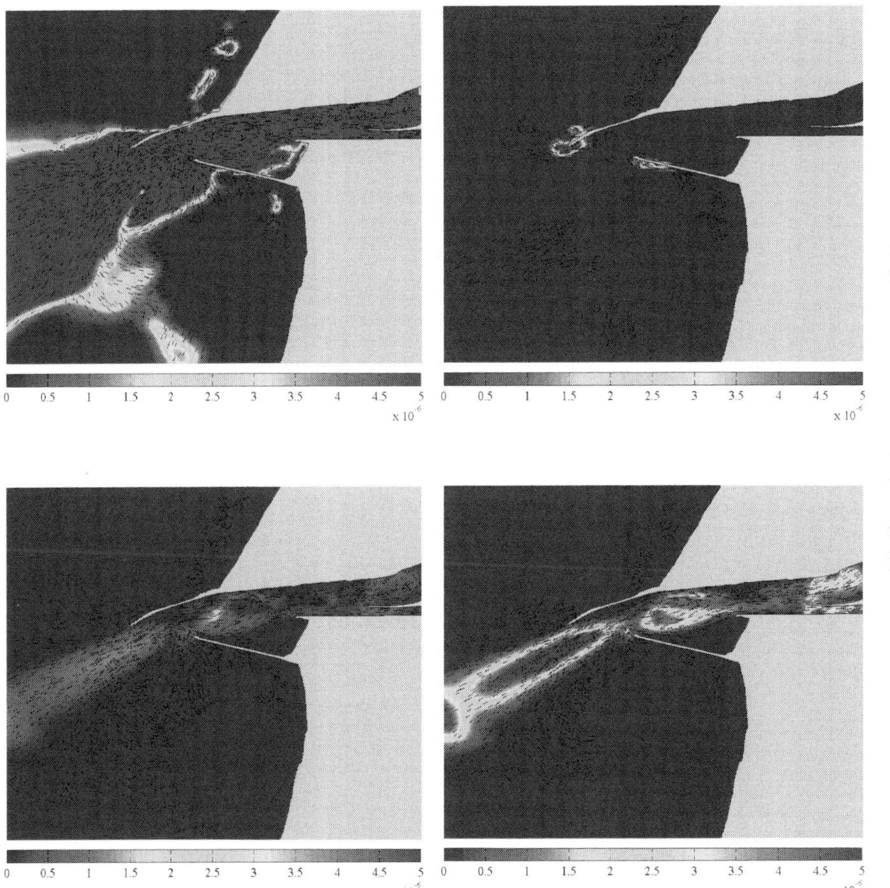

Fig. 4. Residual sediment fluxes (m^2s^{-1}) induced by tidal currents for a coastal lagoon inlet and adjacent nearshore, computed with (a) bha, (b) vR, (b) MPM and (d) AW formulations.

For the simulations forced by tidal currents coupled to a wave regime, the formulations predict similar patterns. Despite this, and for the reason pointed earlier for the case when only tidal currents forcing was considered, the Bi formulation should be used with care. The SvR formulation predicts higher residual fluxes intensities. The more realistic results were obtained with AW formulation, for the case of coupled forcing of tidal currents and wave regime. The residual sediment fluxes obtained for the AW formulation are illustrated in Figure 5.

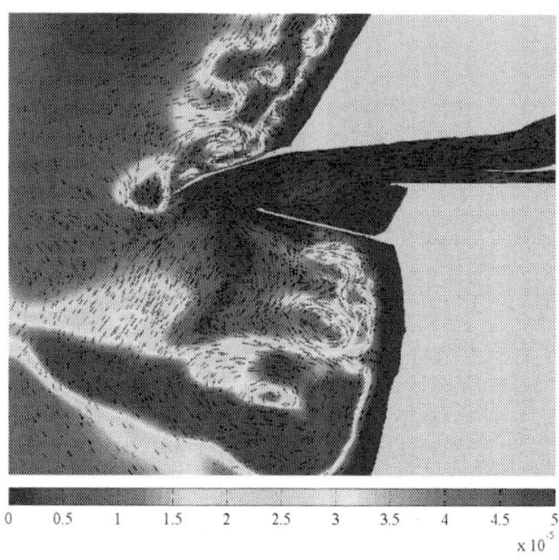

Fig. 5. Residual sediment fluxes (m^2s^{-1}) due to the forcing of tidal currents coupled with a wave regime, for a coastal lagoon inlet and adjacent nearshore, computed with the AW formulation.

When compared with the results obtained only for the tidal current forcing, it is observed that a wave regime induces one order magnitude higher sediment transport rates. In the inner areas of the inlet (inside the lagoon), the sediment transport is ruled by the tidal currents. However, at the adjacent nearshore area only the presence of a wave regime induce transport rates.

Contrasting to the general overall outward fluxes, an inward flux is observed near the south breakwater. This inward flux is induced by the currents observed at downdrift side of several inlets, where breaking waves are turned toward the inlet due to refraction over the outer bar and on breaking, creating currents toward the inlet.

At the nearshore areas higher residual transport rates are observed, predominantly directed north-south. This flux is induced by the longshore currents generated by the wave regime characteristics of the study area. At the inlet the fluxes are dominated by the outward residual tidal currents.

Comparing Figures 4 and 5, is visible that the residual fluxes patterns inside the lagoon and at the inlet are similar. In order to investigate the origin of these patterns, the influence of the sediment size d_{50} and the water depth on the residual sediment patterns is analyzed.

3.2 Dependency on d_{50} and depth

As concluded in Section 2.4, the sediment size d_{50} and the water depth have influence on the sediment transport rates. Therefore, these parameters will also have influence on the residual sediment fluxes and consequently on the bathymetric changes that occurs in the study domain.

To analyze the influence of d_{50} and the water depth on the domain morphology, numerical simulations were performed considering: a constant sediment size distribution (d_{50}); a constant depth at the study area; and the two cases coupled. All cases considered only the tidal currents forcing. Only the results obtained using the AW formulation to compute the sediment transport are presented herein.
In Figures 6 and 7 the residual sediment fluxes are illustrated considering a constant d_{50} distribution (Figure 6) and considering both d_{50} distribution (=0.5 mm) and depth (=10 m) constants (Figure 7). The results obtained considering constant depth and a heterogeneous sediment size distribution are similar to those represented in Figure 7 and are not shown.

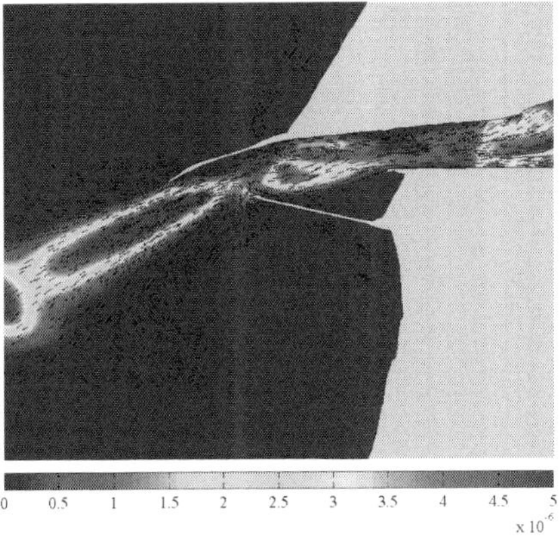

Fig. 6. Residual sediment fluxes (m^2s^{-1}) due to the forcing of tidal currents coupled with a wave regime, for a coastal lagoon inlet and adjacent nearshore, computed with the AW formulation, considering a constant d_{50} distribution.

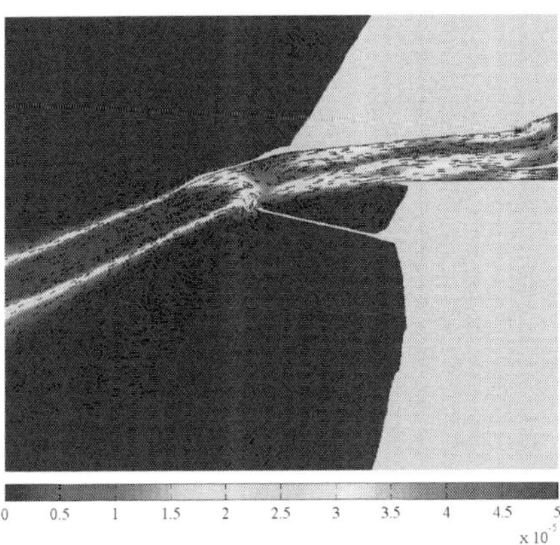

Fig. 7. Residual sediment fluxes (m^2s^{-1}) due to the forcing of tidal currents coupled with a wave regime, for a coastal lagoon inlet and adjacent nearshore, computed with the AW formulation, considering constant d_{50} distribution and constant water depth.

Analyzing Figure 6 it is observed that the residual sediment flux is very similar to that obtained when using a heterogeneous d_{50} distribution (Figure 4(d)). Thus the sediment size distribution, in this study area, is not mandatory on the definition of the residual fluxes pattern.
The results of considering a constant d_{50} distribution coupled with a constant depth set to 10 m, are illustrated in Figure 7. It is observed that the overall pattern of the residual sediment fluxes is similar to the one presented in Figure 4(d). Thus, it can be concluded that for this study area, the residual sediment fluxes are mostly originated by the geometry configuration, depending also, in minor scale, on the bathymetry.
For the study area presented herein, the geometry of the navigation channel and the configuration of the breakwaters that delimit the artificial inlet induces strong residual sediment fluxes near the north breakwater. Large bathymetric changes can result from these patterns, compromising the stability of that structure.

3.3 Trends of erosion and accretion

The residual sediment flux at a domain generates erosion or accretion locally, depending on the pattern and intensity of the flux. From the analysis of these patterns it is possible to have an idea of the bottom morphologic changes expected. In areas where the residual fluxes converge with large intensity, erosion patterns are observed due to the high amount of sediments that are transported.
As example, in Figure 8(a), (b) and (c) are illustrated the final bathymetry and the bathymetric changes induced by the residual sediment flux illustrated in Figure 5, for a one year period simulation. The white areas in figure illustrating the difference between the final and initial bathymetry represent unchanged depth, and the solid lines represent the initial bathymetry.

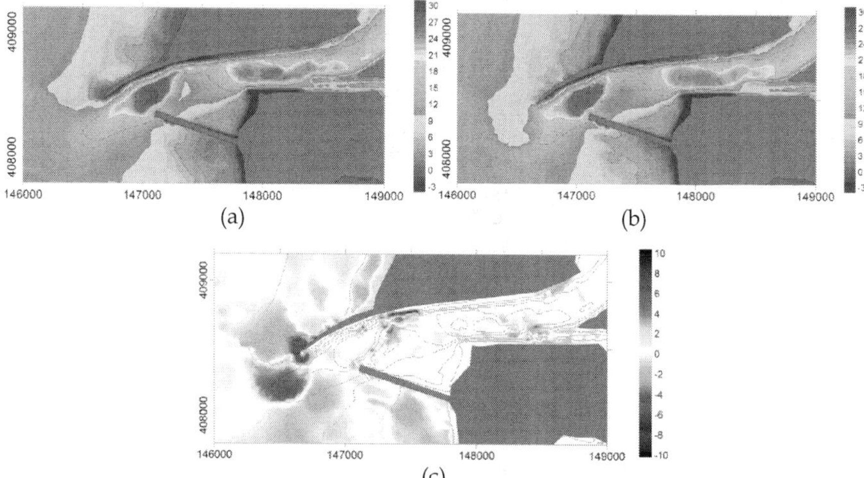

Fig. 8. (a) Initial and (b) final bathymetries (metre) for numerical simulations and (c) difference between the initial and computed final bathymetry (metre). In (c) the solid lines illustrate the initial bathymetry. The negative (positive) values represent erosion (accretion). The values in axes are in metres.

The residual sediment flux patterns at the inlet generates an erosion trend at the downstream side of the deeper zone located between the heads of the breakwaters. At the upstream side of this bottom feature and due to the decrease in the residual sediment flux, an accretion trend is obtained. The high magnitude of the residual sediment flux at the beginning of the channels at the right side of the domain (near the side walls) induces erosion trends. At the downstream side of the bifurcation, the orientation and decrease of the fluxes generates accretion trends.

4. Conclusion

There are several formulations published in literature used to compute the sediment transport rates, either as longshore or as bedload and suspended load. These formulations have different dependencies on lagoon characteristics such as sediment size d_{50}, bottom slope, water depth, current velocities and wave characteristics.
A sensitivity analysis to longshore formulations revealed that the K86 formulation is highly sensitive to the value of d_{50} and bottom slope considered, due to its inverse dependency on the sediment size and linear dependency on the bottom slope *i*. Also, the C2 formulation, with an exponential dependency on d_{50} revealed a strong dependency to this characteristic. The K&I formulation revealed sensible behavior to the value defined to the bottom slope.
The results obtained with the formulations of Bhattacharya *et al.* (2007) and Bijker (1967) over-predict the observed bathymetric changes, while the formulation of Msoeyer-Peter and Muller (1948) under-predicts the bathymetric variations. The results obtained with the formulations of Engelund and Hansen (1967), Ackers and White (1973), Karim and Kennedy (1990), van Rijn (1984a, b, c) and Soulsby-van Rijn (Soulsby 1997) seem to result in predictions more consistent. These conclusions are also illustrated by the spatial distribution of the residual sediment fluxes in a tidal inlet.
The analysis of sensibility performed also illustrates that the dependence of the sediment transport on d_{50} is more important for the fine-medium sediments. It was also concluded that the distribution of the sediment size is not influent on the residual sediment flux patterns. The major influence comes from the geometry of the inlet channel and also, in minor magnitude, from the bathymetric configuration.
Concerning the Ria de Aveiro inlet study area, the sediment fluxes obtained for simulations forced only by tidal currents are restricted to the navigation channel, dominating the long term transport in this zone. At the inlet the fluxes are still dominated by the seaward tidal currents, however near the north side of the inlet, they are affected by the longshore currents. At the inner sections of the lagoon mouth the fluxes induced by the tidal currents are higher than those produced only by the wave regime. At the nearshore area the sediment fluxes are only due to the influence of the wave regime. Therefore, an accurate estimation of the longshore sediment transport should be performed to describe the morphodynamics of this coastal area.
The analysis of the residual sediment transport allows the prediction of the erosion and accretion trends: in areas where the residual fluxes converge with high intensity, erosion patterns are observed due to the high amount of sediments that are transported.
Although these results were obtained for the specific case of Ria de Aveiro inlet, they can be extrapolated to understand and predict other similar systems.

5. Acknowledgement

The first author of this work is supported by the FCT through a PhD grant (FRH/BD/29368/2006). This work has been partly supported by FCT and by European Union (COMPETE, QREN, FEDER) in the frame of the research projects: POCI/ECM/59958/2004: EMERA - Study of the Morphodynamics of the *Ria de Aveiro* Lagoon Inlet; GRID/GRI/81733/2006: G-cast - Application of GRID-computing in a coastal morphodynamics nowcast-forecast system and PTDC/AAC-CLI/100953/2008: ADAPTARia - Climate Change Modelling on Ria de Aveiro Littoral - Adaptation Strategy for Coastal and Fluvial Flooding. The authors would like to thank Profs. A.M. Baptista and Joseph Zhang for the model ELCIRC (www.stccmop.org/CORIE/modeling /elcirc/index.html), the Aveiro Harbor Administration for providing the bathymetric data and the people involved in the EMERA project, in particular, Maria Virgínia Martins, for the sediment characteristics. The authors would like also to thank to André Fortunato, Xavier Bertin and Nicolas Bruneau, from LNEC, for the help provided in the implementation of MORSYS2D to the *Ria de Aveiro*.

6. References

Ackers, P. & White, W.R. (1973). Sediment transport: a new approach and analysis. *J Hydraul Div*, Vol.99, pp. 2041-2060

Aires, J.P.; Nogueira, J. & Martins, H. (2005). Survival of sardine larvae o_ the Atlantic Portuguese coast: a preliminary numerical study. *ICES Journal of Marine Science*, Vol.62, pp. 634-644

Bertin, X.; Fortunato, A.B. & Oliveira, A. (2007). Sensitivity analysis of a morphodynamic modeling system applied to a Portuguese tidal inlet. In: *5th IAHR Symposium on River, Coastal and Estuarine Morphodynamics*, Netherlands

Bertin, X.; Fortunato, A.B. & Oliveira, A. (2009a) A modeling-based analysis of processes driving wave-dominated inlets. *Cont Shelf Res*, Vol.29, No.5-6, pp. 819-834

Bertin, X.; Fortunato, A.B. & Oliveira, A. (2009b). Simulating morphodynamics with unstructured grids: description and validation of a modeling system for coastal applications. *Ocean Model*, Vol.28, No.1-3, pp. 75-87

Bertin, X.; Fortunato, A.B. & Oliveira, A. (2009c). Morphodynamic modeling of the Ancão inlet, South Portugal. *Journal of Coastal Research*, Vol.SI56, pp. 10-14

Bhattacharya, B.; Price, R.K. & Solomatine, D.P. (2007). A machine learning approach to modeling sediment transport. ASCE *J Hydraul Eng*, Vol.133, No.4, pp. 440-450

Bijker, E.W. (1971). Longshore transport computations. *J Waterway, Ports, Harbor, Coastal and Ocean Eng*, Vol.97, No.4, pp. 687-703

Booij, N.; Ris, R. & Holthuijsen, L. (1999). A third-generation wave model for coastal regions. 1. Model description and validation. *Journal of Geophysical Research*, Vol.104, No.7, pp. 649-666

Bruneau, N.; Fortunato, A.B.; Oliveira, A.; Bertin, X.; Costa, M. & Dodet, G. (2010). Towards long-term simulations of tidal inlets: performance analysis and application of a partially parallelized morphodynamic modeling system. In: *XVIII International Conference on Water Resources*. Spain

Carmo, J.S.A. (1995). Contribuição para o Estudo dos Processos Morfodinâmicos em Regiões Costeiras e Estuarinas. PhD dissertation, University of Coimbra, Portugal, pp.238

Cayocca, F. (2001). Long-term morphological modeling of a tidal inlet: the Arcachon Basin, France. *Coastal Engineering*, Vol.42, pp. 115-142

Chonwattana, S.; Weesakul, S. & Vongvisessomjai, S. (2005). 3D Modeling of morphological changes using representative waves. *Coastal Engineering Journal*, Vol.47, No.4, pp. 205-229

Coastal Engineering Manual (2002) US Army Corps of Engineers

Engelund, F. & Hansen, F. (1967). *A Monograph of Sediment Transport in Alluvial Channels*. Teknisk Forlag, Copenhagen

Fortunato, A.B. & Oliveira, A. (2004). A modeling system for long-term morphodynamics. *J Hydraul Res*, Vol.42, No.4, pp. 426–434

Fortunato, A.B. & Oliveira, A. (2006). Uma comparação de métodos para melhorar a stabilidade de um sistema de modelos morfodinâmico. In: *Conferência Nacional de Métodos Numéricos em Mecânica dos Fluidos e Termodinâmica*, Portugal

Fortunato, A.B. & Oliveira, A. (2007). Improving the stability of a morphodynamic modeling system. *J Coastal Res*, Vol.SI50, pp. 486–490

Fortunato, A.B.; Bertin, X. & Oliveira, A. (2009). Space and time variability of uncertainty in morphodynamic simulations. *Coastal Engineering*, Vol.56, pp. 886-894

Fredsoe, J. & Deigaard, R. (1992). *Mechanics of Coastal Sediment Transport*, World Scientific Publishing Co., ISBN 9810208405, Singapore

Grunnet, N.M.; Walstra, D.-J. R. & Ruessink, B. G. (2004). Process-based modelling of a shoreface nourishment. *Coastal Engineering*, Vol.51, pp. 581-607

Guerreiro, M.; Fortunato, A.B.; Oliveira, A.; Bertin, X.; Bruneau, N. & Rodrigues, M. (2010). Simulation of morphodynamic processes in small coastal systems: application to the Aljezur coastal stream (Portugal). In: *Geophysical Research Abstracts*. Vol.12

Kamphuis, J.W. (1991). Alongshore sediment transport rate. *J Waterway Port Coastal and Ocean Eng*, Vol.117, No.6, pp. 624-641

Kamphuis, J. W.; Davies, M. H.; Nairn, R. B. & Sayao, O. J. (1986). Calculation of littoral sand transport rate. *Coastal Engineering*, Vol.10, No.1, pp. 1-21

Karim, M.F. & Kennedy, J.F. (1990). Menu of coupled velocity and sediment discharge relation for rivers. *J Hydraul Eng*, Vol.116, No.8, pp. 978–996

Komar, P.D. & Inman, D.L. (1970). Longshore sand transport on beaches. *Journal of Geophysical Research*, Vol.75, No.30, pp. 5514-5527

Kraus, N.C.; Gingerich, K.J. & Rosati, J.D. (1988). Towards an improved empirical formula for longshore sand transport. In: *Proceedings of the 21st International Conference on Coastal Engineering*, New York, pp. 1183-1196

Larangeiro, S.H.C.D. & Oliveira, F.S.B.F. (2003). Assessment of the longshore sediment transport at Buarcos beach (West coast of Portugal) through different formulations. In: *CoastGis'03*, Genova

Meyer-Peter, E. & Muller, R. (1948). Formulas for bed-load transport. Proc. *3rd Meeting IAHR*, Stockholm, pp. 39–64

Malhadas, M.S.; Silva, A.; Leitão, P.C. & Neves, R. (2009). Effect of the bathymetric changes on the hydrodynamic and residence time in Óbidos lagoon (Portugal). *Journal of Coastal Research,* SI56, pp. 549-553

Neves, R.J.J.; Leitão; P.C.; Braunschweig, F.; Martins, F.; Coelho, H.; Santos, A. & Miranda, R. (2000). The advantage of a generic coordinate approach for ocean modeling. In: *Proceedings of Hydrosoft 2000 Conference,* Lisbon, Portugal.

Nicholson, J.; Broker, I.; Roelvink, J.A.; Price, D.; Tanguy, J. M. & Moreno, L. (1997). Intercomparison of coastal area morphodynamic models. *Coastal Engineering,* Vol.31, No.1, pp. 97-123

Pinto, L.; Fortunato, A.B. & Freire, P. (2006). Sensitivity analysis of non-cohesive sediment transport formulae. *Continental Shelf Research,* Vol.26, pp. 1826-1829

Oliveira, A.; Fortunato, A.B. & Sancho, F.E.P. (2004). Morphodynamic modeling of the Óbidos lagoon. *Coastal Engineering,* Vol.3, pp. 2506-2518

Oliveira, A.; Fortunato, A.B. & Dias, J.M. (2007). Numerical modeling of the Aveiro inlet dynamics. In: *Proceedings of the 30th International Conference on Coastal Engineering.* Vol.4, pp. 3282-3294

Plecha, S.; Rodrigues, S.; Silva, P.; Dias, J.M.; Oliveira, A. & Fortunato, A.B. (2007). Trends of bathymetric variations at a tidal inlet. In: Dohmen-Janssen M, Hulscher S (eds) Proceedings of the *5th IAHR Symposium on River, Coastal and Estuarine Morphodynamics.* Enschede, The Netherlands, pp. 19–23. ISBN: 0415453631

Plecha, S.; Silva, P.A.; Vaz, N.; Bertin, X.; Oliveira, A.; Fortunato, A.B. & Dias,J.M. (2010). Sensitivity analysis of a morphodynamic modeling system appliedto a coastal lagoon inlet. *Ocean Dynamics,* Vol.60, pp. 275-284

Plecha, S.; Silva, P.A.; Oliveira, A. & Dias, J.M. (2011). Establishing the Wave Climate Influence on the Morphodynamics of a Coastal Lagoon Inlet. *Ocean Dynamics,* submitted

Plecha, S.; Rodrigues, S.; Silva, P.; Dias, J.M.; Oliveira, A. & Fortunato, A.B. (2007). Trends of bathymetric variations at a tidal inlet. In: Dohmen-Janssen M, Hulscher S (eds) Proceedings of the *5th IAHR Symposium on River, Coastal and Estuarine Morphodynamics.* Enschede, The Netherlands, pp. 19–23. ISBN: 0415453631

Ramos, M.; Silva, P.A. & Sancho, F. (2005). *Morphological Modelling Using a 2DH Model. SANDPIT - Sand Transport and Morphology of Offshore Sand Mining Pits.* Aqua Publications. The Netherlands

Ranasinghe, R. & Pattiaratchi, C. (2003). The seasonal closure of tidal inlets: causes and effcts. *Coastal Engineering Journal,* Vol.45, No.4, pp. 601-627

Ranasinghe, R.; Pattiaratchi, C. & Masselink, G. (1999). A morphodynamic model todmulate the seasonal closure of tidal inlets. *Coastal Engineering,* Vol. 37, pp. 1-36

Robins, P.E. & Davies, A.G. (2010). Morphological controls in sandy estuaries: the inuence of tidal ats and bathymetry on sediment transport. *Ocean Dynamics,* Vol.60, pp. 503-517

Roelvink, J.A. & Banning, G.K.F.M. (1994). Design and development of Delf3D and application to coastal morphodynamics. *Hydroinformatics,* Vol.94, pp. 451-455

Rosa, J.; Gonçalves, D.; Silva, P.A.; Pinheiro, L.M.; Rebêlo, L.; Fortunato, A.B. & Bertin, X. (2011). Estudo da evolução de uma área de extracção de sedimentos ao largo de Vale de Lobo (Algarve - Portugal) - Comparação entre resultados numéricos e dados batimétricos adquiridos. *Revista de Gestão Costeira Integrada* (submitted)

Silva, A.J.R.; Leitão, J.C. & Abecassis, C. (2004). Improving the navigability and safety conditions in Douro estuary inlet. In: ASCE (Ed.), *Proceedings of International Conference in Coastal Engineering 2004*, pp. 3277-3289

Silva, P.; Bertin, X.; Fortunato, A.B. & Oliveira, A. (2009). Intercomparison of sediment transport computations in current and combined wavecurrent conditions. *J Coastal Res*, SI56, pp. 559-563

Silvio, G.D.; Dall'Angelo, C.; Bonaldo, D. & Fasolato, G. (2010). Long-term model of planimetric and bathymetric evolution of a tidal lagoon. *Continental Shelf Research*, Vol.30, pp. 894-903

Smith, J. M.; Sherlock, A. R. & Resio, D. T. (2001). *Steady-state Wave Model User's Manual for STWAVE. V3.0.* U.S Army Engineer Research and Development Center, Vicksburg

Soulsby, R. (1997). *Dynamics of Marine Sands.* Thomas Telford Publications, London

Tung, T. T.; Walstra, D.-J. R.; van de Graaff, J. & Stive, M. J. (2009). Morphological modeling of tidal inlet migration and closure. *Journal of Coastal Research*, SI56, pp. 1080-1084

Valle, R.; Medina, R. & Losada, M.A. (1993). Dependence of coefficient K on grain size. *J Waterways, Port, Coastal and Ocean Eng*, Vol.119, No.5, pp. 568-574

van de Graaff, J. & van Overeem, J. (1979). Evaluation of sediment transport formulae in coastal engineering practice. *Coastal Engineering*, Vol.3, pp. 1-32

van Rijn, L.C. (1984a). Sediment transport—part 1: bed load transport. *J Hydraul Eng* Vol.110, No.10, pp. 1431-1456

van Rijn, L.C. (1984b). Sediment transport-part 2: suspended load transport. *J Hydraul Eng*, Vol.110, No.11, pp. 1613-1614

van Rijn, L.C. (1984c). Sediment transport-part 3: bed forms and alluvial roughness. *J Hydraul Eng*, Vol.110, No.12, pp. 1733-1754

Vongvisessomjai, S.; Chowdhury, S.H. & Huq, M.A. (1983). *Monitoring of Shoreline and Seabed of Map Ta Phut Port.* Tech. Rep. 262, AIT Research Report, Bangkok, Thailand

Warren, I.R. & Bach, H.K. (1992). MIKE21: a modelling system for estuaries, coastal waters and seas. *Environmental Software*, Vol.7, pp. 229-240

Work, P.A.; Guan, J.; Hayter, E.J. & Elci, S. (2001). Mesoscale model for morphological change at tidal inlet. *Journal of Waterway Port Coastal and Ocean Engineering*, Vol.127, No.5, pp. 282-289

Xie, D.; Wang, Z.; Gao, S. & de Vriend, H. J. (2009). Modeling the tidal channel morphodynamics in a macro-tidal embayment, Hangzhou Bay, China. *Continental Shelf Research*, Vol.29, pp. 1757-1767

Zhang, Y.; Baptista, A.M. & Myers, E.P. (2004). A cross-scale model for 3D baroclinic circulation in estuary-plume-shelf systems: I. Formulation and skill assessment. *Cont Shelf Res*, Vol.110, No.10, pp. 1431–1456

9

Coupling Watershed Erosion Model with Instream Hydrodynamic-Sediment Transport Model: An Example of Middle Rio Grande

Dong Chen and Li Chen*
Desert Research Institute, 755 E. Flamingo RD
Las Vegas
USA

1. Introduction

1.1 Study area and background

The Middle Rio Grande (MRG) is located in Central New Mexico (Figure 1 insert). As one of the most historically documented rivers in the United States, MRG is under constant supervision from regulatory agencies such as the U.S. Bureau of Reclamation (USBR) and U.S. Army Corps of Engineers (USACE) (Albert, 2004). MRG was historically characterized as an aggrading sand bed channel with extensive lateral bank movement, which caused serious flooding problem (Richard et al., 2005; Sixta, 2004). To improve channel conveyance and reduce flood risks, channelization works, levees, and dams were built to control sediment concentrations in the MRG and to inhibit bed aggradation.

The reach of this study is the diversion dam reach of the MRG, which spans from Alameda Blvd bridge to Paseo Del Norte bridge including the Calabacillas Arroyo (Figure 1). The diversion dam has been built about 1,500 feet south of the Alameda Bridge on the river. The city of Albuquerque is diverting water from the Rio Grande with the operation of this diversion dam to supplement the city's drinking water supply. Previously, all of their drinking water needs were supplied by groundwater wells. There are two USGS gauges located at the Alameda Blvd bridge and Paseo Del Norte bridge, respectively. The Calabacillas Arroyo is located in Northwestern Bernalillo and South-central Sandoval counties. The Calabacillas Arroyo is a steep, relatively straight channel with a large width-to-depth ratio and has significant potential for lateral and vertical instability. The channel, consisting of a Bluepoint loamy fine sand, drains a watershed with a total area of approximately 220 square kilometers and enters the MRG about 70 km downstream of Cochiti Dam. The Calabacillas Arroyo watershed is primarily underlain by the sand-rich Santa Fe Formation that is comprised of the basin fill sediments (fluvial, paludal, and lacustrine) of the Rio Grande basin (Simons, Li and Associates, 1983). In addition to the Calabacillas Arroyo watershed, the arroyo also discharges water from portions of the Black's Arroyo watershed due to contributions from the concrete-lined Black Diversion Channel, the only significant tributary of the Calabacillas Arroyo. The drainage area of the Black's Arroyo is approximately 25 square kilometers (Mussetter, 1996).

Flows in the Calabacillas Arroyo are ephemeral, responding only to local rainfall events. In this climatic region, approximately half of the precipitation occurs between July and October, often in heavy thunderstorms. There is potential for significant floods due to the scale and characteristics of the drainage area, and high precipitation events can lead to flash floods and significant sediment transport events (Chen et al., 2009). Swinburne Dam was completed in 1991 at Unser Boulevard across the Calabacillas Arroyo. The structure was constructed as both a stormwater detention facility and to mitigate the recurrence of the sediment plug at the Rio Grande through sediment trapping. Multiple grade control structures have also been placed in the Calabacillas Arroyo to manage channel incision. Currently, many of these structures have been significantly or fully submerged by sediment deposition from upstream. Quantifying sediment transport capacity affected by these structures is a need for local watershed management (personal communication with the Albuquerque Metropolitan Arroyo Flood Control Authority - AMAFCA).

1.2 Purpose of the study

The purpose of this study is to examine how the sediment input from the Calabacillas Arroyo, a tributary to the Middle Rio Grande affects the geomorphology in the main stream. As shown in Figure 2, a huge alluvial fan at the confluence has resulted from sediment outflows from the Calabacillas Arroyo during extreme flood events, which has reduced the main MRG channel width by approximately one half and significantly affected the sediment transport capacity of the stream. The large amount of sediment input breaks the sediment balance in the MRG and may result a number of hydrological and ecological consequences, such as the channel geometry, flood frequency and changes in aquatic habitat. Quantitatively predicting geomorphic changes in the mainstream MRG considering the tributary sediment input will greatly improve the understanding of hydrological processes for the entire stream-watershed system, and provide better guidance for future management practices. Despite the great challenge to the research, the goal was achieved by simulating the watershed, tributary and mainstream as an interconnected system.

2. Models and approach

Output from a physically based watershed erosion model has seldom been used as input to a channel sediment transport model in a previous study (Beven, 2001). Aiming at accurately assessing the sediment transport in forested watershed, Conroy et al. (2006) has coupled the Water Erosion Prediction Project (WEPP) model (Flanagan et al., 1995) with the National Center for Computational Hydrodynamics and Engineering's One-Dimensional (CCHE1D) hydrodynamic-sediment transport model (Wu et al., 2004).

Our approach is to conduct a combined modeling study to investigate the mainstream geomorphic changes with sediment input from the tributary during typical storm events. The whole study will include three major parts: the watershed sediment yield, the tributary sediment transport, and the main stream sedimentation. Firstly, the Kinematic Runoff and Erosion Model (KINEROS2) model was used to estimate sediment yield from the Calabacillas Arroyo watershed during different storm events. Secondly, analysis of sediment transport in the channel was performed using the HEC-RAS program developed by Hydraulic Engineering Center (HEC) of U. S. Army Corps of Engineers. Thirdly, the main stream sedimentation process was modeled with a two-dimensional sediment transport program CCHE2D. Each model will be briefly described as follows.

Fig. 1. Map of the Study Site.

Fig. 2. Aerial photograph of the sediment plug on August 19, 1988 (AMAFCA).

2.1 KINEROS2 model

The KINEROS2 model has undergone long term development in the US Department of Agriculture, Agriculture Research Service (USDA-ARS). It describes the physical processes of interception, infiltration, surface runoff and erosion from small agricultural and urban watersheds and uses physically-based approach to simulate dynamics of short duration rainfall-runoff processes in watersheds. In this model, a watershed is represented by a series of planes (hill-slopes) and channels. Interception, infiltration and overland flows are simulated for planes to generate runoff, and channel routing is simulated to transport runoff to the outlet of the watershed with the consideration of transmission loss. Infiltration on planes as well as in channels is modeled using a three-parameter general infiltration model (Parlange et al., 1982). Technical details can be found from the website http://www.tucson.ars.ag.gov/kineros/. The model is included in the Automated Geospatial Watershed Assessment Tool (AGWA) package and is made available to public as an ArcGIS extension to simplify the data processing and modeling process (http://www.tucson.ars.ag.gov/agwa/).

2.2 HEC-RAS model

HEC-RAS program was developed by Hydraulic Engineering Center (HEC) of U. S. Army Corps of Engineers. The latest HEC-RAS model provides a module for sediment transport analysis. This model was designed for modeling one-dimensional sediment transport, and can simulate trends of scour and deposition typically over periods of years or alternatively, for single flow events. For unsteady flow events, it segments the hydrograph into small time periods and simulates the channel flow for each time interval assuming a steady state flow in the whole channel. The non-equilibrium sediment transport approach included in the module makes the sediment transport process more realistic. The sediment transport potential is computed by grain size fraction so that the non-uniform sediment can be represented more accurately. The model can be used for evaluating sedimentation in fixed channels and estimating maximum scour during large flood events among other purposes (Hydrologic Engineering Center, 2008). The HEC-RAS sediment transport module provides the option of several different sediment transport functions, thus users can select the most appropriate function according to the site conditions. This module also has the ability to limit degradation to specified elevations/depths at individual cross-sections which allows for the representation of Grade Control Structures (GCS) in the arroyo for the modeling study.

2.3 CCHE2D model

CCHE2D is an integrated software package for two-dimensional simulation for analysis of river flows, non-uniform sediment transport, morphologic processes, coastal processes, pollutant transport and water quality developed at the National Center for Computational Hydro-Science and Engineering at the University of Mississippi. These processes in the model are solved using the depth integrated Reynolds equations, transport equations, sediment sorting equation, bed load and bed deformation equations. The model is based on Efficient Element Method, a collocation approach of the Weighted Residual Method. Internal hydraulic structures, such as dams, gates and weirs, can be formulated and simulated synchronously with the flow. A dry and wetting capability enables flow simulation on complex topography. There are three turbulence closure schemes in the model, depth-averaged parabolic, mixing length eddy viscosity models and k-ε model. The numerical scheme can handle subcritical, supercritical, and transitional flows (NCCHE, 2009). The applicability of CCHE2D for the study reach has been proved by Chen et al. (2007).

2.4 Modeling procedure

The modeling system described in this article uses an aggregated approach to model watershed sediment yield, stream sediment transport, and bedform change. As shown in figure 3, the HEC-1 and KINEROS2 results provide the flow and sediment boundary conditions (BCs) for HEC-RAS Calabacillas Arroyo channel model, respectively. The HEC-RAS model result is further used as the tributary flow and sediment boundary conditions (BCs) for the CCHE2D main stream model. The CCHE2D model for the main stem MRG also needs the input data of flow and sediment from the main stream. Using this interconnected system, we are able to examine how the sediment input from the tributary affects the geomorphology in the main stream of the MRG.

3. Data & model prepration

3.1 Calabacillas arroyo watershed sediment yield

For the watershed sediment yield analysis we need the topographic data, soil data, and land cover data. All of these data sets can be downloaded from the public available internet sites. A ten-meter digital elevation model (DEM) data used for watershed delineation and 2001 land cover data were obtained from the U. S. Geological Survey (USGS) via internet. SSURGO soil data was obtained from the U. S. Department of Agriculture Soil Data Mart site. Using the topographic data, we can delineate the watershed via the AGWA extension in the ArcGIS environment. The soil data and land cover data were used to parameterize the model, which can be done within the AGWA interface. Meteorological data for different storm events were shapefiles of precipitation-frequency grids from NOAA. These grids are based on high-resolution (~800 m) NOAA Atlas 14 precipitation frequency estimates that were calculated from the analysis of partial duration series (see hdsc.nws.noaa.gov/hdsc/pfds/pfds_gis.html). The computed sediment yield totals from KINEROS2 were used as the sediment boundary conditions in the HEC-RAS analysis.

3.2 Calabacillas arroyo channel sediment transport

Arroyo channel hydraulics and sediment transport conditions were simulated with HEC-RAS. The first step of this study is to generate cross sections to represent the channel geometry. This was done using the high resolution (2-ft contour) digital terrain model (DTM) data provided by the Albuquerque Metropolitan Arroyo Flood Control Authority (AMAFCA) within the HEC-GeoRAS software for ArcGIS environment developed by HEC, USACE (http://www.hec.usace.army.mil/software/hec-ras/hec-georas.html). The study reach in Calabacillas Arroyo ranges from Swinburne Dam to the confluence of the arroyo at the MRG. A total of 114 cross sections were extracted from the DTM data (see Figure 4). Among others, 17 cross sections designate the approximate locations of existing grade control/soil cement structures. Locations of these structures were estimated using engineering drawings provided by AMAFCA. The elevations of the bases of the grade control structures were determined from engineering drawings and entered into HEC-RAS as minimum elevations (below which the channel bed will not be eroded at those cross sections).

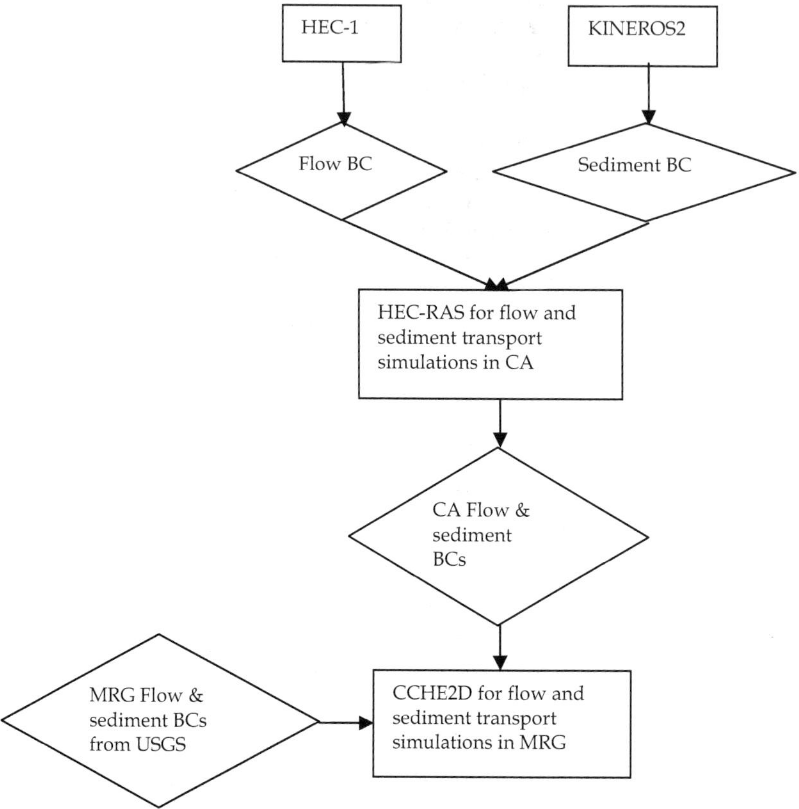

Fig. 3. Flowchart for modeling system.

HEC-1 hydrograph data for existing conditions from a previous study for AMAFCA was used to create the upstream flow boundary condition in HEC-RAS. The flow data from HEC-1 was on five-minute intervals. During the modeling study, we have used both the original data and the small interval high density data down to 30 second interpolated by HEC-RAS to test the sensitivity of the HEC-RAS sediment module to time step. The peak flow values for the 100-year, 25-year, and 10-year 24-hour storms were calculated as 12,302 cfs, 7,921 cfs, and 5,391 cfs, respectively. The downstream flow boundary condition was set to normal depth, which was based on the friction slope (0.0105 ft/ft) calculated from the cross sections in the most downstream 400 feet of the channel.

Sediment bed gradation characteristics for the channel bed were based on sediment sampling data provided in the Calabacillas Arroyo Prudent Line Study and Related Work (Mussetter Engineering, Inc., 1998). Figure 5 shows the gradation curves of five different samples along the Calabacillas Arroyo channel. While the locations are not provided here, the samples are numbered starting from most downstream to most upstream. Based on Figure 5, about 60-90% bed particles belong to sand (smaller than 2mm).

Fig. 4. Image from HEC-GeoRAS showing the cross sections (green lines) and channel centerline (blue line) drawn over the study area. The green and blue contours in the bottom-right show the DTM elevation data processed within ArcGIS.

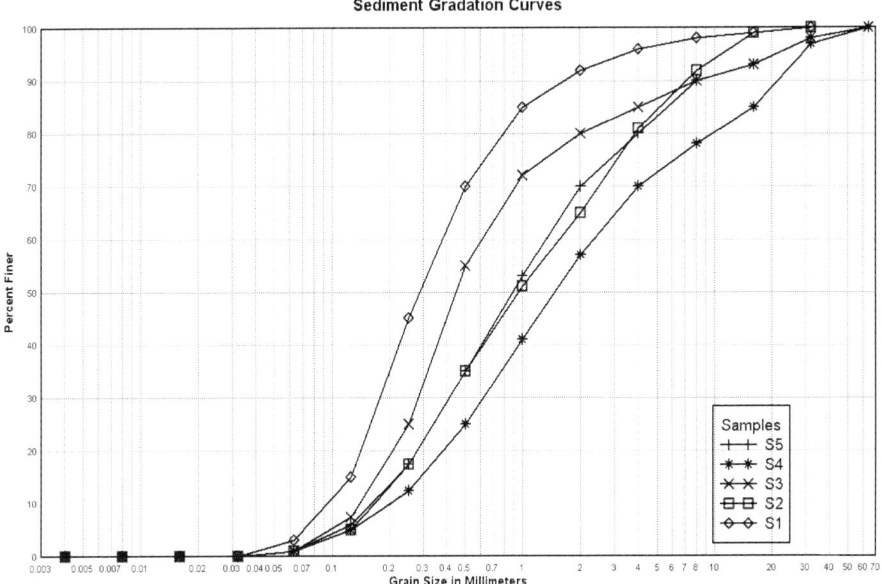

Fig. 5. Sediment particle size distribution curves used in the Calabacillas Arroyo sediment transport study. Based on sediment sampling from Mussetter Engineering, 1998.

It is very difficult to select the most appropriate sediment transport function without field data verification. In the study, we made the choice based on literature and our modeling experiences. We initially selected three transport functions: Engelund-Hansen (EH), Meyer-Peter-Muller (MPM), and Yang(Chien and Wan, 1999). The EH function calculates the total sediment load and is most appropriate for sandy rivers with substantial suspended load. The EH function was developed from flume data and has been found to fairly consistent with field data when tested. The MPM function is a bedload transport method. In the MPM function, transport rate is proportional to the difference between the mean shear stress acting on a sediment grain and the critical shear stress. This method has been widely accepted for coarse sediment rivers. The Yang method is based on the theory that unit stream power is the dominant factor for sediment concentration or unit sediment discharge. The method has been supported by both field measurements and flume experiments in many studies.

3.3 Main stream sediment transport

A 2-D computational mesh was generated based on channel topographic data extracted from a digital contour map which was produced combining the cross section data from HEC-RAS model (downstream of the diversion dam) and measurement by AMAFCA (arroyo) and University of New Mexico (upstream of the diversion dam). Bed material gradations were obtained from textural analyses of in-Channel Cored Samples taken by Sandia National Laboratories (SNL). The original data include bed material gradation in five layers at 27 sampling sites along the reach. For the simplification and requirement by CCHE2D, we reduced it to three layers and applied an averaged bed material gradation for each layer.

The final computational mesh covered the entire study reach with 400 cross sections in the main stem of MRG and 40 cross sections in the tributary (Calabacillas Arroyo). Each cross section in the main stem has 45 computational nodes, while each tributary cross section has 7 nodes.

4. Modeling results

4.1 Calabacillas arroyo watershed sediment yield

We used KINEROS2 model to calculate the sediment yield of the watershed for 100-year, 25-year, and 10-year recurrence interval 24-hour storms. The model calculated sediment yield at each element and obtained the total sediment yield at the watershed outlet for the whole watershed (with an area of 20077.16 hectares) by calculating transport of sediment from all elements through channel network. The result is summarized in the following table.

Event	Yield (kg/hectare)	Yield (kg)
10year-24hour	0.0036	73.24
25year-24hour	0.6886	13,825.14
100year-24hour	333.41	6,693,968.04

Table 1. Sediment yield estimates from KINEROS2 for the entire watershed.

Figure 6 shows an example of the calculated spatial distribution of sediment yield from KINEROS2 model for the 100-year event. The results disclose the spatial distribution and the major source of the flow and sediment generated from the watershed during a large storm event. The simulation may provide useful information for future watershed management. Those sediment yield results were used as the input sediment load for HEC-RAS simulations.

4.2 Calabacillas Arroyo channel sediment yransport

Table 2 shows the cumulative sediment load near the confluence of the Calabacillas Arroyo from different design storm events from the HEC-RAS simulations. The cumulative sediment load at the second most downstream cross section was recorded for each simulation. The most downstream cross section was held at a constant elevation (zero bed change) in the boundary condition file. Model runs were made with the selected transport functions, design storm events, and for two different conditions, with and without the presence of grade control structures (GCS). The presence or absence of GCS did not lead to large changes in the totals. In some cases, the presence of GCS led to slightly larger sediment totals as compared to runs without GCS.
In the HEC-RAS sediment transport analysis of the Calabacillas Arroyo, three transport functions were used: Engelund-Hansen (EH), Meyer-Peter-Muller (MPM), and Yang. The choice of transport function strongly impacted the totals. The EH transport equation provided the highest totals and the MPM equation provided the lowest totals. The highest total produced, 284,259 tons (US short tons), was with the EH transport function with GCS for a 24-hour, 100-year storm. The lowest total, 2,430 tons, was produced with the MPM transport function without GCS for a 24-hour, 10-year storm. For the sake of save channel design, EH function with GCS was adopted for the current study. Figure 7 shows the Calabacillas Arroyo channel bed change after 100-year event when using EH transport function with GCS. Alternative erosion and sediment deposit were simulated along the Calabacillas Arroyo channel. Since the upstream sediment boundary was far below the equilibrium condition, serious bed scouring was calculated at the Calabacillas Arroyo inlet.

Function		100year	25year	10year
EH	No GCS	275,928	172,754	120,384
	With GCS	284,259	175,318	118,929
Yang	No GCS	164,184	111,935	86,323
	With GCS	167,549	109,421	87,073
MPM	No GCS	4,490	3,094	2,430
	With GCS	4,489	3,097	2,440

Table 2. Cumulative sediment load at the most downstream cross section in US short tons.

4.3 Main stream bedform change

The CCHE2D model for the main stem Rio Grande needs the input data of flow and sediment from both the main stream and the tributary Calabacillas Arroyo. In this modeling work, the equilibrium sediment boundary was adopted for the main stream MRG, while the sediment input from the tributary Calabacillas Arroyo was calculated by HEC-RAS model using EH function with GCS (see Section 4.2).

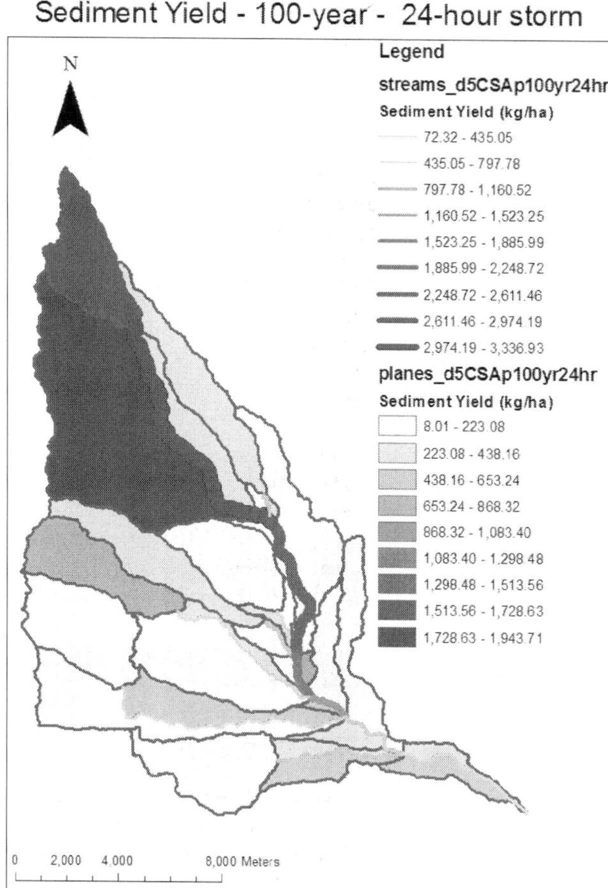

Fig. 6. Sediment yield results from KINEROS2 model for 100-year event.

Figure 8 shows the bed elevation changes at two extreme scenarios: (a) 100-year flood occurs in the MRG with synchronous 10-year storm in the tributary; (b) 100-year storm occurs in the tributary with based flow in the MRG. Figure 8a represents a favorable condition for river maintenance: although the whole channel system was an aggrading system in the past (Chen et al., 2007), sedimentation was minimal under those conditions. In contrast, Figure 8b represents the worst case: a heavy rainstorm has been restricted to the upper watershed of the Calabacillas Arroyo which results in a 100-year storm in the tributary and huge sediment load, while the flow in the main stem of MRG remains around 25m³/s (base flow). In the case of Figure 8b, a great amount of sediment has settled around the confluence area which indicates that base flow in the main stem does not have enough hydraulic power to transport the deposited sediment to downstream reaches. Most serious deposition (around 1m) is located a bit downstream of the confluence, while depositing depths elsewhere are less than 0.5m. In addition, it could be observed that sedimentation in Figure 8b traces toward the upstream along the right bank for several hundred meters, which is believed due to the ponding effect produced by excessive downstream aggradation.

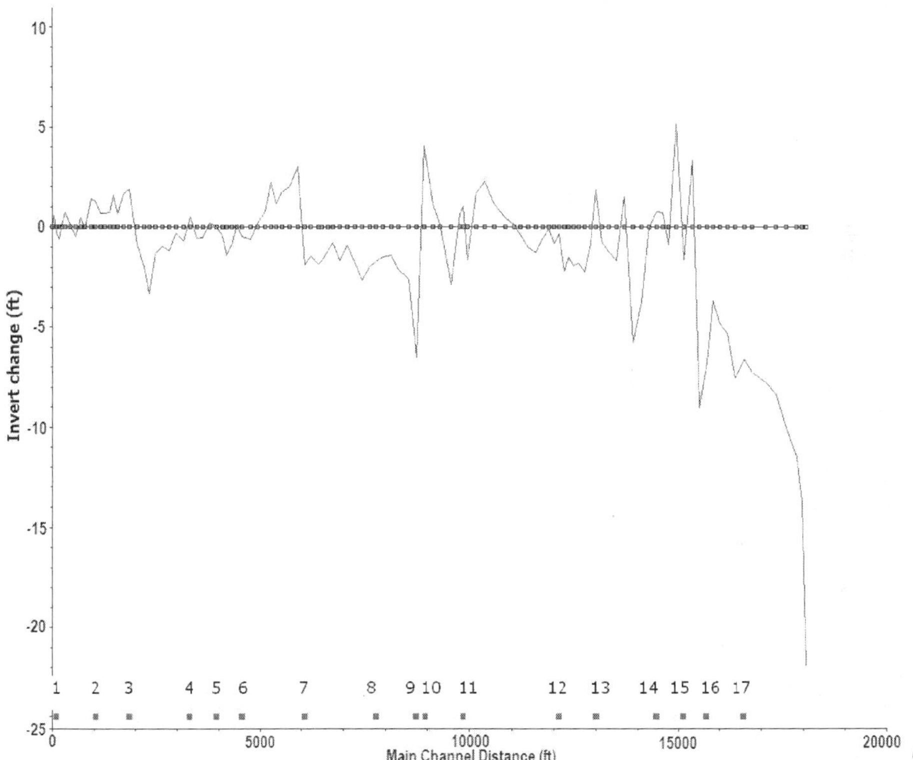

Fig. 7. Calabacillas Arroyo channel bed elevation change - 100-year event, EH transport function, with GCS.

Figure 9 shows the bed elevation changes at three in-phase scenarios: (a) 100-year flood occurs in the MRG with synchronous 100-year storm in the tributary; (b) 25-year flood occurs in the MRG with synchronous 25-year storm in the tributary; (c) 10-year flood occurs in the MRG with synchronous 10-year storm in the tributary. Based on Figure 9 (a) - (c), it can be found that the amount of bed change is mainly determined by sediment input from the Calabacillas Arroyo. For instance, flood of 100-year storm in the arroyo will bring the most amount of sediment into the main stem of MRG and cause serious deposition problem. The crucial role of sediment source from the tributary can also be revealed when comparing Figure 8 (b) with Figure 9 (a), although the 100-year flood in MRG does mitigate, to a certain extent, the deposition in the main channel. The channel is wide and shallow and the whole channel system is in aggrading.

5. Discussion

To better understand the impact of Calabacillas Arroyo inflow sediment on the evolution of main stream geomorphology, scenarios with equilibrium sediment boundary conditions at Calabacillas Arroyo inlet was also simulated to provide a most extreme conditions.

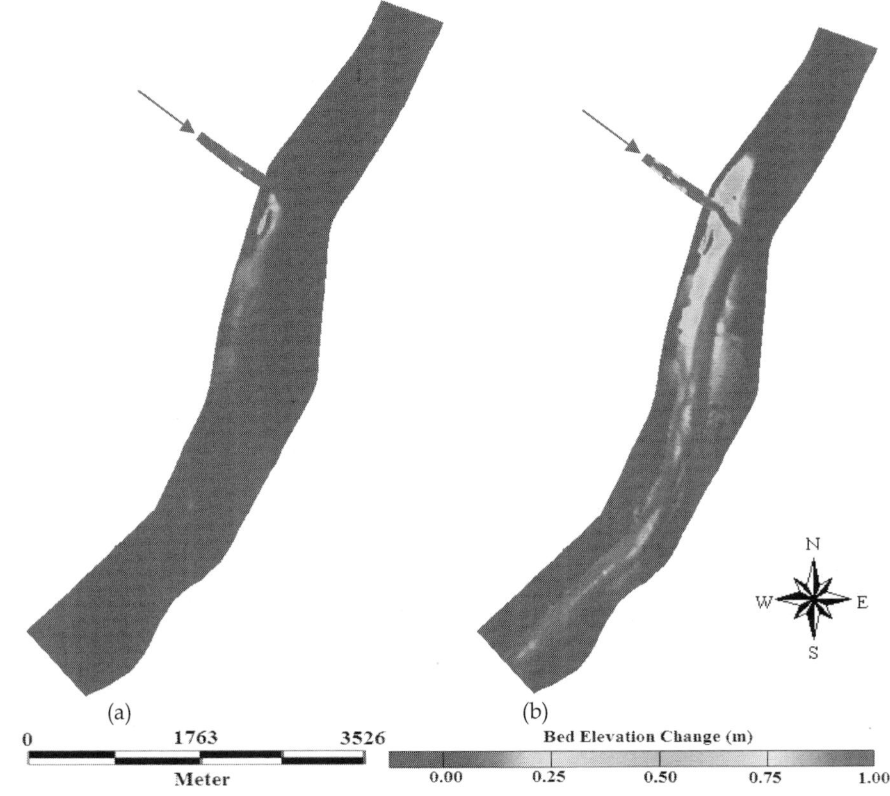

Fig. 8. Bed elevation change after combined flood events: (a) 10-year flood in the tributary with synchronous 100-year flood occurs in the MRG; (b) 100-year flood occurs in the tributary with based flow (25m3/s) in the MRG

Figure 10 shows the bed elevation changes at two scenarios: (a) 100-year flood occurs in the tributary (equilibrium sediment inflow at the Calabacillas Arroyo inlet) with the based flow (25 m3/s) in the MRG; (b) 100-year flood occurs in the MRG with synchronous 100-year flood in the tributary (equilibrium sediment inflow at the upstream boundary of the tributary). The crucial role of sediment source from the tributary can also be revealed when comparing Figure 8b with Figure 10a. There is more inflow sediment from the Calabacillas Arroyo under the equilibrium boundary scenarios which will bring more serious deposition in the main stream. Even the 100-year flood in the MRG has limited ability to flush away the sand bar formed along the right bank. Unlike the non-equilibrium sediment boundary scenarios, the most serious deposition area located just upstream the confluence which indicates stronger ponding effects.

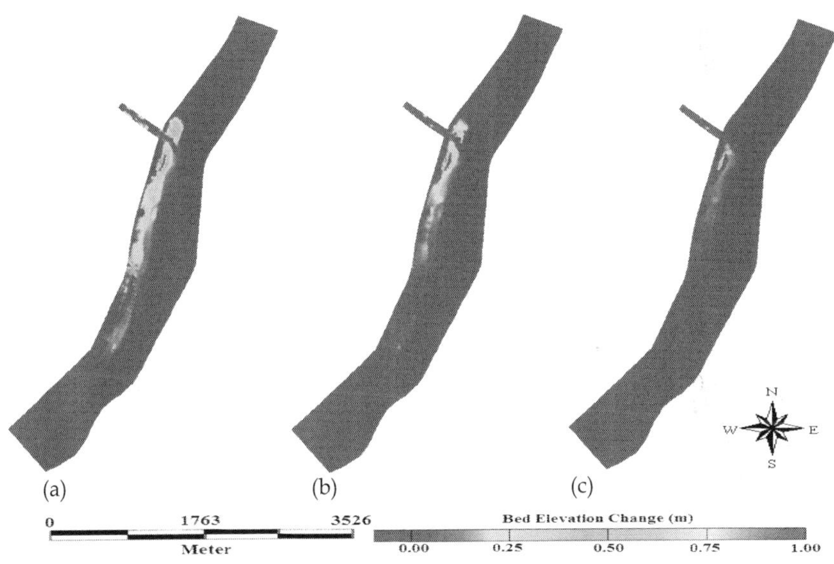

Fig. 9. Bed elevation change after combined flood events: (a) 100-year flood occurs in the MRG with synchronous 100-year flood in the tributary; (b) 25-year flood occurs in the MRG with synchronous 25-year flood in the tributary; (c) 10-year flood occurs in the MRG with synchronous 10-year flood in the tributary

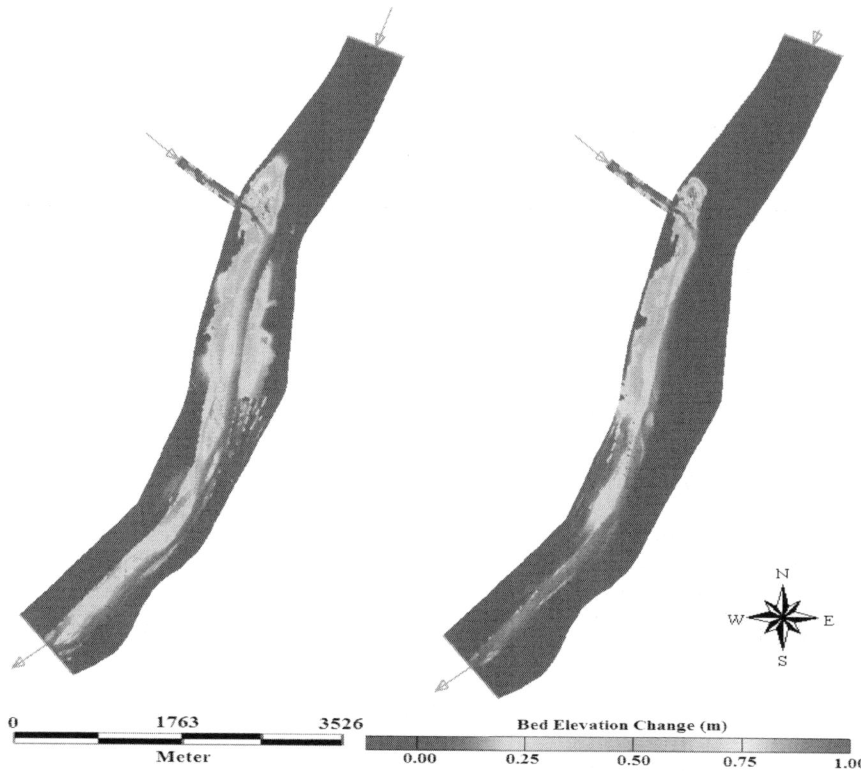

Fig. 10. Bed elevation change after combined flood events: (a) 100-year flood occurs in the tributary (equilibrium sediment inflow at the inlet) with based flow (25 m3/s) in the MRG; (b) 100-year flood occurs in the MRG with synchronous 100-year flood in the tributary (equilibrium sediment inflow at the upstream boundary of the tributary)

6. Conclusions

The Middle Rio Grande (MRG) in Central New Mexico was suffering severe bed aggradation in the history which caused serious flooding problem. Among others, inflow from the Calabacillas Arroyo, a tributary of MRG downstream of the Albuquerque Diversion Dam delivers a large amount of sediment, manifested by the delta formation and the narrowing of the main channel Rio Grande. It implies significant erosion and sediment yield from the surrounding watershed. The goal of this research was to investigate the impact of tributary sediment on the channel geomorphology of main stream MRG.

Three models have been conjunctively applied in this study to conduct a combined investigation of watershed sediment yield, tributary sediment transport, and main channel sedimentation modeling. The following conclusions can be drawn from the study:

1. Significant amount of sediment (7,378 US tons) can be eroded from the watershed and enter the Calabacillas Arroyo channel during a 100-year, 24-hour storm event. However, this portion of the sediment is relatively small compared with the amount of sediment that can be transported to the MRG through Calabacillas Arroyo ;
2. Large amount of sediment (ranging 167,549 - 519,967 US tons) will be scoured from the Calabacillas Arroyo channel to supply the main stream MRG during a 100-year storm event. The GCSs will have limited impact on the sediment transport process under the current situation;
3. According to findings in literature and the current study, EH and Yang sediment transport functions are better applicable for the Calabacillas Arroyo site. MPM function should be used with care because it is limited to bedload prediction;
4. The 25-year and 10-year storms will produce about 60% and 40% of sediment transported by a flood produced by a 100-year storm event, respectively, which are also significant amounts of sediment to the MRG;
5. The sediment input from the Calabacillas Arroyo channel can significantly affect the geomorphic features in the main stream MRG. Serious deposition will happen in the main stream mainly in the downstream reach of the confluence and also in a short reach upstream. The worst situation happens when large events occur in the tributary coincident with low flows in the main stream.

7. Acknowledgement

This study is supported by Urban Flood Demonstration Program (UFDP). We are grateful to Darrell Eidson of US Army Corps of Engineers, Albuquerque and John Kelly of AMAFCA who provided valuable data and assistance to this study.

8. References

Albert, J.M. (2004). *Hydraulic Analysis and Double Mass Curves of the Middle Rio Grande From Cochiti to San Marcial, New Mexico*, Thesis, Colorado State University, Fort Collins, 207 pp.

Beven, K. J. (2001). *Rainfall-Runoff Modelling: The Primer*. John Wiley and Sons Ltd. 347pp.

Chen, L., Chen, D., and Forsee, W. (2009). *Impact of Sediment Supply from Calabacillas Arroyo on Stream Morphology of Middle Rio Grande.* Prepared for U.S. Army Corps of Engineering, Technical Report of Desert Research Institute.

Chen, C., Stone, M., and Acharya, K. (2007). *Hydrodynamic and sediment transport modeling in the Middle Rio Grande.* Prepared for U.S. Army Corps of Engineering, Technical Report of Desert Research Institute.

Chien, N. and Wan, Z. (1999). *Mechanics of Sediment Transport.* ASCE Press.

Conroy, W.J., Hotchkiss, R.H., and Elliot, W.J. (2006). A coupled upland-erosion and instream hydrodynamic-sediment transport model for evaluating sediment transport in forested watersheds. *American Society of Agricultural and Biological Engineers*, 49(6),1713-1722.

Flanagan, D.C., Ascough, J.C., Nicks, A.D., Nearing M.A., and Laflen, J.M. (1995). *Chapter 1: Overview of the WEPP erosion prediction model. In Technical Documentation: USDA-Water Erosion Prediction Project (WEPP).* Flanagan D.C. and Nearing, M.A., eds. West Lafayette, Ind. USDA-ARS national soil erosion research laboratory.

Hydrologic Engineering Center, 2008: *HEC-RAS River Analysis System, User's Manual*. U.S. Army Corps of Engineers, Davis, CA.

Mussetter Engineering, Inc., (1998) *Calabacillas Arroyo Prudent Line Study and Related Work. Evaluation of Existing Erosion-Risk Limits between Coors Road and Swinburne Dam*. Fort Collins, CO.

Mussetter Engineering, Inc. (1996) *Calabacillas Arroyo Prudent Line Study and Related Work. Sediment Sampling and Field Reconnaissance*. Fort Collins, CO.

National Center for Computational Hydroscience and Engineering (NCCHE), (2009). University of Mississippi, http://www.ncche.olemiss.edu/

Parlange, J.Y., Lisle, I., Braddock, R.D. (1982). The three-parameter infiltration equation. *Soil Science*, 133(6), 337-341

Richard, G.A., Julien, P. and Baird, D.C. (2005). "Case Study: Modeling the Lateral Mobility of the Rio Grande below Cochiti Dam, New Mexico," *Journal of Hydraulic Engineering*, 131(11), 931-941.

Simons, Li, & Associates (1983) *Erosion Study to Determine Boundaries For Adjacent Development - Calabacillas Arroyo Bernalillo County, New Mexico*. Fort Collins, CO.

Sixta, M.J. (2004). *Hydraulic Modeling and Meander Migration of the Middle Rio Grande, New Mexico.*, Thesis, Colorado State University, Fort Collins.

Wu, W., Vieira, D.A., and Wang, S.Y. (2004). One-dimensional numerical model for nonuniform sediment transport under unsteady flows in channel network. *J. Hydraulic Eng*. 130(9), 914-923.

10

Coastal Morphological Modeling

Yun-Chih Chiang[1] and Sung-Shang Hsiao[2]
[1]*Tzu Chi University,*
[2]*National Taiwan Ocean University*
Taiwan

1. Introduction

The coastal zone is the area where the action of waves and wave-driven-currents on the seabed is very intense, and where the bed level and sediment are almost always in motion. If the mainly seasonal climate changes, the wind and wave conditions will also change. A new waves and wave-driven-currents conditions will change the rate of sediment transport and the beach topography. Natural beaches are significant characterized by an annual cycle of seasonal erosion and accretion. In other words, natural beaches are generally in dynamic equilibrium from the balance of sediment transport budge for specific area on the beach. We can utilize the conservation law of sediment transport to describe the coastal morphodynamic evolution. It is the same way to predict beach evolution due to changes in wave conditions or caused by coastal structures. Therefore, we can utilize numerical models to predict the change of bottom topography from the spatial distribution of the alongshore and offshore sediment transport rates in real coastal area.
Coastal morphological models are indispensable and powerful tools that allow harbour and hydraulic engineers to predict nearshore topography, to analyze the impact of coastal structures, and to verify the planning and design of harbours and coastal defences. Morphological models are based on various sub-models for waves, tidal currents, nearshore currents, and sediment transport, coupled with the sediment transport model. The sediment transport model solves the sediment conservation equation to calculate bed-level evolution. The local sediment transport is first calculated by wave and current sub-models, and the bed form evolution is then computed based on the conservation of sediment and its continual redistribution in time. The aim of this chapter is to describe the theories, techniques, applications and robust algorithms for computing bed level change which is flexible enough to handle the nonlinearity present in the sediment conservation equations describing bed evolution in a complex coastal area.

1.1 Classification of prediction methods for morphological evolution

The changes of coastal topography were the complex and irreversible morphodynamic processes, influenced by local bathymetry, weather, wave, tide, and coastal structures, etc. The technology of prediction morphological evolution caused by coastal structures or varied with monsoon, typhoon, and climate changes, etc. is needed. In order to predict morphological evolutions, the experience analysis with many survey data in similar cases and the results of hydraulic model tests are considered in the past.

The empirical morphodynamic analysis was based on observed trends of coastal evolution with a series of sequential bathymetry survey data or present trends under similar geographic and natural conditions. This prediction method should take a lot of times and resources to complete. However, the prediction with empirical analysis cannot be applied widely with general coastal areas subjecting to specific condition in similar case.

Hydraulic modeling analysis is a well established and widely accepted fact among the coastal engineering professionals that the physical hydraulic scale model test is not only a choice but an reliable tool for testing coastal structures before its construction due to its ability to solve complex hydraulic problems which otherwise cannot be solved analytically or experience morphological analysis. However, model tests for coastal morphodynamic evolution involve scaling problems such as sediment particle hardly scaling down and similitude law for coastal movable bed tests not well established. Furthermore, hydraulic model tests usually have been studied with expensive facilities, laboratory resource and time.

In recent years, computer technology has made remarkable progress and a computer has become an indispensable tool for coastal morphodynamic analysis. Upon the deficiencies of the empirical morphological analysis and hydraulic model tests, the numerical model simulation is a convenient, economical and efficient tool for analyzing coastal morphodynamic evolution. In order to predict coastal morphological evolution, the coastal processes must be understood, appropriately simplified, and mathematically modelling. Numerical simulations for long-term (years to decades) periods and wide coastal area should be different from short-term (hours to days) and medium-term (weeks, months to years). According to the objective and simplified mathematic formulation of numerical modelling, the coastal morphological models can be classified into two general groups: the shoreline models (single-line or multi-line model) and three-dimensional models.

The shoreline model is a numerical prediction model based on the cross-shore-section sediment continuity equation and an equation for the longshore sediment transport rate. It is also called the one-line theory for the prediction of beach position changes, where the 'one-line' refers to the shoreline. The two-line and multi-line models have also been developed to calculate the movement of selected contours. The shoreline models, which have been well developed and applied to many practical problems, simplified the actual phenomena, and hence require only a relatively shore computation time, but they cannot be used to predict local changes in the bottom topography.

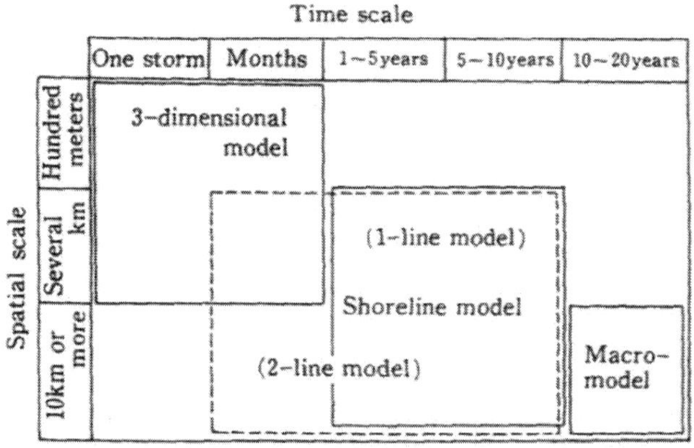

Fig. 1.1 Application ranges of morphological evolution predictive models. (Horikawa, 1986)

The three-dimensional models are used to predict the changes of bed level and bottom topography from spatial distributions of the sediment transport rates in cross-shore and longshore direction, which are estimated with the results of nearshore wave and current simulation. Compared with the shoreline models, the three-dimension models of coastal topography changes don't require more simplification and idealizations. On the other hand, the three-dimension models could be applied to analyze local coastal topography changes and therefore they have wider applicability, but a long computation time is required. However, the deficiencies of much computation time have be overcame with remarkable progress in computer techniques. Therefore, the three-dimension models of coastal topography changes could be applied with predicting local bed level changes in bottom topography over short-term and middle-term time interval.
According to Horikawa (1986), Figure 1.1 indicates criteria of application ranges of coastal morphological evolution models in terms of time scales and spatial scales. The macro-models in the figure denote the more simplified empirical morphological analysis based on similar evolution trend experience and a lot of local survey data, and they are effective for a qualitative analysis but not applicability to quantitative analysis.

1.2 The main concern of coastal morphological models in present chapter

Coastal morphological models are indispensable and powerful tools that allow harbour and hydraulic engineers to predict nearshore topography, to analyze the impact of coastal structures, and to verify the planning and design of harbours and coastal defences. Morphological models are based on various sub-models for waves, tidal currents, nearshore currents, and sediment transport, coupled with the sediment transport model. The sediment transport model solves the sediment conservation equation to calculate bed-level evolution. The local sediment transport is first calculated by wave and current sub-models, and the bed form evolution is then computed based on the conservation of sediment and its continual redistribution in time. In the last twenty years, two dimensional depth-averaged coastal morphological models have been developed, and these models have been applied in the short-term (hours to days) and medium-term (weeks, months to years) (Coeffe' and Pe'chon, 1982; Yamaguchi and Nishioka, 1984; Watanabe, 1986; Anderson et al., 1991; Wang et al., 1992; de Vriend et al., 1993b; Sato et al, 1995; Nicholson et al., 1997)

2. Modelling coastal morphological evolution

2.1 Conservation of sediment transport

The change of bed form in local bottom elevation z_b can be computed by solving the conservation equation for sediment mass. In two dimensions, this can be written as:

$$\frac{\partial}{\partial t}\left[(1-n)z_b + \int_{z_b}^{\eta} c dz\right] + \left(\frac{\partial q_x}{\partial x} + \frac{\partial q_y}{\partial y}\right) = 0 \quad (1)$$

where z_b is the bed level elevation, defined as positive up from a fixed datum, x and y are horizontal space coordinates, t is time, n is the bed porosity, η is the free surface elevation, c is the suspended sediment concentration in the water column per unit area, and q_x and q_y are the total volumetric sediment transport rates (unit: m^3/sec) in the x- and y- directions, as shown in Figure 2.1, respectively. The sediment transport rates are expressed in terms of the effective volume of sediment passing through the vertical cross-section of unit width in unit time. The effective volume means the volume consisted of sediment particles included voids by the changes of bed level in the unit area. If the income of sediment transport rates is larger than outcome, the sediment effective volume will settle on the bottom and increase the bed level. If the outcome of sediment transport rates is larger than income, the bed form will be washed out to follow mass conservation law. Therefore, in applying Eq. (1) it should be noted that the changes in bed level is caused by net averaged sediment transport affected with the wave and current field.

In general, the suspended load contribution of sediment can be consisted in sediment flux, and the sediment transport conservation equation can be reduced to:

$$\frac{\partial z_b}{\partial t}+\frac{1}{1-n}\left(\frac{\partial q_x}{\partial x}+\frac{\partial q_y}{\partial y}\right)=0 \tag{2}$$

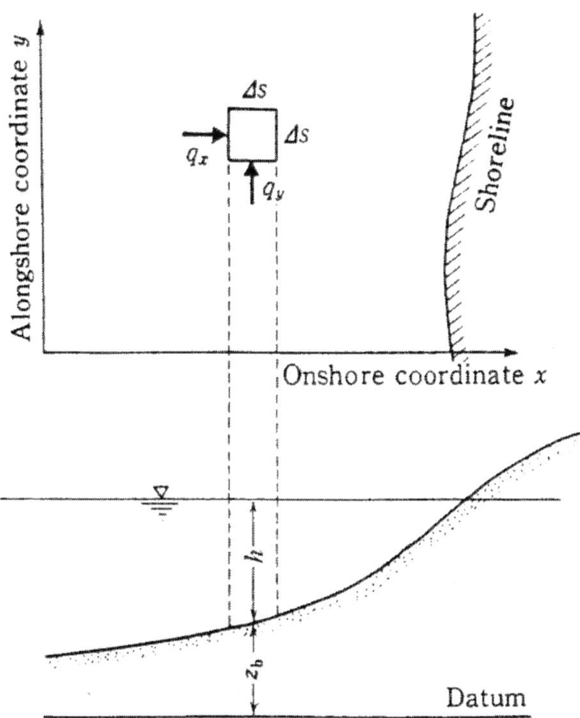

Fig. 2.1 The definition of coordination.

The sediment transport rates q_x, and q_y are complex functions of several parameters, including waves, currents, water depth, density, and sediment properties (including grain size and porosity). Here, we assume the sediment properties and water level are fixed in every time step. Under these assumptions, the sediment transport rates caused by waves and currents can be calculated using the formula obtained through experimentation or theory. At present, no established and widely accepted formulas are available for estimating local sediment transport rates from local coastal quantities while general wave and current conditions.

Therefore, the most important problem is how to quantify the sediment transport rates caused by waves and currents. There are many studies on sediment transport formulas have been reviewed in previous chapters. However, there is no universally acceptable and reliable sediment transport formulas have been found for accurately evaluating the local sediment transport rates everywhere without local data verification. For following simulation,examples, the experimental results of alongshore and offshore transport rate for following simulation examples suggested by Chiang et al. (1996), which is applicable for Taiwan coastal areas are employed:

$$q_x = q_c\left(u + U_r\right) \tag{3}$$

$$q_y = q_c\left(v + V_r\right) \tag{4}$$

$$q_c = \left\{A_1 f_c\left[\left(u + U_r\right)^2 + \left(v + V_r\right)^2\right] + A_2\left(U_w{}^2 - U_{wc}{}^2\right)\right\} / g \tag{5}$$

$$U_w = \sqrt{\frac{\tau_m}{\rho}} = \sqrt{\frac{f_w}{2}} \times U_{w\max},\ U_{w\max} = \frac{\pi H}{T\sinh(kh)} \tag{6}$$

$$U_{wc} = 8.41 \times d_{50}{}^{11/32} \tag{7}$$

$$f_c = g / C_c{}^2 \tag{8}$$

$$f_w = \begin{cases} 0.00251 \times \exp\left(5.21 \times \exp\left(\dfrac{A}{k_s}\right)^{-0.19}\right), \dfrac{A}{k_s} > 1.57 \\ 0.3, \dfrac{A}{k_s} \le 1.57 \end{cases} \tag{9}$$

$$A = \frac{H}{2}\sinh(kh) \tag{10}$$

where u and v are the depth integrated average current velocity in the x- and y- directions, respectively; U_r and V_r are the average equivalent river flow velocity in the x- and y- directions; A_1 and A_2 are the coefficients of sediment transport due to currents and waves; f_c and f_w are the friction factors for mean current and wave orbital fluid motion; U_w and $U_{w\max}$ are the shear velocity and its maximum velocity due to wave motion; U_{wc} is the critical shear velocity for the inception of particle motion; g is the gravitational acceleration; C_c is the non-dimensional Chezy coefficient; d_{50} is the median grain diameter; A is the semi-orbital excursion; H is the wave height; T is the wave period; k is the wave number; and h is the water depth.

2.2 Sample computation and discussions

In the example of coastal morphological models, the morphological evolution in the vicinity of a shore parallel breakwater is investigated from Johnson and Zyserman (2002) using

MIKE 21 CAMS (The same governing equation and similar sediment transport formula as below). The test conditions are described below.

A shore parallel breakwater 310 m long is placed at a distance of 360 m from the shoreline. The initial bathymetry is characterised by a 1:50 plane beach profile to a depth of 14 m, and a 1:20 plane beach slope to the boundary of the model area (depth of 20 m). Irregular unidirectional waves with H_{rms} = 1.98 m and T_p = 8.0 s propagate towards the coast from 167.5°N at a water depth of 20 m. The normal to the beach points towards the south (180°N). The flow of water along the coast is driven only by wave action.

The sand characteristics are specified as D_{50} = 0.20 mm and $(D_{84}/D_{16})^{0.5}$ = 1.40 everywhere along the coast. The porosity of the bed material is taken as 40%. The model area covers an area of 2700×840 m (alongshore extent × offshore extent). This model setup is the same as test KM3 of Zyserman et al. (1998), with different sediment properties.

In this example, a maximum morphological time step of 1 hour is specified. The wave field is updated every 3 hours (or every third morphological step), and the morphological simulation carried out for 14 days. The computed bathymetry is shown in Fig. 2-2.

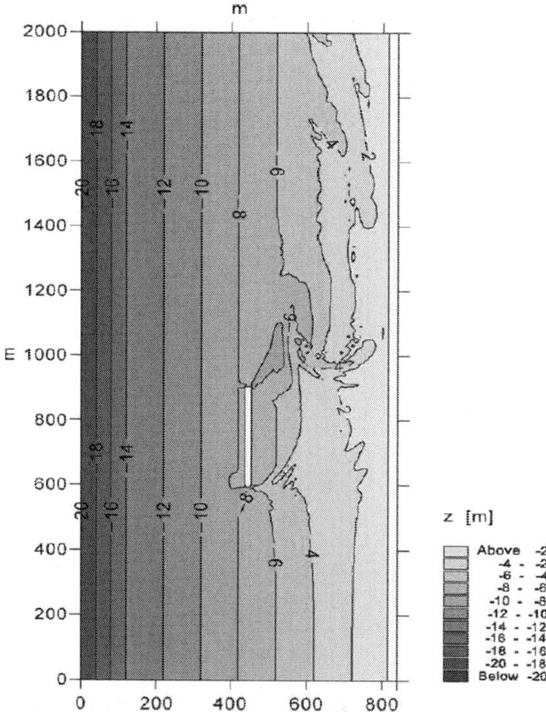

Fig. 2.2 Morphological evolution in the vicinity of a detached breakwater. Computed bathymetry after 14 days. (Johnson and Zyserman, 2002)

Fig. 2.2 shows the general features of the morphological evolution after 14 days. Some oscillatory salient had formed in the lee of the breakwater. A tendency for unsteady erosion

at the down drift edge of the breakwater can also be observed. In addition, high wave number spatial oscillations of the bottom contours can also be observed at several sections.
In the example of Johnson and Zyserman (2002), the morphological simulation is completed by MIKE 21 CAMS, the wave simulation is carried out using MIKE 21 PMS, the flow simulations using MIKE 21 HD and the sand transport calculations using MIKE 21 ST. However, MIKE PMS and HD are one of the state-of-art numerical model for evaluating wave and current field, and perform well in many studies. Why the unsteady results appeared? According to Johnson and Zyserman (2002), the origin of spatial oscillations in numerical morphological models is traced to the dependence of the bed celerity with bed levels, which is due to the non-linear relationship between sediment transport and bed levels.
In these morphodynamic systems as shown in sec. 1.2, the governing equations are a nonlinear function of bed level. The sediment transports, which are caused by waves and currents, are also calculated from complicated nonlinear hydrodynamic systems. The nonlinear couplings and numerical scheme errors in these sub-models can generate unstable and inaccurate numerical results, whose natures are still poorly understood. Even though the results of all sub-models are accurate and robust, their combination in the sediment conservation equation also leads to numerical oscillations and instabilities (Jensen et al. 1999). They generally fail to accurately predict the bed form evolution in surf zone and in the areas around coastal structures for long term simulations (Johnson and Zyserman 2002).
Several techniques to improve accuracy and stability for coastal morphological modeling have been developed in the last decades. The main concern of these studies is to stabilize the solutions when solving the sediment conservation equation. As reviewed in Long et al. (2008), many state-of-the-art models introduce oscillation controlling schemes for bed form evolution modeling. The Delft Hydraulics model Delft2D-MOR (Roelvink et al. 1994 and 1998) utilizes a forward time, central space explicit scheme with a corrected sediment transport rate to offset the negative diffusion terms in the scheme. The University of Liverpool model (O'Connor and Nicholson, 1995, Nicholson et al. 1997) uses a two-step Lax-Wendroff scheme considering the effects of gravity on the bed slope for sediment transport rates. Vincent and Caltagirone (1999) also use a modified Lax-Wendroff scheme with the Total Variation Diminishing scheme (LW-TVD) and a slope limiter. Cayocca (2001) uses a forward-time upwind scheme with the transport rate corrected due to the effect of the bed form slope (de Vriend, 1987a,b) and an input filtering technique (de Vriend et al., 1993a,b)[10, 11] to prevent oscillation. Johnson and Zyserman (2002) illustrate that the bed form slope plays a principal role in the instability of morphological numerical schemes and expands a second order Taylor-Series of the bed level in time: the first order time derivative term is composed of the sediment conservation equation calculated with the Lax-Wendroff scheme, and the second order time derivative term is regarded as the diffusion terms of advection equation. Their morphological scheme is also modified with a low-pass filter to dissipate spurious high frequency oscillations (suggested by Jensen et al., 1999). Saint-Cast (2002) applied the high-resolution NOC (Non-Oscillating Centered) scheme (based on Jiang and Tadmor, 1998) and bedform slope updating technique (suggested by Watanabe, 1988) to solve the sediment conservation equation without a filter or limiter. Shao et al. (2004), Long et al. (2008) and Chiang et al. (2010) utilized the WENO (Weighted Essentially Non-Oscillatory) algorithm from the Computational Fluid Dynamics scheme (based on Liu et al., 1994 and Jiang et al., 1999) to solve the 1D and 2D sediment conservation equation without any filters or limiters.

In order to perform well to calculate the change in coastal topography for long term simulations under waves and wave-driven currents, the model should not only be able to control the oscillations in space, but also improve the accuracy in time. The model should also take into account the effects caused by the discontinuity of the contours and beach slopes.
There are two points to be noted. First, the simulating oscillations of coastal morphological evolution should be recognized. The second point is, which techniques of controlling oscillations perform well? We will discuss in next sub-section and next section.

2.3 Analysis of oscillations for morphological scheme

As mentioned in the Introduction, several works have focused on controlling oscillations of coastal morphological schemes. Watanabe (1988), Cayocca (2001), Johnson and Zyserman (2002) suggested modifying the conservation equation to consider bottom elevation changes. Following Watanabe (1988), it is apparent that the sediment grains tend to move downward due to the distribution force of gravity, while the local slope becomes steep; the effect of bottom slope should be taken into account. Although the simulation of the wave-current field by hydraulic models varies with the beach transformation, the change of the sediment transport flux alone cannot be expected to completely suppress the creation of a jagged bed form profile. Modification of sediment transport rate was suggested as (Watanabe, 1988):

$$q_x = q_{x0} - \varepsilon_s |q_{x0}| \frac{\partial z_b}{\partial x} \tag{11}$$

$$q_y = q_{y0} - \varepsilon_s |q_{y0}| \frac{\partial z_b}{\partial y} \tag{12}$$

Where ε_s is a positive diffusivity constant; the value of which will be determined empirically. The second subscript "0" indicates sediment transport on a flat bottom.
It is noted that Watanabe (1988) did not consider the influence of the cross-slope terms in the alongshore direction, but they should be included for complex bathymetries. Johnson and Zyserman (2002) suggested including the cross-slope terms:

$$q_x = q_{x0} - \varepsilon_{xx} |q_{x0}| \frac{\partial z_b}{\partial x} - \varepsilon_{xy} |q_{x0}| \frac{\partial z_b}{\partial y} \tag{13}$$

$$q_y = q_{y0} - \varepsilon_{yy} |q_{y0}| \frac{\partial z_b}{\partial y} - \varepsilon_{yx} |q_{y0}| \frac{\partial z_b}{\partial x} \tag{14}$$

where ε_{xx}, ε_{xy}, ε_{yy}, and ε_{yx} are also positive diffusivity constants.
We substitute Eqs. (12), (13) into Eq. (2), to obtain:

$$\begin{aligned} &\frac{\partial z_b}{\partial t} + \frac{1}{1-n}\left(\frac{\partial q_{x0}}{\partial z_b}\frac{\partial z_b}{\partial x} + \frac{\partial q_{y0}}{\partial z_b}\frac{\partial z_b}{\partial y}\right) = \frac{\partial}{\partial x}\left[\frac{\varepsilon_{xx}|q_{x0}|}{(1-n)}\frac{\partial z_b}{\partial x}\right] + \\ &+ \frac{\partial}{\partial y}\left[\frac{\varepsilon_{yy}|q_{y0}|}{(1-n)}\frac{\partial z_b}{\partial y}\right] + \frac{\partial}{\partial x}\left[\frac{\varepsilon_{xy}|q_{x0}|}{(1-n)}\frac{\partial z_b}{\partial y}\right] + \frac{\partial}{\partial y}\left[\frac{\varepsilon_{yx}|q_{y0}|}{(1-n)}\frac{\partial z_b}{\partial x}\right] \end{aligned} \tag{15}$$

We can rewrite Eq. (15) as an advection-diffusion equation of bed form elevation:

$$\frac{\partial z_b}{\partial t}+C_{x0}\frac{\partial z_b}{\partial x}+C_{y0}\frac{\partial z_b}{\partial y}=\frac{\partial}{\partial x}\left[\left(\frac{\varepsilon_{xx}\left|q_{x0}\right|}{1-n}\right)\frac{\partial z_b}{\partial x}\right]+\frac{\partial}{\partial y}\left[\left(\frac{\varepsilon_{yy}\left|q_{y0}\right|}{1-n}\right)\frac{\partial z_b}{\partial y}\right]+$$
$$+\frac{\partial}{\partial x}\left[\left(\frac{\varepsilon_{xy}\left|q_{x0}\right|}{1-n}\right)\frac{\partial z_b}{\partial y}\right]+\frac{\partial}{\partial y}\left[\left(\frac{\varepsilon_{yx}\left|q_{y0}\right|}{1-n}\right)\frac{\partial z_b}{\partial x}\right] \quad (16)$$

where

$$C_{x0}=\frac{1}{1-n}\frac{\partial q_{x0}}{\partial z_b}\quad C_{y0}=\frac{1}{1-n}\frac{\partial q_{y0}}{\partial z_b}$$

The right-hand side of Eq. (16) represents the diffusion terms. These terms show that the effect of bed slope and the value of transport rates play an important role as diffusion parameters. The stability of the morphological scheme is based on these diffusion terms under control. Consequently, these diffusivity constants ε_{xx}, ε_{xy}, ε_{yy}, and ε_{yx} should be chosen carefully and sediment transport rates in Eqs. (13) and (14) should be updated in every time step after the new bed elevation is computed.

Long et al. (2008), Hsu and Hanes (2004), and Henderson et al. (2004) address the appearance of wave phase-resolving sediment transport models for nearshore applications by waves and wave-driven nearshore currents. Because the wave orbital fluid motion is oscillatory in space and time, the resulting sediment transport fluxes should also be oscillatory. Unfortunately, there are no schemes that can be applied to solve the advection-diffusion conservation equation while removing the effect of the oscillation by wave orbit motion without any limiters or artificial viscosities. Chiang et al. (2010) recommend the application of numerical oscillation removal techniques with the simulation of bed form evolution by solving the conservation equation of sediment under wave motions and wave-driven currents, such as bed-slope feedback, oscillation and the removal in spatial and temporal discretization. The details of these techniques are shown in the following sections.

3. Modification of coastal morphological models

More recently, the advantages and disadvantages of these controlling oscillatory morphological models were reviewed. Several numerical schemes are reviewed by Callaghan et al. (2006), including a first order upwind scheme, two Lax-Wendroff schemes (Vincent and Caltagirone, 1999 and Johnson and Zyserman, 2002) and the NOC scheme (Saint-Cast, 2002). Long et al. (2008) review two Lax-Wendroff schemes (based on the Richtmyer scheme and the MacCormack scheme) and three WENO schemes (the TVD-RK-WENO scheme by Shao et al., 2004, the Euler-WENO scheme by Long et al., 2008 andBed-level updated two-steps three-time-levels WENO scheme by Chiang et al., 2010). According to these reviewed results, Lax-Wendroff schemes or modified Lax-Wendroff schemes for morphodynamic system are not stable for long term simulation of bed level evolution. The filter, limiter, or artificial viscosity should be added to prevent numerical oscillation in these schemes. They also found that it is difficult to determine the phase celerity of the bed form, which is the most important parameter for the scheme's stability, when these weakened morphological schemes are applied to complex bathymetry. These powerful oscillations removal techniques will be introduced after.

3.1 Feedback by bed-slope updating scheme

As mentioned before, the morphodynamic system to be calculated when coupling the hydrodynamic (waves and wave-driven currents) and sediment transport into the morphological models (governed by conservation equation of sediment) is inherently unstable. This highly non-linear system will lead to diffusion if the effect of bed-slope variations in time is not taken into account (Watanabe, 1988; de Vriend et al., 1993a, b. In the morphological model, the linear stability analysis of bed-level and quasi-steady conditions for waves and currents are assumed. The waves and currents are assumed to remain unchanged during the entire calculation period, while the bathymetry does vary. Under this assumption, the sediment transport rates obtained from waves and currents remains unchanged over the same time duration. Even though the wave orbital velocity and current velocity remain unchanged, the bed level changes at every time step. This inconsistency implies that some quantities related to bed level, such as the friction factor and the sediment transport direction factor will also change.

In order to modify sediment transport with bed form slope, the down slope gravitational transport rate is the most commonly utilized while the bed-level changes are greater than a threshold value (Watanabe 1988; Maruyama and Takagi, 1988; de Vriend et a., 1993a, and b; Cayocca 2001; Anunes do Carmo and Seabra-Santos, 2002). Consequently, the conservation equation of sediment mass coupled with the bed-slope updated terms in two-dimensions can be rewritten as:

$$\frac{\partial z_b}{\partial t}+\frac{\partial}{\partial x}\left(q_{x0}-\varepsilon_{xx}\left|q_{x0}\right|\frac{\partial z_b}{\partial x}-\varepsilon_{xy}\left|q_{x0}\right|\frac{\partial z_b}{\partial y}\right)+\frac{\partial}{\partial y}\left(q_{y0}-\varepsilon_{yy}\left|q_{y0}\right|\frac{\partial z_b}{\partial y}-\varepsilon_{yx}\left|q_{y0}\right|\frac{\partial z_b}{\partial x}\right)=0 \quad (17)$$

Chiang et al. (2010) recommend that the modified bed-slope feedback should be processed at every time step. It is very effective for the removal oscillations for long term simulations.

3.2 Controlling oscillations in spatial discretization

In order to control the oscillations in space for the morphological model, there are several numerical schemes as reviewed by Callaghan et al. (2006) and Long et al. (2008). Among these oscillation removal schemes, the NOC and WENO scheme are more stable finite-difference schemes than the others.

3.2.1 NOC (Non-Oscillating Centered) scheme

NOC scheme is a general non-oscillating scheme originally developed by Nessyahu and Tadmor (1990) and extended to 2D by Jiang and Tadmor (1998). It solves multidimensional hyperbolic fluid conservation equations using a staggered grid method, again directly utilizing fluxes. It is similar to be applied with sediment conservation equation, and directly using sediment fluxes is good for avoiding the estimation of characteristic bedform velocity. The examples of NOCS applications for coastal area include linear oblique advection, the two-dimensional Burger and Euler equations and the one-dimensional shallow water wave equations (Williams and Peregrine, 2002). Their approach uses two staggered grids, which are moving the calculated results from one grid to another half grid with every time step. However, the staggered grid system, while minimizing scheme diffusion is impractical for morphological models consisting of a looped arrangement of individual components for sediment transport caused by waves and currents. According to Saint-Cast (2002), there is an extension of the original scheme where the new solution was initially determined on the

staggered grid and then reconstructed on the original grid (Jiang et al., 1998). This extension is easily integrated into looped morphological models for the NOC scheme.
The NOC scheme was first applied for coastal morphological modeling by Saint-Cast (2002) with the purpose of explaining ridge and runnel formation on sand beaches exposed to consistent north Atlantic swells. The work by Saint-Cast (2002) included initial testing against analytical solutions and simulations of an idealized crescent shape barred beach evolutions. Callaghan et al. (2006) utilize the NOC scheme with an idealized river entrance, perpendicular to a straight beach with constant slope, under waves and currents and show its stability by comparison with other schemes.

3.2.2 WENO (Weighted Essentially Non-Oscillatory) scheme

The WENO schemes are based on the essentially non-oscillatory (ENO) schemes, which were first developed by Harten et al. (1987) in the form of finite volume schemes and were later improved by Shu and Osher (1988). The ENO schemes are generalizations of the total variation diminishing (TVD) schemes of Harten (1983). The TVD schemes are designed so that the total variation of specific quantity in space remains constant or only decrease in time. During the solution process, there will be no new extrema generated. In other words, the TVD schemes typically degenerate to first-order accuracy at locations with smooth extrema, while the ENO schemes maintain high-order accuracy. The key idea of the ENO schemes is to use the smoothest stencil among several candidates to approximate the fluxes at the cell boundaries to high order and at the same time to avoid spurious oscillations near shocks and discontinuities. The WENO schemes process one step further by taking a weighted average of all candidates, and the weights are adjusted by the local smoothness.
The first version of WENO schemes was developed by Liu et al. (1994) for one-dimensional conservation laws of fluid mechanics. Jiang and Shu (1996) applied the scheme to multi-dimensional cases with a new weighting procedure to obtain optimized accuracy. Later, Jiang and Wu (1999) extended a high-order (5th) accurate WENO finite difference scheme, which has successfully attained comparable accuracy with fewer time-steps in computations. Shao et al. (2004) first applies the WENO scheme to solve the one-dimensional conservation equation of sediment mass and study the evolution of periodic sand bars in the presence of waves at the resonant Bragg frequency. Long et al. (2008) use the Euler-WENO schemes, based on first order explicit time discretization with the WENO scheme, to study the evolution of periodic alternating sand bars in a rectangular open channel with gravity flow. Chiang et al. (2010) applied the bed-slope feedback, 3 levels 2 time-steps, WENO morphological scheme to calculate topography changes of complex coastal area under waves and currents, and found that the stability was performed very well. The full detail of the WENO scheme applied with coastal morphological modeling can be found in Chiang et al. (2010).
The key of inaccuracy of numerical morphological modeling is to obtain the characteristic phase velocity of bedform variation to modified sediment fluxes. Fortunately, another advantage of WENO scheme is that split sediment fluxes only involve the sign of characteristic bedform velocity instead of accurate calculations. Based on the accuracy of algorithm, the fifth order WENO scheme is recommended in this chapter.

3.2.3 Controlling oscillatory morphological models with WENO scheme

In this subsection, we briefly present the finite difference version for the sediment conservation equation (Eq. (2)) using the WENO schemes. Following the numerical algorithms of Jiang et al. (1998) and the assumption of Long et al. (2008), we will describe the one-dimensional problem first, and then extend to two-dimensions. To achieve numerical stability and to avoid entropy-violating solutions, upwinding and sediment flux splitting approaches are used. The sediment transport rate can be split into two parts associated with bedform propagation in the positive and negative x (offshore) directions. This can be written as:

$$q(C) = q^{+}(C) + q^{-}(C) \tag{18}$$

$$q^{+} = (1-n)\int_{0}^{z_b} C^{+}(z)dz \tag{19}$$

$$q^{-} = (1-n)\int_{0}^{z_b} C^{-}(z)dz \tag{20}$$

where C is phase velocity of the bedform and C^+ and C^- are the phase velocities of the bedform propagating in the positive and negative *x*-directions, respectively, i.e. $C^+=\max(C,0)$, $C^-=\min(C,0)$. Thus,

$$\frac{dq^{+}(C)}{dC} \geq 0, \text{ for } C = C^{+} \tag{21}$$

$$\frac{dq^{-}(C)}{dC} \leq 0, \text{ for } C = C^{-} \tag{22}$$

Following Jiang and Wu (1999), the WENO scheme uses a conservative approximation to the spatial derivatives,

$$\frac{\partial z_b}{\partial t} = -\frac{\partial q}{\partial x} = -\frac{\hat{q}_{i+1/2} - \hat{q}_{i-1/2}}{\Delta x} \tag{23}$$

where $\hat{q}_{i+1/2}$ and $\hat{q}_{i-1/2}$ are the approximations of the sediment transport rate at grid locations (i+1/2) and (i-1/2), respectively, for the three stencil system of the WENO scheme (Fig. 3.1).

We then apply the WENO approximation procedure, as was given in Eq. (18), to obtain two numerical fluxes, $\hat{q}^{\pm}_{i+1/2}$, and sum them to obtain the numerical flux $\hat{q}_{i+1/2}$:

$$\hat{q}_{i+1/2} = \hat{q}^{-}_{i+1/2} + \hat{q}^{+}_{i+1/2} \tag{24}$$

The left-biased-flux $\hat{q}^{-}_{i+1/2}$ of Eq. (24) is calculated by the WENO scheme with approximations of three sub-stencils in a five point stencil system (grid positions from *i*-2 to *i*+2 in Fig. 1). The idea of the WENO scheme is to properly weight the three sub-stencils for the five points. This is written as:

$$\hat{q}^{-}_{i+1/2} = \begin{cases} \omega_0 q^{0}_{i+1/2} + \omega_1 q^{1}_{i+1/2} + \omega_2 q^{2}_{i+1/2}, & C_{i+1/2} \geq 0 \\ 0, \quad C_{i+1/2} < 0 \end{cases} \tag{25}$$

where ω_s (s = 0, 1, or 2) are the positive weights, and $q^s_{i+1/2}$ (s=0, 1, or 2) are the approximations of the sub-stencils.

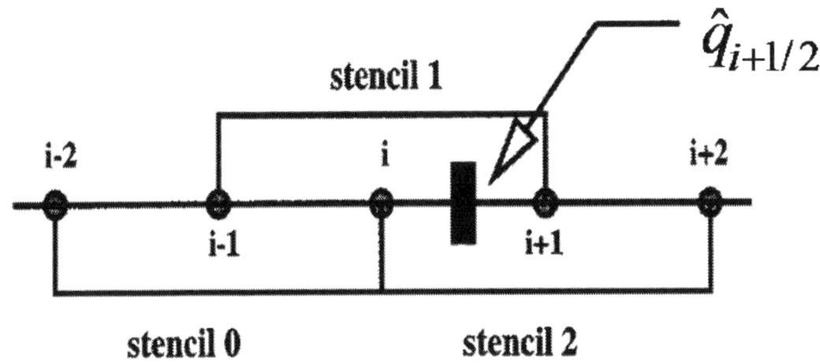

Fig. 3.1 The three sub-stencils with five points of WENO approximation

In each sub-stencil s (s=0, 1, or 2), the 3rd-order accurate approximation $q^S_{i+1/2}$ is obtained by the Taylor series expansion as:

$$q^0_{i+1/2} = \frac{1}{3}q_{i-2} - \frac{7}{6}q_{i-1} + \frac{11}{6}q_i \tag{26}$$

$$q^1_{i+1/2} = -\frac{1}{6}q_{i-1} + \frac{5}{6}q_i + \frac{1}{3}q_{i+1} \tag{27}$$

$$q^2_{i+1/2} = \frac{1}{3}q_i + \frac{5}{6}q_{i+1} - \frac{1}{6}q_{i+2} \tag{28}$$

The WENO scheme procedure of three sub-stencils with the five point system possesses the following properties in Eq. (25): (1) the approximation $\hat{q}_{i+1/2}$ at grid position i+1/2is accurate to the fifth-order; and (2) no Gibbs phenomena occur (i.e., spurious oscillations), while $\hat{q}_{i+1/2}$ is discontinuous near i+1/2.
In accordance with the above two properties, Jiang and Shu (1996) suggested the calculation of weights as:

$$\omega_0 = \frac{\alpha_0}{\alpha_0 + \alpha_1 + \alpha_2} \tag{29}$$

$$\omega_1 = \frac{\alpha_1}{\alpha_0 + \alpha_1 + \alpha_2} \tag{30}$$

$$\omega_2 = \frac{\alpha_2}{\alpha_0 + \alpha_1 + \alpha_2} \tag{31}$$

where the coefficients α_0, α_1, α_2 are calculated by Long et al. (2008) :

$$\alpha_0 = \frac{0.1}{(\varepsilon + IS_0)^2} \tag{32}$$

$$\alpha_1 = \frac{0.6}{(\varepsilon + IS_1)^2} \tag{33}$$

$$\alpha_2 = \frac{0.3}{(\varepsilon + IS_2)^2} \tag{34}$$

In which $\varepsilon \approx 10^{-20}$ is used to prevent the denominators of Eqs. (32) - (34) from becoming zero, and the smoothness measurements are written as:

$$IS_0 = \frac{13}{12}(a-b)^2 + \frac{1}{4}(a-3b)^2 \tag{35}$$

$$IS_1 = \frac{13}{12}(b-c)^2 + \frac{1}{4}(b+c)^2 \tag{36}$$

$$IS_2 = \frac{13}{12}(c-d)^2 + \frac{1}{4}(3c-d)^2 \tag{37}$$

Where

$$a = q_{i-2} - q_{i-1} \tag{38}$$

$$b = q_{i-1} - q_i \tag{39}$$

$$c = q_i - q_{i+1} \tag{40}$$

$$d = q_{i+1} - q_{i+2} \tag{41}$$

Similarly, we can give the right-biased-flux $\hat{q}^{+}_{i+1/2}$ of Eq. (24) in the same procedure.

$$\hat{q}^{+}_{i+1/2} = \begin{cases} \tilde{\omega}_0 \tilde{q}^0_{i+1/2} + \tilde{\omega}_1 \tilde{q}^1_{i+1/2} + \tilde{\omega}_2 \tilde{q}^2_{i+1/2}, & C_{i+1/2} < 0 \\ 0, \quad C_{i+1/2} \geq 0 \end{cases} \tag{42}$$

with

$$\tilde{q}^0_{i+1/2} = -\frac{1}{6} q_{i-1} + \frac{5}{6} q_i + \frac{1}{3} q_{i+1} \tag{43}$$

$$\tilde{q}^1_{i+1/2} = \frac{1}{3} q_i + \frac{5}{6} q_{i+1} - \frac{1}{6} q_{i+2} \tag{44}$$

$$\tilde{q}^2_{i+1/2} = \frac{11}{6} q_{i+1} - \frac{7}{6} q_{i+2} + \frac{1}{3} q_{i+3} \tag{45}$$

$$\tilde{\omega}_0 = \frac{\tilde{\alpha}_0}{\tilde{\alpha}_0 + \tilde{\alpha}_1 + \tilde{\alpha}_2} \tag{46}$$

$$\tilde{\omega}_1 = \frac{\tilde{\alpha}_1}{\tilde{\alpha}_0 + \tilde{\alpha}_1 + \tilde{\alpha}_2} \tag{47}$$

$$\tilde{\omega}_2 = \frac{\tilde{\alpha}_2}{\tilde{\alpha}_0 + \tilde{\alpha}_1 + \tilde{\alpha}_2} \tag{48}$$

$$\tilde{\alpha}_0 = \frac{0.1}{\left(\varepsilon + I\tilde{S}_0\right)^2} \tag{49}$$

$$\tilde{\alpha}_1 = \frac{0.6}{\left(\varepsilon + I\tilde{S}_1\right)^2} \tag{50}$$

$$\tilde{\alpha}_2 = \frac{0.3}{\left(\varepsilon + I\tilde{S}_2\right)^2} \tag{51}$$

$$I\tilde{S}_0 = \frac{13}{12}(b-c)^2 + \frac{1}{4}(b-3c)^2 \tag{52}$$

$$I\tilde{S}_1 = \frac{13}{12}(c-d)^2 + \frac{1}{4}(c+d)^2 \tag{53}$$

$$I\tilde{S}_2 = \frac{13}{12}(d-e)^2 + \frac{1}{4}(3d-e)^2 \tag{54}$$

$$b = q_{i-1} - q_i \tag{55}$$

$$c = q_i - q_{i+1} \tag{56}$$

$$d = q_{i+1} - q_{i+2} \tag{57}$$

$$e = q_{i+2} - q_{i+3} \tag{58}$$

At the grid location i-1/2, the left-biased-flux $\hat{q}_{i-1/2}^{-}$ can be repeated using Eqs. (24) - (58) by simply shifting i backward one grid. Consequently, the spatial discretization of WNEO finite difference scheme for one-dimensional sediment conservation problem is complete. With a simple 1st-order forward temporal discretization, Eq. (2) can be solved by the Euler-WENO scheme as:

$$\frac{z_{bi}^{n+1} - z_{bi}^{n}}{\Delta t} + \frac{1}{1-n}\frac{\hat{q}_{i+1/2} - \hat{q}_{i-1/2}}{\Delta x} = O\left(\Delta t, \Delta x^5\right) \tag{59}$$

The extension of the Euler-WENO scheme for Eq. (2) in two-dimensions can be easily given as:

$$\frac{z_{bi,j}^{n+1}-z_{bi,j}^{n}}{\Delta t}+\frac{1}{1-n}\frac{\hat{q}_{x+1/2,j}^{n}-\hat{q}_{x-1/2,j}^{n}}{\Delta x}+\frac{1}{1-n}\frac{\hat{q}_{y,j+1/2}^{n}-\hat{q}_{y,j-1/2}^{n}}{\Delta y}=O\left(\Delta t,\Delta x^{5},\Delta y^{5}\right) \tag{60}$$

3.2.4 Phase celerity of the bedform

There is only one undetermined parameter for the bed-level updated WENO scheme; namely the phase celerity of the bed form. Eq. (2) can be written as:

$$\frac{\partial z_b}{\partial t}+\frac{1}{1-n}\left(\frac{\partial q_x}{\partial z_b}\frac{\partial z_b}{\partial x}+\frac{\partial q_y}{\partial z_b}\frac{\partial z_b}{\partial y}\right)=0 \tag{61}$$

Since the phase velocity of bedforms can be assumed as $C_x=1/(1-n)\partial q_x/\partial z_b$, $C_y=1/(1-n)\partial q_y/\partial z_b$, Eq. (61) becomes:

$$\frac{\partial z_b}{\partial t}+C_x\frac{\partial z_b}{\partial x}+C_y\frac{\partial z_b}{\partial y}=0 \tag{62}$$

In order to calculate the phase celerity C_x and C_y, Eq. (66) can be written in vector form and Eq. (2) is substituted into it:

$$\vec{C}=\frac{-\frac{\partial z_b}{\partial t}}{\left|\nabla z_b\right|^2}\vec{\nabla}z_b=\frac{\vec{\nabla}\cdot\vec{q}}{(1-n)\left|\nabla z_b\right|^2}\vec{\nabla}z_b \tag{63}$$

where vector $\vec{C}=\left(C_x,C_y\right)$, $\vec{q}=\left(q_x,q_y\right)$.

For one-dimensional case, Eq. (63) is reduced to:

$$C\left(z_b\right)=\frac{\frac{\partial q}{\partial x}}{(1-n)\frac{\partial z_b}{\partial x}} \tag{64}$$

We can calculate the phase celerity of the bedform by solving Eq. (63) and (64) for one- and two-dimensional problems. Hudson et al. (2005) suggested a central difference scheme to calculate the phase celerity for the one-dimensional case:

$$C_i=\frac{q_{i+1}-q_{i-1}}{(1-n)\left(z_{bi+1}-z_{bi-1}\right)} \tag{65}$$

There are some disadvantages when applying Eq. (65) to estimate phase celerity in a real coastal area. It fails when $z_{bi+1}=z_{bi-1}$, and causes significant errors when the central difference spatial grid spans sand bars, dunes, or ripples. Fortunately, the WENO scheme only requires the sign of the bedform phase celerity for split sediment transport. Long et al. (2008) suggested a simple formula to calculate the sign of the phase celerity:

$$sign(C_i) = sign((q_{i+1} - q_i)(z_{bi+1} - z_{bi})) \tag{66}$$

3.3 Controlling oscillations in temporal discretization

As mentioned before, the inaccuracy of bed-level simulation in every time-step is caused by discretization errors and oscillation factors from the sub-models of waves, currents, or sediment transport rates. They will lead to diffusions and dispersions in the long term. There are three typical techniques for controlling oscillations in temporal discretization as following.

3.3.1 Explicit two-step, three-time-level finite difference scheme

In this subsection, the explicit two-step, three-time-level finite difference scheme with stagger grid is introduced (Chiang et al., 2010). This technique can be easily implemented in time discretization and does not lead to a significant CPU time increases. The gradient of sediment fluxes is expanded at half grid position and the bed-level is calculated at original grid position. It is convenient to apply the operator of WENO scheme with approximated sediment fluxes. For one-dimensional sediment conservation problem, Eq. (2) becomes:
Step 1.

$$\frac{z_{bi}^{n+1/2} - z_{bi}^{n}}{\Delta t / 2} + \frac{1}{1-n} \frac{\hat{q}\left(q_0, z_{bi}^{n}\right)_{i+1/2} - \hat{q}\left(q_0, z_{bi}^{n}\right)_{i-1/2}}{\Delta x} = 0 \tag{67}$$

Step 2.

$$\frac{z_{bi}^{n+1} - z_{bi}^{n}}{\Delta t} + \frac{1}{1-n} \frac{\hat{q}\left(q_0, z_{bi}^{n+1/2}\right)_{i+1/2} - \hat{q}\left(q_0, z_{bi}^{n+1/2}\right)_{i-1/2}}{\Delta x} = 0 \tag{68}$$

where the $\hat{q}(\ )$ is the operator of WENO scheme. This scheme gives 2nd order accuracy $O(\Delta t^2)$ in time.

3.3.2 TVD-Runge-Kutta scheme

TVD (Total Variation Diminishing) schemes are designed such that the total variance of the solution TV= $\int_{-\infty}^{+\infty} \left|\partial z_b / \partial x\right| dx$ will remain constant or only decrease in time. During the solution process, there will be no new extrema generated. TVD scheme were often proposed based on existing schemes. Shu and Osher (1988) and Shao et al. (2004) applied a TVD-Runge-Kutta (TVD-RK) scheme for 3rd order time integration of Eq. (2). The TVD-RK scheme can be summarized as an algorithm of 5 steps:
The 1st step consists of an Euler forward step to get time level n+1:

$$\frac{1}{\Delta t}\left(z_b^{n+1} - z_b^{n}\right) + \frac{\partial}{\partial x}\left(\frac{1}{1-n} q\left(z_b^{n}\right)\right) = 0 \tag{69}$$

The 2nd step uses a second forward step to time level n+2:

$$\frac{1}{\Delta t}\left(z_b^{n+2} - z_b^{n+1}\right) + \frac{\partial}{\partial x}\left(\frac{1}{1-n} q\left(z_b^{n+1}\right)\right) = 0 \tag{70}$$

The 3rd step uses an averaging step to obtain an approximation solution at n+1/2:

$$z_b^{\,n+1/2} = \frac{3}{4} z_b^{\,n} + \frac{1}{4} z_b^{\,n+2} \tag{71}$$

The 4th step uses a 3rd Euler step to get time level n+3/2:

$$\frac{1}{\Delta t}\left(z_b^{\,n+3/2} - z_b^{\,n+1/2}\right) + \frac{\partial}{\partial x}\left(\frac{1}{1-n} q\left(z_b^{\,n+1/2}\right)\right) = 0 \tag{72}$$

the 5th step uses another averaging step finally to get solution at time level n+1:

$$z_b^{\,n+1} = \frac{1}{3} z_b^{\,n} + \frac{2}{3} z_b^{\,n+2/3} \tag{73}$$

3.3.3 Predictor-corrector method

As mention before, instability problems appear to originate from the explicit discretization of the sediment conservation equation, Fortunato and Oliveira (2007) recommend a predictor-correct scheme was implemented and performed well. It is shown as below:
Predictor step: an estimate of depth at time n+1 is first calculated as:

$$h^{(p)} = h^n + \frac{1}{1-n} \nabla \int_n^{n+2} q\left(u(t), \eta(t), h^n\right) dt \tag{74}$$

Corrector step: a fully or semi-implicit scheme is applied with the correction step as:

$$h^{n+1} = h^n + \frac{1}{1-n} \nabla \int_n^{n+2} q\left(u(t), \eta(t), h^*\right) dt \tag{75}$$

where h is depth, $h^* = ah^{(p)} + (1\text{-}a)h^n$, a is the implicitness parameter in [0,1]. Eqs. (11) and (12) can be repeated iteratively for a user-specified number of correction cycles. This scheme gives 2nd order accuracy $O(\Delta t^2)$ in time.

4. The controlling oscillatory coastal morphological models

Consider the computing efficiency, stability and accuracy of numerical schemes, we recommend a bed-slope updating, 2 steps with 3-time-levels (2nd order accuracy), WENO morphodynamic scheme (5th order accuracy) to be applied with complex coastal estuary area. The hydraulic modeling system for waves and currents is suggested by Lin et al. (1996), and the sediment transport modeling is suggested by Chiang et al. (1996). It is briefly describe as following. The procedure of the coastal morphological model system is shown as Figure 4.1.

4.1 Coastal morphological system

The bed-slope feedback updating, 2 steps with 3-time-levels, WENO morphological scheme with accuracy $O\left(\Delta t^2, \Delta x^5, \Delta y^5\right)$ is shown as below:

Step 1.

$$\frac{z_{bi,j}^{n+1/2}-z_{bi,j}^{n}}{\Delta t/2}+\frac{1}{1-n}\frac{\hat{q}_x\left(q_{x0},z_{bi}^{n}\right)_{i+1/2,j}^{n}-\hat{q}_x\left(q_{x0},z_{bi}^{n}\right)_{i-1/2,j}^{n}}{\Delta x}+\frac{1}{1-n}\frac{\hat{q}_y\left(q_{y0},z_{bi}^{n}\right)_{i,j+1/2}^{n}-\hat{q}_y\left(q_{y0},z_{bi}^{n}\right)_{i,j-1/2}^{n}}{\Delta y}=0 \quad (76)$$

Step 2.

$$\frac{z_{bi,j}^{n+1}-z_{bi,j}^{n}}{\Delta t}+\frac{1}{1-n}\frac{\hat{q}_x\left(q_{x0},z_{bi}^{n+1/2}\right)_{i+1/2,j}^{n+1/2}-\hat{q}_x\left(q_{x0},z_{bi}^{n+1/2}\right)_{i-1/2,j}^{n+1/2}}{\Delta x}+\frac{1}{1-n}\frac{\hat{q}_y\left(q_{y0},z_{bi}^{n+1/2}\right)_{i,j+1/2}^{n+1/2}-\hat{q}_y\left(q_{y0},z_{bi}^{n+1/2}\right)_{i,j-1/2}^{n+1/2}}{\Delta y}=0 \quad (77)$$

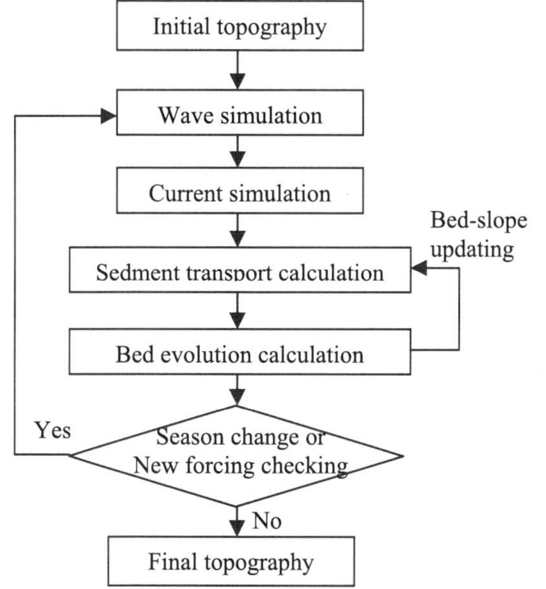

Fig. 4.1 The procedure of the coastal morphological modelling system

4.2 Stability condition analysis

The stability requirement of most morphological schemes is the Courant number $|C_i\Delta t / \Delta x| \le 1$, and the large Courant numbers observed in the simulation of morphological evolutions suggest that reducing the time steps could improve the stability of the numerical schemes. However, this will increase the computational demand. Fortunately, if the diffusion terms of Eq. (3) could be properly eliminated, the limit of Courant number could be great than unity. This can be done by carefully giving the values of the diffusivity constants (ε_{xx}, ε_{xy}, ε_{yx}, and ε_{yy}). Some diffusivity constants have been suggested for real coastal environments. Watanabe (1988) suggested that the values are determined empirically through experiments. Struiksma et al. (1985) and Cayocca (2001) used $\varepsilon_{xx} = \varepsilon_{yy} = 4$, and Kuroiwa and Kamphuis (2003) suggest $\varepsilon_{xx} = \varepsilon_{yy} = 2$. Chiang et al. (2010) recommend $\varepsilon_{xx} = \varepsilon_{yy} = 4$ and $\varepsilon_{xy} = \varepsilon_{yx} = 2$ for complex coastal area. The diffusivity constants were set to $\varepsilon_{xx} = \varepsilon_{yy} = 4$ and $\varepsilon_{xy} = \varepsilon_{yx} = 2$ in following simulation test.

5. Numerical examples and discussions

5.1 Example 1: Miao-Li coastal area in Western Taiwan

5.1.1 Environmental conditions

The developed model is first applied for the complex topography in the Miao-Li county coastal area of western Taiwan. The simulation area (Fig. 5.1) is a sandy beach, which is 7.0 km long in the alongshore direction and 3.5 km wide in the on-off shore direction. This amounts to a maximum depth of around 35.0 m. The median diameter of the beach sand is $d_{50} = 0.25$ mm. The area also has complex beach slopes (from 1/10 to 1/150) and depth contours. The shoreline orientation is in the NNE to NW direction. The tide is semidiurnal with a mean range of around 3.0 m. Since tidal current is negligible in the shore area, waves are the main factors in the coastal dynamics. The dominant wave is the winter northerly monsoon waves between September and March. The significant wave height $H_{1/3}$ is 2.5 m and the significant wave period $T_{1/3}$ is 7.8 sec.

5.1.2 Numerical conditions

In example 5.1, we analyze the morphodynamic evolution results by FTCS (forward time central space), Euler-WENO, and our bed-slope updated 2-step, 3-time-level WENO schemes using the same numerical conditions. The selected sediment coefficients are: A_1=1.5 and A_2=2.5. The spatial grid sizes of $\Delta x = \Delta y = 10$ m are used in all models (including the sub-models for waves and currents). The time step interval of $\Delta t = 1$ s is used for the nearshore current sub-model and Δt=60 sec for the morphological model.

5.1.3 Results and discussions

Because we utilize a single directional regular wave to represent all random sea waves, it is very difficult to accurately examine two topography surveys at different times. In order to demonstrate the oscillation removal performance, we compare our results with other schemes that have been reviewed previously. The morphodynamic results after 90 days using the FTCS and Euler-WENO schemes are shown in Figs. 5.2 and 5.3. The results of present morphological scheme after 30, 60, and 90 days are shown in Figs. 5.4, 5.5 and 5.6. We can easily see the significant differences between these three schemes. Present morphological scheme is more stable than the other schemes. As shown in Fig. 5.5, the numerical dispersion, diffusion and oscillations are clearly seen for the FTCS scheme. The results of the Euler-WENO scheme are more stable than FTCS scheme, but the diffusions still occur around -5.0m ~ -2.0m. Similar situations are also found in the wave field and wave-driven current flow field: wave breaking and maximum velocities occur in the same region, which is the steepest area of the beach. The wave breaking and the instability of the numerical scheme (such as the results of FTCS and Euler-WENO) will cause "shockwaves" in the local area. It is unreasonable for the bed-slope exceeding the rest angle of sand under water. The significance of the bed-slope updating technique and the nonlinear coupling sub-models for morphodynamic system can be clearly identified. This shows that the two-step, three-time-level method can improve the stability in the steep slopes and sharp gradients of beaches in a real coastal area.

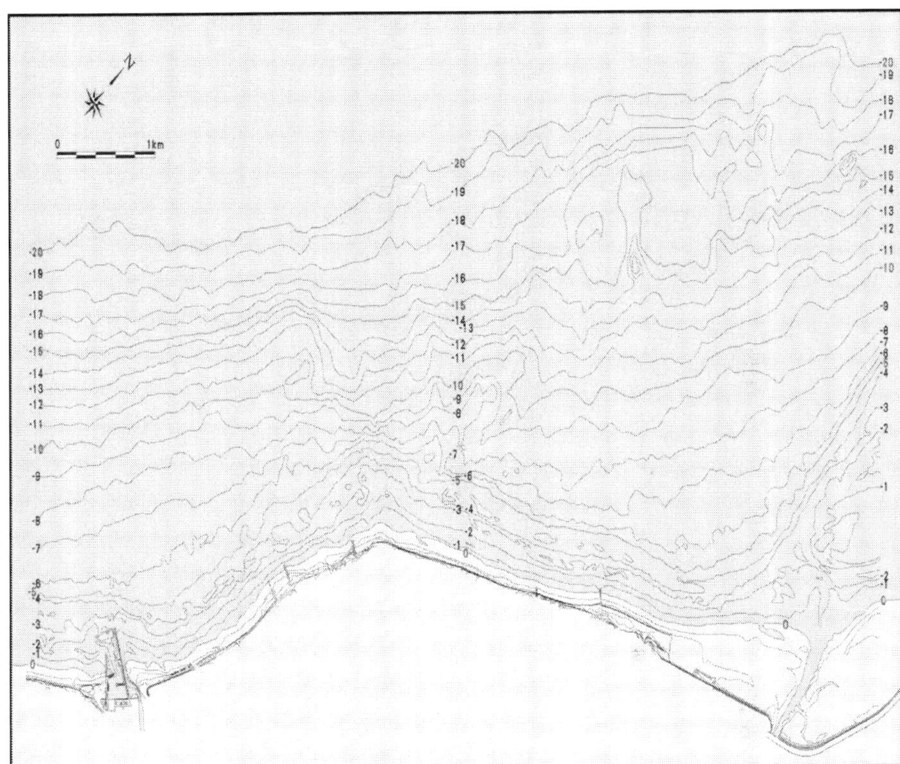

Fig. 5.1 The modeling area of example 1 at Miao-Li county coastal area of western Taiwan.

5.2 Example 2: Tai-Dong coastal area in Eastern Taiwan

In example 2, the Fu-Guon coastal area in Tai-Dong County in eastern Taiwan is considered (Fig. 5.7). The area has a steep and convex topography, which causes wave energy to concentrate. At this site, the tide is semidiurnal with a mean tidal range of less than 1 m. The dominant wave is the summer southern monsoon waves with a significant wave height $H_{1/3}$ = 1.5 m and significant wave period $T_{1/3}$ = 7.0 sec. The numerical conditions are A_1=1.7, A_2=2.6. The spatial grid interval is used uniformly with Δx = Δy = 5.0 m in all models (including sub-models of wave and current). The time interval for the nearshore current sub-model is Δt = 1 s and Δt = 60 sec for the morphological model.

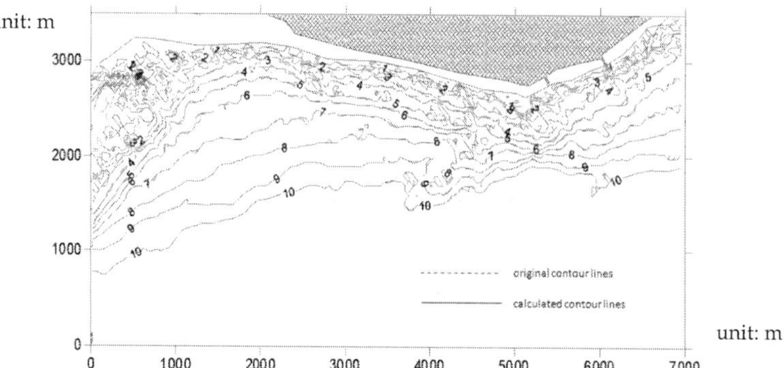

Fig. 5.2 The result of FTCS scheme after 90 days.

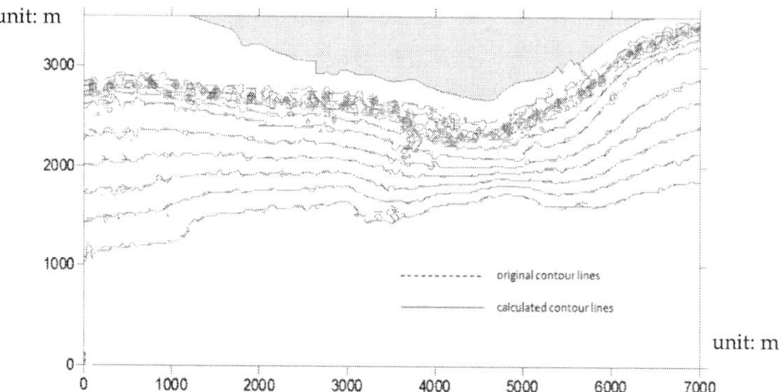

Fig. 5.3 The result of Euler-WENO scheme after 90 days.

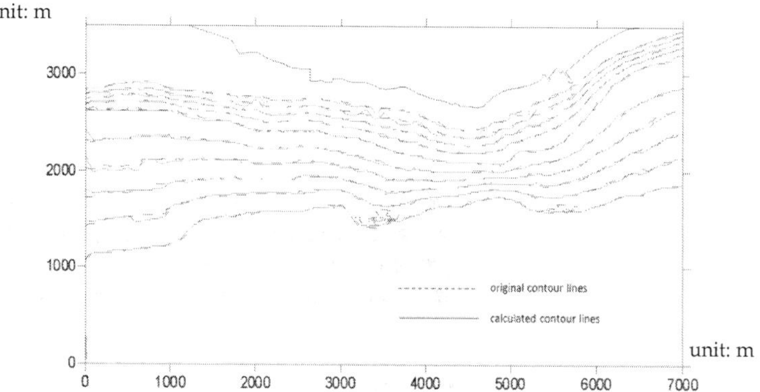

Fig. 5.4 The result of bed-slope updated 2-steps 3-time-levels WENO scheme after 30 days.

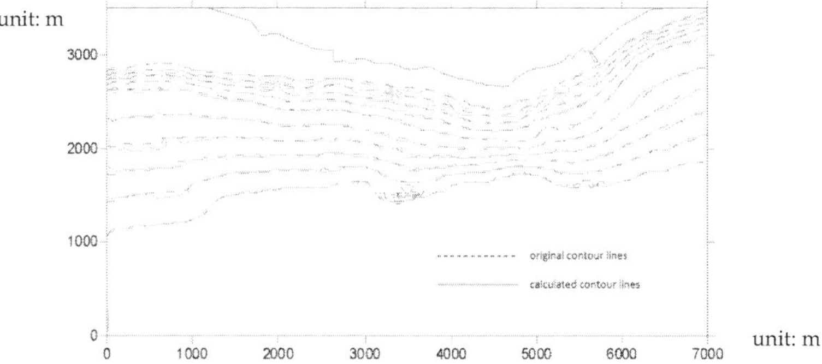

Fig. 5.5 The result of bed-slope updated 2-steps 3-time-levels WENO scheme after 60 days.

unit: m

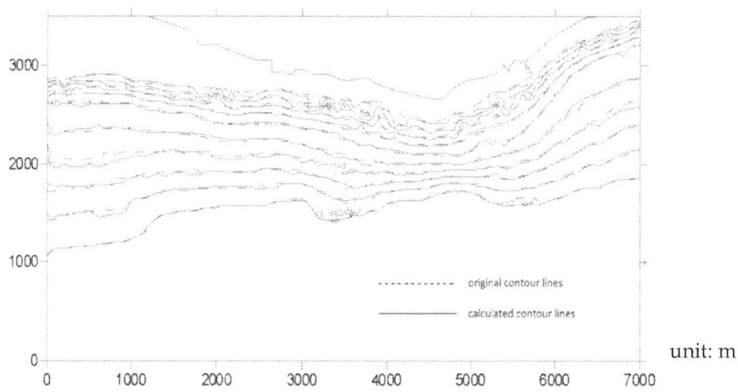

Fig. 5.6 The result of bed-slope updated 2-steps 3-time-levels WENO scheme after 90 days

The morphodynamic results of the Euler-WENO and our schemes are shown in Figs. 5.8 and 5.9. Because of the existence of underwater rigs in this area, the wave energy tends to converge at some locations and diverge at others. Consequently there are oscillations in the numerical simulation of the bed-level evolution, and our schemes are more stable than the others. This can be seen in the topography change results where wave energy was concentrated in Figs. 5.8 and 5.9.

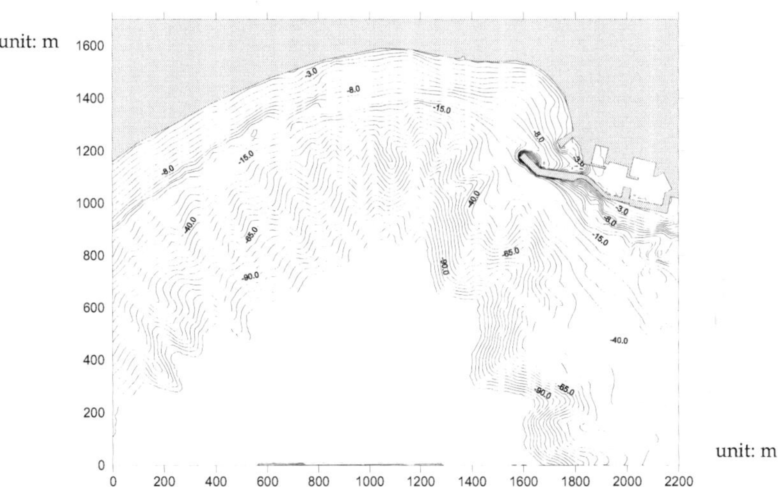

Fig. 5.7 The modeling area of example 2 at Ti-Dong county coastal area of eastern Taiwan.

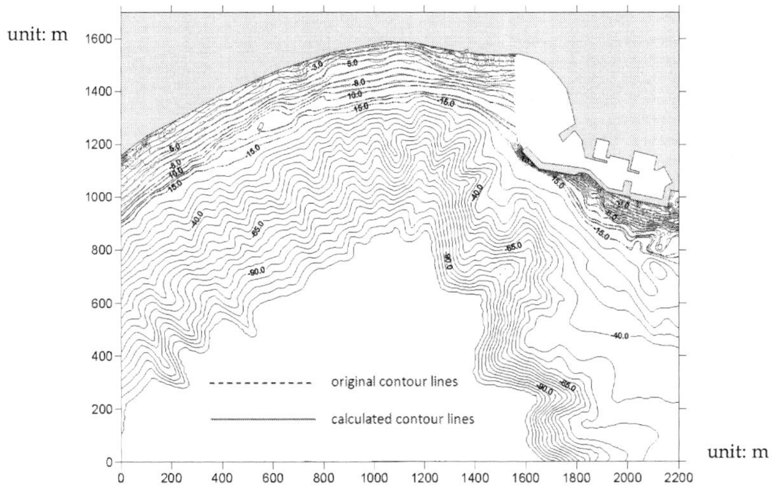

Fig. 5.8 The result of Euler-WENO scheme after 90 days.

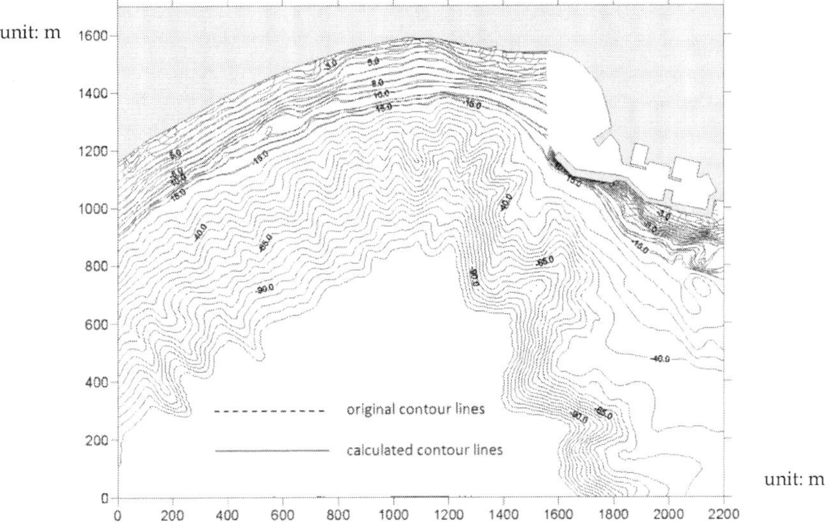

Fig. 5.9 The result of bed-slope updated 2-steps 3-time-levels WENO scheme after 90 days.

6. Conclusion

The evolution of morphodynamic schemes for oscillation removal over the past decade are summarized, the instability of a morphological system is discussed, and numerical solutions of morphodynamic evolution for complex topographies are presented. The bed-slope updated 2-step, 3-time-level WENO scheme has performed well for a real coastal area subject to waves and wave- driven currents. It is sufficiently demonstrated that this scheme provides a significant improvement for shockwave capture and stability in different Δt based on two schematized examples. Results from these two examples suggest the following: the effect of diffusions, dispersions, and oscillations from coupling the sub-models of morphodynamic systems have been improved significantly. The multi-time-level schemes for temporal discretization can improve the stability in examples with steep slopes and sharp gradients of beaches. With carefully selected diffusivity constants, the Courant number can generally exceed unity with reduced time steps efficiently for long-term simulations.

7. References

Andersen, O.H., Hedegaard, I.B., Deigaard, R., de Girolamo, P., Madsen, P., 1991. Model for morphological changes under waves and current, Preprints IAHR, Symposium Suspended Sediment Transport, Florence, Italy, pp. 327–338.

Antunes Do Carmo, J.S., Seabra-Santos, F.J., 2002. Nearshore sediment dynamics computation under the combined effects of waves and currents, Advances in Engineering Software, 33(1), 37-48.

Callaghan, D. P., Saint-Cast, F., Nielsen, P., Baldock, T. E., 2006. Numerical solutions of the sediment conservation law; a review and improved formulation for coastal morphological modeling, Coastal Engineering, 53, 557-571.

Cayocca, F., 2001. Long-term morphological modeling of a tidal inlet: the Arcachon Basin, France, Coastal Engineering, 42(2), 115-142.

Chiang, Y. C., Lin, M. C., Liou, J. Y., 1996. A Model Formula for Estimation of the Coastal Sediments in West Coast of Taiwan. Proc. 18th Conf. on Ocean Engineering in Republic of China Nov. 1996, pp.619-626.

Chiang Y.C., Hsiao S.S., Lin M.C., 2010. Numerical solutions of coastal morphodynamic evolution for complex topography. Journal of Marine Science and Technology, Vol. 18, No. 3, pp. 333-344.

Coeffe´, Y., Pe´chon, P., 1982. Modelling of sea-bed evolution under wave action. Proceedings 18th Conference ICCE, Capetown.

de Vriend, H.J., 1987a. 2DH Mathematical modelling of morphological evolution in shallow water. Coastal Eng. 11, 1–27.

de Vriend, H.J., 1987b. Analysis of horizontally two-dimensional morphological evolution in shallow water. J. Geophys. Res. 92 _C4., 3877–3893.

de Vriend, H.J., 1994. Two-dimensional horizontal and weakly three-dimensional models of sediment transport due to waves and currents. In: Abbott, M.B., Price, W.A._Eds.., Coastal, Estuarial and Harbour Engineers' Reference Book. E&FN. Spon, London, pp. 215–238.

de Vriend, H.J., Copabianco, M., Chesher, T., De Swart, H.E., Latteux, B., Stive, M.J.F., 1993a. Long term modeling of coastal Morphology. Coastal Eng. 21, 225–269.

de Vriend, H.J., Zyserman, J., Nicholson, J., Roelvink, J.A., Pe´chon, P., Southgate, H.N., 1993b. Medium term 2DH coastal area modeling. Coastal Eng. 21, 193–224.

Harten, A., 1983. High resolution schemes for hyperbolic conservation laws. J. Comput. Phys. 49, 357.

Harten, A., Engquist, B., Osher, S., Chakravarthy, S., 1987. Uniformly high-order accurate essentially non-oscillatory schemes, III. J. Comput. Phys. 71, 231. pp. 231-303.

Henderson, S.M., Allen, J.S., Newberger, P.A., 2004. Nearshore sandbar migration predicted by an eddy-diffusive boundary layer model. J. Geophys. Res. 109, C06024. doi:10.1029/2003JC002137.

Hsu, T.-J., Hanes, D.M., 2004. Effects of wave shape on sheet flow sediment transport. J. Geophys. Res. 109, C05025. doi:10.1029/2003JC002075.

Hudson, J., Damgaard, J., Dodd, N., Chesher, T., Cooper, A., 2005. Numerical approaches for 1D morphodynamic modelling. Coast. Eng. 52, 691–707.

Jensen, J.H., Madsen, E.Ø., Fredsøe, J., 1999. Oblique flowover dredged channels: II. Sediment transport and morphology. J. of Hydraul. Eng. 125 (11),1190–1198.

Jiang, G.-S., Shu, C.-W., 1996. Efficient implementation of weighted ENO schemes. J. Comput. Phys. 126, 202.

Jiang, G.-S., Wu, C.-C., 1999. A high-order WENO finite difference scheme for the equations of ideal magnetohydrodynamics. J. Comput. Phys. 150, 561–594.

Jiang, G.-S., Levy, D., Lin, C.-T., Osher, S., Tadmor, E., 1998. High-resolution nonoscillatory central schemes with nonstaggered grids for hyperbolic conservation laws. SIAM Journal on Numerical Analysis 35 (6), 2147–2168.

Johnson, H.K., Zyserman, J.A., 2002. Controlling spatial oscillations in bed level update schemes. Coast. Eng. 46, 109–126.

Lin, M. C., Kuo, J. C., Chiang, Y. C., Liou, J. Y., 1996. Numerical Modeling of Topography Changes in Sea Region. Proc. 18th Conf. on Ocean Engineering in Republic of China Nov. 1996, pp.627-637.

Liu, X.-D., Osher, S., Chan, T., 1994. Weighted essentially non-oscillatory schemes. J. Comput. Phys. 115, 200.

Long, W., Kirby, J. T., Shao Z., 2008. A numerical scheme for morphological bed level calculations, Coastal Engineering, Vol. 55 167–180 .

Maruyama, K., Takagi, T., 1988. A simulation system of nearshore sediment transport for the coupling of the sea-bottom topography, waves and currents. Proc. IAHR, Copenhagen.

Nicholson, J., Broker, I., Roelvink, J.A., Price, D., Tanguy, J.M., Moreno, L., 1997. Intercomparison of coastal area morphodynamic models. Coast. Eng. 31, 97–123.

Nairn, R.B., Southgate, H.N., 1993. Deterministic profile modelling of nearshore processes. Part 2. Sediment transport and beach profile development. Coastal Engineering 19 (1–2), 57–96.

O'Connor, B.A., Nicholson, J., 1989. Modelling changes in coastal morphology, Sediment Transport Modeling. In: Wang, S.S.Y. (Ed.), ASCE, pp. 160–165.

Roelvink, J.A., van Banning, G.K.F.M., 1994. Design and development of Delft3D and application to coastal morphodynamics. In: Verwey, Minns, Babovic, Maksimovic (Eds.), Hydroinformatics '94. Balkema, Rotterdam, pp. 451–455.

Roelvink, J.A., Walstra, D.J.R., Chen, Z., 1998. Morphological modelling of Keta lagoon case. Proc. 24th Int. Conf. on Coastal Engineering. ASCE, Kobe, Japan.

Roelvink, J.A., 2006. Coastal morphodynamic evolution techniques. Coast. Eng. 53, 277–287.

Saint-Cast, F., 2002. Modelisation de la morphodynamique des corps sableux en milieu littoral (Modelling of Coastal Sand Banks Morphodynamics). University Bordeaux I, Bordeaux. 245 pp.

Saint-Cast, F., Caltagirone, J.P., Bonneton, P., 2001. On the splitting of the sediment fluxes balance: a new formulation for the sand waves equation. Coastal Engineering 2001, Computer Modelling of Seas and Coastal Regions, Rhodes, Greece, pp. 2–12.

Sato, K., Shuto, N., Tanaka, H., 1995, Numerical simulation of thesand spit flushing at a river mouth. In: Chinese Hydraulic Engineering Society and International Research and Training Center on Erosion and Sedimentation (Ed.), Adv. In Hydro-Science and Eng., vol. II, Part B, Tsinghua University Press, Beijing, pp. 1399-1406.

Shao, Z.Y., Kim, S., Yost, S.A., 2004. A portable numerical method for flow with discontinuities and shocks. Proceedings of 17th Engineering Mechanics Conference, ASCE, June 13-16. Paper, vol. 65. University of Delaware, Newark, DE, USA (on CD).

Shu, C.-W., 1997, Essentially non-oscillatory and weighted essentially non-oscillatory schemes for hyperbolic conservation laws, lecture notes. ICASE Report No. 97-65, NASA/CR-97-206253. November.

Shu, C.-W., Osher, S., 1988. Efficient implementation of essentially non-oscillatory shock-capturing schemes. J. Comput. Phys. 77, 439–471.

Struiksma, N., Olewesen, K.W., Flokstra, C., de Vriend, H.J., 1985. Bed deformation in curved alluvial channels. J. Hydraul. Res. 23 (1).

Vincent, S., Caltagirone, J.P., 1999. Efficient solving method for unsteady incompressible interfacial flow problems. International Journal For Numerical Methods In Fluids 30 (6), 795–811.

Wang, Z.B., 1992. Theoretical analysis on depth-integrated modeling of suspended sediment transport. J. Hydrol. Res. 30 (3).

Watanabe, A., 1982. Numerical models of near-shore currents and beach deformation. Coastal Eng., Jpn. 25, 147–161.

Watanabe, A., Maruyuma, K., Shimizu, T., Sakakiyama, T., 1986. Numerical prediction model of three-dimensional beach deformation around a structure. Coastal Eng., Jpn. 29.

Watanabe, A., 1988, Modeling of sediment and beach evolution. In: Horikawa, K. (Ed.), Nearshore Dynamics and Coastal Processes. University of Tokyo Press, Tokyo, Japan, pp. 292-302.

Yamaguchi M., Nishioka Y., 1984, Numerical simulation on the change of bottom topography by the presence of coastal structures, Proc. 19th Int. Conf. on Coastal Engineering, ASCE, Houston, pp. 1732-1748.

Part 3

River, Delta and Lake Sediment Processes

11

Computation of Lake or Reservoir Sedimentation in Terms of Soil Erosion

Vlassios Hrissanthou
Democritus University of Thrace
Greece

1. Introduction

Lake or reservoir sedimentation, which decreases the useful life of the lake or reservoir, is closely associated with soil and stream bed erosion in the corresponding basin. Sediment inflowing into the lake or reservoir, through the river which feeds the lake or reservoir, originates mainly from the products of soil and stream bed erosion in the corresponding basin. Therefore, the computation of lake or reservoir sedimentation requires, in a previous step, the computation of soil erosion and stream sediment transport in the lake or reservoir basin. The computation of soil erosion, in turn, requires, in a previous step, the computation of runoff due to rainfall, because both rainfall and runoff cause soil erosion.

According to the above mentioned process chain, the physical processes, that should be quantified, are: runoff resulting from rainfall, soil erosion due to rainfall and runoff, inflow of eroded particles into streams, sediment transport in streams, sediment inflow into the lake or reservoir, and sediment deposition in the lake or reservoir. Since the centre of gravity of the present chapter falls into the quantification of soil erosion and stream sediment transport, an overview and classification of soil erosion models and stream sediment transport (total load) models will be given in Sections 2 and 3, respectively.

The physical processes mentioned above could be included in a mathematical simulation model, which will compute the inflowing sediment quantity into a lake or reservoir from the surrounding basins. For the verification of the mathematical model, usually, no systematic, long-term sediment yield measurements of the rivers, streams or torrents, discharging their water into the lake or reservoir, are available. Contrarily, rainfall and, in general, meteorological data, which serve as input data of the mathematical models, can be found. Additionally, model parameters, which serve also as input data, can be estimated by means of topographic and geologic maps. However, the lack of output data, e.g. sediment yield measurements, leaves the computational results unchecked.

In the present chapter, four case studies, regarding the computation of the mean annual sediment inflow into two reservoirs and two natural lakes, will be described. The mathematical simulation model, used in all cases, consists of three submodels:

- a hydrologic rainfall-runoff submodel
- a soil erosion submodel
- a stream sediment transport submodel

In the following paragraph, the names of the lakes (artificial or natural), the submodels used and the available output data are given:

Forggensee Reservoir (Bavaria, Germany)
- Rainfall-runoff submodel of Lutz (1984)
- "Universal Soil Loss Equation" (USLE)
- Stream sediment transport submodel of Yang & Stall (1976)

Available output data: values of sediment yield (suspended load) at the reservoir inlet for 12 years (1966 - 1977).

Yermasoyia Reservoir (Cyprus)
- Rainfall-runoff submodel (Giakoumakis et al., 1991)
- Soil erosion submodel of Schmidt (1992)
- Stream sediment transport submodel of Yang & Stall (1976)

Available output data: mean annual rate of soil erosion in the corresponding basin.

Kastoria Lake (Greece)
- Rainfall-runoff submodel (Giakoumakis et al., 1991)
- Soil erosion submodel of Schmidt (1992)
- Stream sediment transport submodel of Yang & Stall (1976)

No available output data.

Vistonis Lake (Greece)
- Rainfall-runoff submodel (Giakoumakis et al., 1991)
- Soil erosion submodel of Schmidt (1992)
- Stream sediment transport submodel of Yang & Stall (1976)

Available output data: topographic maps with isobath contours of Vistonis Lake for the years 1949 and 1970.

2. Classification of soil erosion models

Two broad categories of mathematical soil erosion models are the deterministic and the stochastic ones (e.g. Smith et al., 1977). In this chapter, a deterministic approach to the soil erosion, as well as to the other physical processes, is made. Therefore, a further classification of the deterministic erosion models is given below:
- empirical models (e.g. Universal Soil Loss Equation, USLE, Wischmeier & Smith, 1978; Santos et al., 1977; Johnson & Julien, 2000; De Vente et al., 2005)
- physically-based models (e.g. Hairsine & Rose, 1992a, 1992b; Flanagan & Nearing, 1995; De Roo et al., 1996; Lukey et al., 2000; Bathurst, 2002; De Aragão et al., 2005)
- models based on the concept of unit sediment graph (e.g. Das & Agarwal, 1990; Sharma et al., 1993)

The empirical models are simple, but they do not go into the mechanisms of the physical processes. In the physically-based models, the soil surface is subdivided into rills and interrill areas, while the soil erosion process is also decomposed into physical subprocesses. Gully erosion is not quantified in the above models. The physically-based models require, apart from rainfall data and physiographic characteristics of the basin, numerous other data, i.e. field measurements, for the determination of the parameters (constants, exponents) in the corresponding equations; therefore, these models are difficult to apply to large basins. By contrast, the empirical models require mostly rainfall data and maps (topographic, soil, vegetation etc.).

RUSLE (Revised Universal Soil Loss Equation, Renard et al., 1996) is the newest improved version of USLE and could be considered as a transition from the empirical to the physically-based models. However, the detailed information required for estimating the

individual factors of RUSLE, with the exception of the topographic factor, renders the application of RUSLE to large basins practically impossible.
The unit sediment graph, as well as the unit hydrograph in hydrology, is a system function describing the sediment behaviour of a basin. It enables the conversion of the effective rainfall or soil detachment (system input) in the basin into sediment yield (system output). Generally, models based on a system function have the possibility to follow the time variation of sediment yield.

3. Classification of stream sediment transport models

The last physical process in the process chain simulated by the mathematical model is the stream sediment transport. In concrete terms, a total load model (bed load plus suspended load) is used for the quantification of sediment transport capacity by streamflow. Therefore, a classification of total load models is made below (Vetter, 1992):

- stochastic and regression models (e.g. Einstein, 1950; Karim & Kennedy, 1983)
- energy and power models (e.g. Engelund and Hansen, 1967; Yang, 1973)
- shear stress models (e.g. Zanke, 1982; van Rijn, 1984a, 1984b)

Additionally, Einstein and van Rijn calculate bed load and suspended load separately.
Sediment transport is considered by Einstein (1950) as probability problem. The probability that a grain will be lifted, is related with two dimensionless parameters: the intensity of bed load transport and the shear intensity of the flow. The Total Load Transport Model (TLTM) of Karim and Kennedy (1983) contains two multiple regression equations that compute the dimensionless total load transport and the dimensionless mean flow velocity as functions of different auxiliary dimensionless parameters.
Engelund and Hansen (1967) compare the rate of energy needed in lifting the grains with the rate of work being done by the resistance forces to the transported grains in the same time. Finally, they result to a relationship between the dimensionless total load transport and a dimensionless mobility parameter. According to Yang (1973), the "unit stream power", namely the mean flow velocity - energy slope product, is the basic parameter for the description of sediment transport. By means of a multiple regression, he results to a relationship between the total sediment concentration and the dimensionless effective unit stream power, among other dimensionless parameters. It is noted that the effective unit stream power is the difference between the dominant unit stream power and its critical value characterizing the incipient motion.
According to Zanke (1982), bed load transport and suspended load transport are functions of the difference between the existing shear velocity and its critical value characterizing the incipient motion. Particulary for the suspended load transport, the critical shear velocity for the lifting of the grains to the suspension zone is taken additionally into account. In the model of van Rijn (1894), two dimensionless parameters characterizing the grain diameter and the incipient motion, respectively, dominate. The incipient motion is designated by the difference between the existing shear velocity and its critical value.

4. Mathematical simulation model

As mentioned before, the mathematical simulation model consists of three submodels: (a) a rainfall-runoff submodel, (b) a soil erosion submodel, and (c) a stream sediment transport submodel.

In order to realize the connection of the submodels with each other, the input and output of each submodel are summarized below:

- The input of the rainfall-runoff submodel is the rainfall depth in each sub-basin, while the output is the runoff depth in each sub-basin.
- The input of the soil erosion submodel is the rainfall depth and the runoff depth in each sub-basin, while the output is the soil erosion amount in each sub-basin and the inflowing sediment quantity from the surrounding sub-basin into the main stream of the sub-basin considered.
- The input of the stream sediment transport submodel is the inflowing sediment quantity into the main stream of the sub-basin considered from the surrounding sub-basins, while the output is the sediment yield at the outlet of the sub-basin considered.

In the following sections, the submodels used in the four case studies are described shortly.

4.1 Hydrological submodels

4.1.1 Rainfall-runoff submodel of Lutz

The rainfall-runoff submodel of Lutz (1984) predicts rainfall excess for a given storm by using region-dependent and event-dependent parameters. Region-dependent parameters are the land use and the hydrological soil group which reflects the infiltration rate.
The model of Lutz is expressed mathematically by the following equation:

$$h_o = (N - A_v)c + \frac{c}{k}[e^{-k(h_o - A_v)} - 1] \tag{1}$$

where
h_o : daily rainfall excess (mm)
N : daily rainfall depth (mm)
A_v : initial abstraction consisting mainly of interception, infiltration and surface storage and depending on the land use (mm)
c : maximum end runoff coefficient expected for a rainfall depth of about 250 mm and depending on the land use and the hydrological soil group
k : proportionality factor (1/mm) which is given by the following equation:

$$k = P_1 e^{-2.0/WZ} e^{-2.0/q_B} \tag{2}$$

where
P_1 : region-dependent parameter
WZ : week number which designates the season
q_B : baseflow rate which designates the antecedent moisture condition [l/(s km²)]
Another typical parameter required to estimate soil erosion resulting from runoff is the peak runoff rate. The following formula developed by the US Soil Conservation Service (SCS) is used to determine the peak runoff rate from a sub-basin (Huggins & Burney, 1982):

$$q_p = 0.278 \frac{A_E h_o}{T_A} \tag{3}$$

where

q_p : peak runoff rate (m^3/s)
A_E : sub-basin area (km^2)
h_o : rainfall excess (mm)
T_A : time of rise of the hydrograph (hr)
This formula is based upon the SCS triangular hydrograph analysis procedure for approximating the manner in which an incremental volume of rainfall excess is translated into a time distribution of runoff at the sub-basin's outlet.

4.1.2 Rainfall-runoff submodel of water balance

The hydrologic submodel described in this section is a simplified water balance model (Giakoumakis et al., 1991), in which the variation of soil moisture due to rainfall, evapotranspiration, deep percolation and runoff is considered. The basic balancing equation is:

$$S_n' = S_{n-1} + N_n - E_{pn} \tag{4}$$

where
S_{n-1} : available soil moisture for the time step $n-1$ (mm)
N_n : rainfall depth for the time step n (mm)
E_{pn} : potential evapotranspiration for the time step n (mm)
S_n' : auxiliary variable (mm)
The direct runoff depth h_{on} (mm) and the deep percolation IN_n (mm) for the time step n can be evaluated as follows:
If $S_n' < 0$, then $S_n = 0$, $h_{on} = 0$ and $IN_n = 0$
If $0 \le S_n' \le S_{max}$, then $S_n = S_n'$, $h_{on} = 0$ and $IN_n = 0$
If $S_n' > S_{max}$, then $S_n = S_{max}$, $h_{on} = k(S_n' - S_{max})$ and $IN_n = k'(S_n' - S_{max})$,
where k and k' are proportionality coefficients ($k' = 1 - k$).
According to the structure of the above hydrologic submodel, h_o represents the sum of surface runoff depth and interflow depth. The auxiliary variable S_n' includes the quantitative influence of rainfall and potential evapotranspiration for the considered time step n on the available soil moisture for the preceding time step $n-1$. S_n' is compared with the maximum available soil moisture S_{max} in order to estimate the runoff and the deep percolation for the considered time step.
The maximum available soil moisture S_{max} (mm) is estimated by the following relationship of US Soil Conservation Service (SCS, 1972):

$$S_{max} = 25.4(\frac{1000}{CN} - 10) \tag{5}$$

where CN is the curve number depending on the soil cover, the hydrologic soil group and the antecedent soil moisture conditions ($0 < CN < 100$).
Finally, the potential evapotranspiration E_p is estimated by the radiation method improved by Doorenbos & Pruitt (1977). For this purpose, the following meteorological data are required: mean daily temperature (°C), sunlight hours per day (hr/day), mean daily relative humidity (%) and mean daily wind velocity (m/s).
The rainfall-runoff submodel described above is applicable on a long-term basis, e.g. on a monthly basis, because it is based on a water balance equation. Performing the calculations

on a monthly time basis may be justified by the fact that the rainfall in most rain days of a year in some countries (e.g. Cyprus, Greece) is not particularly high. Furthermore, it is usual to express the evapotranspiration in monthly or annual values, because the evapotranspiration in this case can also be calculated empirically with a good approximation (Hrissanthou, 2002).

4.2 Soil erosion submodels

4.2.1 Universal Soil Loss Equation (USLE)

The classical form of the USLE (Wischmeier & Smith, 1978) is:

$$A = RK(LS)CP \tag{6}$$

where

A : soil loss due to surface erosion (t/ha)
R : rainfall erosivity factor (N/hr)
K : soil erodibility factor [(t/ha)/(N/hr)]
LS : topographic factor
C : crop management factor
P : erosion control practice factor

The USLE is intended to estimate average soil loss over an extended period, e.g. mean annual soil loss (Foster, 1982). However, only raindrop impact is taken into account in this equation to estimate soil loss. An improved erosivity factor was introduced by Foster et al. (1977) to take also into account the runoff shear stresses effect on soil detachment for single storms:

$$R = 0.5R_{st} + 0.5R_R = 0.5R_{st} + 0.5\alpha h_o q_p^{\,0.33} \tag{7}$$

where

R : modified erosivity factor (N/hr)
R_{st} : rainfall erosivity factor (N/hr)
R_R : runoff erosivity factor (N/hr)
h_o : runoff volume per unit area (mm)
q_p : peak runoff rate per unit area (mm/hr)
α : a constant depending on the units ($\alpha = 0.70$)

4.2.2 Soil erosion submodel of Schmidt

The soil erosion submodel of Schmidt (1992) is based on the assumption that the impact of droplets on the soil surface and the surface runoff are proportional to the momentum flux contained in the droplets and the runoff, respectively.

The momentum flux exerted by the falling droplets, φ_r (kg m/s^2), is given by:

$$\varphi_r = Cr\rho A_E u_r \sin\alpha \tag{8}$$

where

C : soil cover factor
r : rainfall intensity (m/s)
ρ : water density (kg/m^3)

A_E : sub-basin area (m^2)
u_r : mean fall velocity of the droplets (m/s)
α : mean slope angle of the soil surface (°)
The original relationship of Schmidt for the momentum flux exerted by the droplets is valid for bare soils. Therefore, an additional factor is necessary to express the decrease of the momentum flux because of the vegetation. It is believed that the dimensionless crop management factor C of the USLE is appropriate to express the vegetation influence.
The fall velocity of the droplets, u_r (m/s), is a function of the rainfall intensity r (mm/hr) according to the following equation (Schmidt, 1992):

$$u_r = 4.5r^{0.12} \tag{9}$$

The momentum flux exerted by the runoff, φ_f (kg m/s^2), is given by:

$$\varphi_f = q\rho bu \tag{10}$$

where
q : direct runoff rate per unit width [m^3/(s m)]
b : width of the sub-basin area (m)
u : mean flow velocity (m/s)
The mean flow velocity u can be obtained from the well-known Manning formula, while the runoff rate per unit width q can be calculated from the mean flow velocity u and the runoff depth h_o, which is assumed to be uniformly distributed over a sub-basin.
The structure of Equations (8) and (10) indicates that the model of Schmidt is based on fundamental physical concepts. Nevertheless, the basic variable u_r is evaluated by the empirical Equation (9).

4.3 Estimate of sediment inflow into the main stream of a sub-basin

Sediment from the soil erosion transported to the main stream of a sub-basin is computed by the concept of overland flow sediment transport capacity. At this point, it must be noted that only the main stream of the sub-basin is considered because large amounts of unavailable data for the geometry and hydraulics of the entire stream system would otherwise be required.
The amount of sediment due to soil erosion transported to the main stream of a sub-basin, ES, is estimated by means of the following controls: If the available sediment in a sub-basin, q_{rf}, exceeds overland flow sediment transport capacity q_t, deposition occurs, and the sediment transported to the main stream of the sub-basin equals sediment transport capacity. If the available sediment in a sub-basin is less than overland flow sediment transport capacity and if the flow's erosive forces exceed the resistance of the soil to detachment by flow, detachment occurs; in this case, sediment transported to the main stream of the sub-basin equals the available sediment. It is symbolized by the following relationships:

$$ES = q_t, \quad \text{if } q_{rf} > q_t \tag{11}$$

$$ES = q_{rf}, \quad \text{if } q_{rf} \le q_t \tag{12}$$

However, sediment from the preceding sub-basin, FLI, is also transported to the sub-basin under consideration. The total sediment transported to the main stream of the sub-basin, ESI, is therefore:

$$ESI = ES + FLI \tag{13}$$

4.3.1 Relationships of Beasley et al.

The following relationships of Beasley et al. (1980) are used to compute the overland flow sediment transport capacity in a sub-basin:

$$q_t = 146sq^{1/2} \quad \text{for } q \le 0.046 \ \text{m}^3/(\text{min m}) \tag{14}$$

$$q_t = 14600sq^2 \quad \text{for } q > 0.046 \ \text{m}^3/(\text{min m}) \tag{15}$$

where

q_t : overland flow sediment transport capacity [kg/(min m)]

s : mean slope gradient

q : flow rate per unit width [m³/(min m)]

The first equation is valid for laminar flow and the second for turbulent flow. The relationships of Beasley et al. (1980) are based upon the equation developed by Yalin (1963), who assumed that the mechanism of sediment transport by a shallow flow, e.g. by the overland flow, is similar to the mechanism of bed load transport in channels and that a critical shear stress exists acting on the soil at the beginning of sediment transport.

Since the relationships of Beasley et al. (1980) are combined with USLE, the quantity q_{rf} is computed on the basis of the soil erosion amount A.

4.3.2 Relationships of Schmidt

The available sediment discharge per unit width, q_{rf} [kg/(s m)], due to rainfall and runoff, in a sub-basin is given by (Schmidt, 1992):

$$q_{rf} = (1.7E - 1.7)10^{-4} \tag{16}$$

where

$$E = \frac{\varphi_r + \varphi_f}{\varphi_{cr}} \qquad (E > 1) \tag{17}$$

φ_{cr} : critical momentum flux (kg m/s²)

The critical momentum flux φ_{cr}, which designates the soil erodibility, can be calculated from:

$$\varphi_{cr} = q_{cr}\rho bu \tag{18}$$

where q_{cr} [m³/(s m)] is the direct runoff rate per unit width at initial erosion.

The critical runoff rate q_{cr} is determined from the critical erosion velocity depending on soil roughness. Equation (17) suggests the concept of critical situation characterizing the initiation of sediment motion on the soil surface.

The sediment transport capacity by overland flow, q_t [kg/(s m)], is computed as follows (Schmidt, 1992):

$$q_t = c_{max}\rho_s q \tag{19}$$

where

c_{max} : concentration of suspended particles at transport capacity (m^3/m^3)

ρ_s : sediment density (kg/m^3)

The concentration c_{max} results from the equation (Schmidt, 1992; Hrissanthou, 2002):

$$c_{max} = \frac{(\varphi_r + \varphi_f)\sin\alpha}{\rho_s A_E w^2} \tag{20}$$

where w (m/s) is the terminal fall velocity of suspended particles.

Equation (20) is obtained from the equilibrium condition between the vertical component of the total momentum flux [$(\varphi_r + \varphi_f)\sin\alpha$] and the critical momentum flux of the suspended particles $(c_{max}\rho_s A_E w^2)$. The critical momentum flux results by multiplying the mass rate of settling particles $(c_{max}\rho_s A_E w)$ by the settling velocity w. If the critical momentum flux is exceeded, the particles do not remain in suspension. The equilibrium condition is valid when the sediment transport capacity is achieved

The sediment supply ES to the main stream of a sub-basin is estimated by means of Equation (11) or (12), as described above.

4.4 Estimate of sediment yield at the outlet of a sub-basin

The sediment yield at the outlet of a sub-basin, FLO , reflects the same basic controls as the sediment supply ES from soil erosion:

$$FLO = q_{ts}, \quad \text{if } ESI > q_{ts} \tag{21}$$

$$FLO = ESI, \quad \text{if } ESI \le q_{ts} \tag{22}$$

where q_{ts} is the sediment transport capacity by streamflow. In the first case, if the available sediment in the main stream of the considered sub-basin, ESI, exceeds sediment transport capacity by streamflow, q_{ts}, deposition occurs. In the second case, if the available sediment ESI is less than sediment transport capacity by streamflow, q_{ts}, bed detachment may occur.

4.4.1 Relationships of Yang & Stall

The following relationships of Yang & Stall (1976) are used to compute sediment transport capacity by streamflow:

$$\log c_t = 5.435 - 0.286\log\frac{wD_{50}}{\nu} - 0.457\log\frac{u_*}{w} + \\ +(1.799 - 0.409\log\frac{wD_{50}}{\nu} - 0.314\log\frac{u_*}{w})\log(\frac{us}{w} - \frac{u_{cr}s}{w}) \tag{23}$$

$$\frac{u_{cr}}{w} = \frac{2.5}{\log(u_* D_{50}/\nu) - 0.06} + 0.66, \quad \text{if } 1.2 < \frac{u_* D_{50}}{\nu} < 70 \tag{24}$$

$$\frac{u_{cr}}{w} = 2.05, \quad \text{if } \frac{u_* D_{50}}{\nu} \ge 70 \tag{25}$$

where
c_t : total sediment concentration (ppm)
w : terminal fall velocity of suspended particles (m/s)
D_{50} : median grain diameter of the bed material (m)
ν : kinematic viscosity of the water (m^2/s)
u_* : shear velocity (m/s)
u : mean flow velocity (m/s)
u_{cr} : critical mean flow velocity (m/s)
s : energy slope
Equation (23) was determined from the concept of unit stream power (rate of potential energy expenditure per unit weight of water, us) and dimensional analysis. The variable u_{cr} in Equation (23) suggests that a critical situation is considered at the beginning of sediment particle motion, as in most sediment transport equations. But the relationship of Yang and Stall has the advantage, in contrast to other published equations, that it was verified in natural rivers.

5. Application of the mathematical model to artificial or natural lakes

5.1 Application to the basin of Forggensee Reservoir

The mathematical model, consisting of the rainfall-runoff submodel of Lutz, the USLE and the stream sediment transport submodel of Yang & Stall, was applied to the 1500 km^2 basin of the Forggensee Resrvoir (Bavaria, Germany). The storage capacity of the reservoir is 168 x 10^6 m^3. The largest part of the basin is in Austria and the main stream is the Lech River. The basin consists mainly of forest (36%), of meadow (49%), and of rock (11%) over 2000 m in altitude (Fig. 1, Hrissanthou, 1990). Information about the soil texture class was available only for a small part of the basin where it consists of loamy sand, sandy loam, and clay loam.
The following data were available:
- daily rainfall amounts from five rainfall stations in the basin for 12 years (1966-1977);
- suspended load at the outlet of the basin for these same 12 years, on a daily basis.

The quantification of the runoff, erosion and sediment transport processes can be made more exact, if it is applied to small land areas. The specification of the size of the sub-areas is the result of a compromise between the following conflicting criteria:
- the precision of the calculations and results;
- the effort and time available for the performance of the calculations.

The smaller the area and time unit the more exact the calculations and results will be and the greater the effort and time that will be required for the performance of the calculations. According to the above considerations, the study basin was divided into 88 natural sub-basins (Fig. 2), about 25 km^2 in area.
Sediment yield at the outlet of the basin was computed by the model on a daily basis because the rainfall amounts (input data) were available on a daily basis. Apart from this, the selection of the "day" as the time unit in the calculations is a very good approximation of the sediment delivery problem of a large basin.
It was assumed that uniform conditions exist in each sub-basin and that steady conditions exist throughout each day for the runoff and erosion processes. Daily rainfall occurrences were treated as individual storm events.

The application of the mathematical model requires the use of a sediment transport routing plan, which specifies the sediment motion from sub-basin to sub-basin.
Regarding the application of USLE to the basin considered, the tables of Schwertmann (1981), which are valid especially for Bavaria, were used for the evaluation of the factors K, C and P.
The rainfall erosivity factor is defined as the product of two rainstorm characteristics: kinetic energy and the maximum 30-min intensity. The computation of this factor for a rainfall event requires a continuous record of rainfall intensity. Only daily rainfall amounts were, however, available for the example basin. A regression analysis was therefore used to estimate the factor R_{st} as a function of the daily rainfall amount. Data for the rainfall erosivity factor and rainfall amount were available from a small basin ($\approx$ 14 km^2) in the neighbouring state of Baden-Württemberg.
The topographic factor LS was evaluated as a function of the slope gradient and slope length (Wischmeier & Smith, 1978). In addition to the tables of Schwertmann, soil, topographic and vegetation maps were required for estimating the K, LS and C factors.
The USLE was developed for small agricultural fields; therefore, the application of this equation to large areas, e.g. the sub-basins of this example, results in only a rough estimate of soil erosion. Moreover, USLE and any equation for classical erosion due to rainfall and runoff are not appropriate to be applied to rock areas without vegetation. As mentioned before, a part (11%) of the basin area consists of rock without vegetation, over 2000 m in altitude.
In Fig. 2, the mean annual erosion amount (t/ha) in each sub-basin is given.
Sediment yield at the outlet of the basin considered was computed by the mathematical model on a daily basis. The daily values of sediment yield were added to produce the annual value of sediment yield at the outlet of the basin. These computed annual values of sediment yield due to soil and stream bed erosion were compared with the measured values of "annual suspended load" plus "annual bed load" at the outlet of the basin. Annual bed load was assumed to be 20% of the annual suspended load (Schröder & Theune, 1984).
The ratios of the computed annual values of sediment yield, associated with soil and stream bed erosion, to the measured values of sediment yield at the outlet of the whole basin are presented in Table 1 (Hrissanthou, 1988).
A sensitivity analysis showed that rainfall amount and sub-basin area strongly affect daily sediment yield at the outlet of the whole basin.

Year	Measured value (t)	Computed value/ Measured value	Year	Measured value (t)	Computed value/ Measured value
1966	585 600	1.46	1972	79 046	5.30
1967	351 600	2.45	1973	408 352	1.23
1968	374 400	1.78	1974	324 037	2.48
1969	246 000	1.74	1975	745 586	0.91
1970	1 165 200	0.88	1976	315 772	1.36
1971	326 052	1.59	1977	312 025	2.18

Table 1. Ratio of computed to measured values of sediment yield

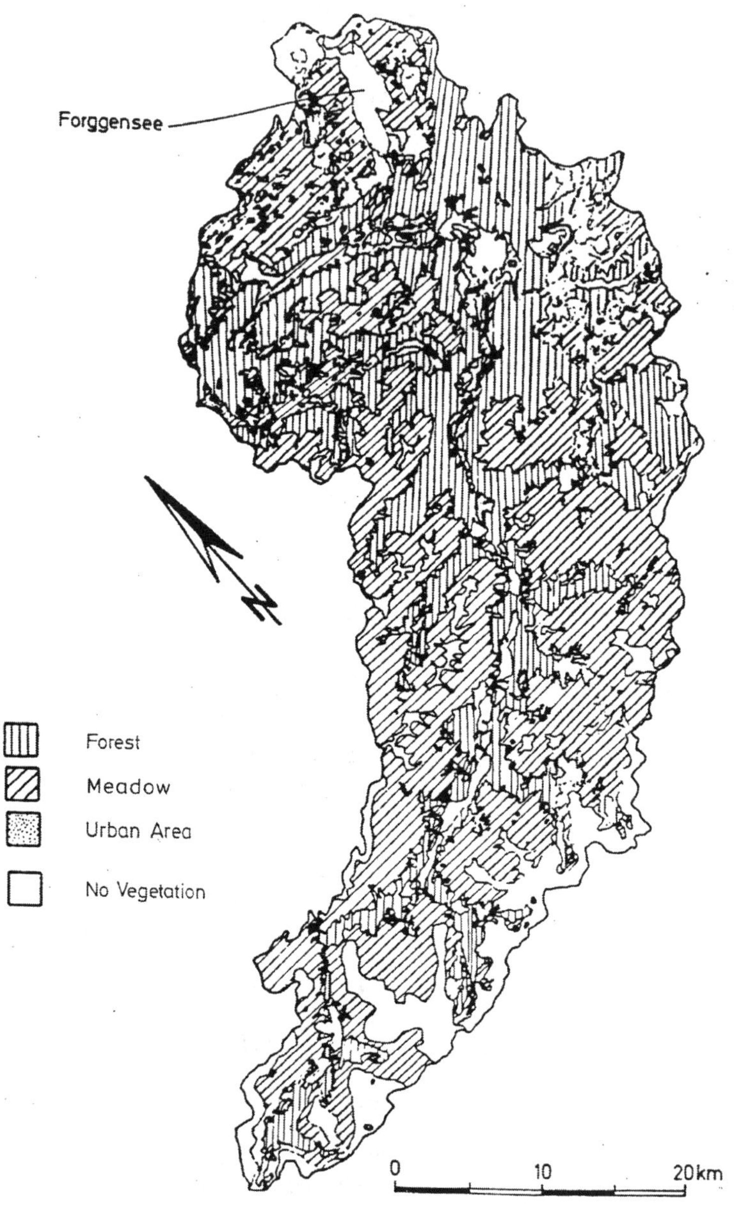

Fig. 1. Soil cover map of the basin of Forggensee Reservoir

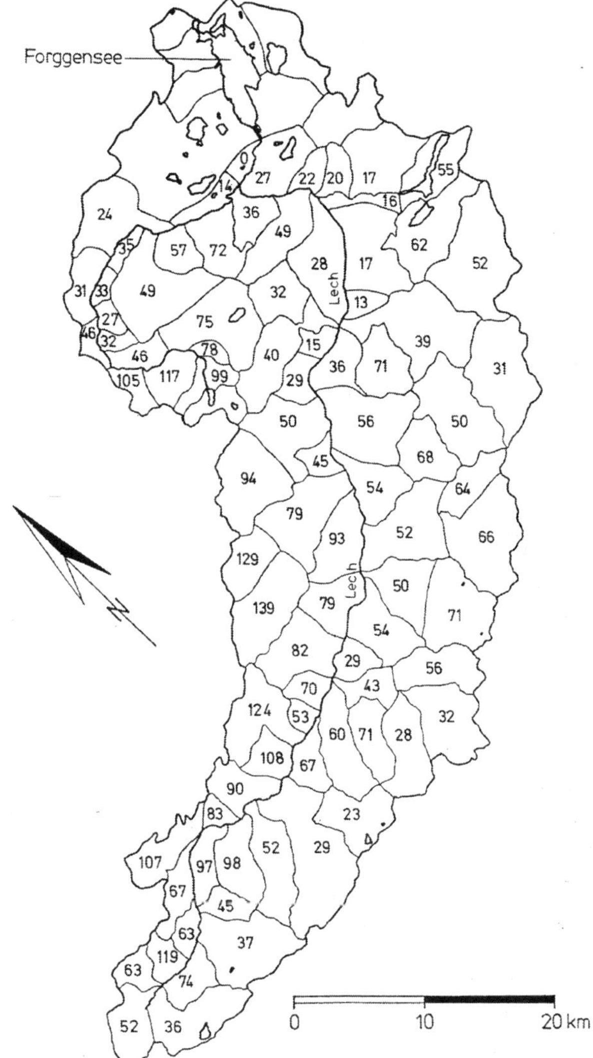

Fig. 2. Mean annual erosion amount (t/ha) in each sub-basin (years: 1966-1977)

5.1.1 Discussion - conclusions

The arithmetic results are satisfactory considering the large basin area and the fact that the computation was performed on a daily basis and that no rainfall, runoff or sediment yield data were available for the sub-basins. These results are much better than those obtained for the same example when only soil erosion was taken into account and the whole basin was subdivided into sub-areas by means of a quadrangular grid (Hrissanthou, 1986).

The rainfall, runoff, soil detachment, deposition and transport processes in the large basin area were not considered in detail, but rather in a macroscopic way. It implies, for example, that soil deposition occurs, according to the model, in sub-basins with very low slope gradients or in streams with very low bed slopes. In the other sub-basins or streams, with greater slopes, all the detached soil mass is transported to the main streams of the sub-basins or to the next sub-basins.

Gully and bank erosion, as well as mass movement were ignored because no information was available about these kinds of erosion. Snowmelt runoff, glacial and snow erosion were not quantified because the research was focused on the classical soil erosion due to rainfall and runoff.

Finally, in this example, the comparison between computed and measured values of sediment yield at the outlet of the entire basin was made on an annual basis, although the calculations were performed on a daily basis. The following reasons are given for using an annual basis for the comparison:

- the very long sediment travel times from the outlets of the most sub-basins to the outlet of the whole basin;
- the problems of using a total daily rainfall amount;
- the lack of runoff and sediment yield data in the sub-basins.

These reasons render the precise computation of daily sediment yield at the outlet of the whole basin difficult. The addition of the daily values of sediment yield at the outlet of the basin causes a decrease in the differences between computed and measured values of sediment yield.

5.2 Application to the basin of Yermasoyia Reservoir

The mathematical model, consisting of the rainfall-runoff submodel of water balance, the soil erosion submodel of Schmidt and the stream sediment transport submodel of Yang and Stall, was applied to the basin of Yermasoyia Reservoir. The Yermasoyia Reservoir is located northeast of the town of Limassol, Cyprus. The storage capacity of the reservoir is 13 x 10^6 m^3. The Yermasoyia River drains a basin that, upstream of the reservoir, amounts to 122.5 km^2. The length of the main stream of the basin is about 25 km, and the highest altitude of the basin is about 1400 m. The basin which consists of forest (57.7%), bush (33.7%), cultivated land (5.8%), urban area (1.8%) and an area with no significant vegetation (1%) (Fig. 3), was divided into four natural sub-basins for more precise calculations (Fig. 4, Hrissanthou, 2006). The sub-basin areas vary between 14 and 44 km^2.

The soils of the basin belong to the following types: Calcaric Cambisols, Eutric Cambisols, Eutric Regosols, Calcaric Lithosols, and Eutric Lithosols. These soil types were further divided into three categories: permeable (Calcaric Cambisols, Eutric Cambisols), semi-permeable (Eutric Regosols) and impermeable (Calcaric Lithosols, Eutric Lithosols) soils, because the above distinction is necessary for the estimate of the curve number in the hydrologic submodel.

Daily rainfall data for four years (1986-1989) from three rainfall stations were available. The mean annual rainfall at these stations amounts to 662 mm. Additionally, mean daily values of air temperature and relative air humidity and daily values of sunlight hours for the above four years were available from a meteorological station. Mean daily values of wind velocity only for one year (1988) were obtained from the same meteorological station.

Monthly runoff volumes for the years 1986-1989 were available from a water gauging station, named "Phinikaria" and located at the outlet of sub-basin 3 (Fig. 4). The basin area corresponding to the gauging station amounts to 108 km^2.

Finally, the distribution of mean annual erosion rates over the island of Cyprus was obtained from the Water Development Department (Nicosia, Cyprus). According to this authority, the erosion rates have been deduced and assigned to the various geomorphologic areas of Cyprus on the basis of existing, randomly obtained, suspended sediment samples and mainly on the basis of estimates derived by surveying three dams.

The mathematical model was applied to each sub-basin separately and on a monthly time basis for a certain year. The monthly values of sediment yield at the basin outlet resulting from the model for a given year were added to produce the annual value of sediment yield YA due to soil and stream bed erosion. The annual soil erosion amount for the whole basin is symbolized with YD. The ratio of YA to YD is called the sediment delivery ratio DR. The computational results for YA, YD and DR for the years 1986-1989 are shown in Table 2. Table 2 contains also the measured and computed annual values of runoff volume VO at the outlet of sub-basin 3 (Fig. 4), as well as the ratio of computed to measured annual values VR. The computed values of runoff volume result from the hydrologic submodel.

Year	YD (t)	YA (t)	DR (%)	Measured VO (10^6 m^3)	Computed VO (10^6 m^3)	VR
1986	113 000	32 000	28	8.0	9.7	1.2
1987	673 000	224 000	33	19.6	24.6	1.3
1988	618 000	238 000	38	24.7	25.4	1.0
1989	108 000	30 000	28	16.3	9.3	0.6
Mean value	378 000	131 000	32	17.2	17.3	1.0

Table 2. Computational results for YD, YA and DR; annual values of computed and measured runoff volume

The mean annual value of YD, 378 000 t, is transformed into the mean annual rate of soil erosion, 1.16 mm. The latter value is 1.7 times higher than the corresponding estimated value of 0.70 mm (Water Development Department, Nicosia, Cyprus). This estimated value is assigned to areas with igneous rocks, steep slopes and rainfall rates of the order of 600-800 mm/year, covered by forest, brush and with little cultivation. These climatic and physiographic conditions are fulfilled by the basin of Yermasoyia Reservoir.

According to the classical diagram of Brune (1953), the trap efficiency of Yermasoyia Reservoir is 100%. This means that all of the sediment yield at the basin outlet is deposited in the reservoir. Considering the storage capacity of the reservoir (13.6 x 10^6 m^3), its useful life thus amounts to 193 years.

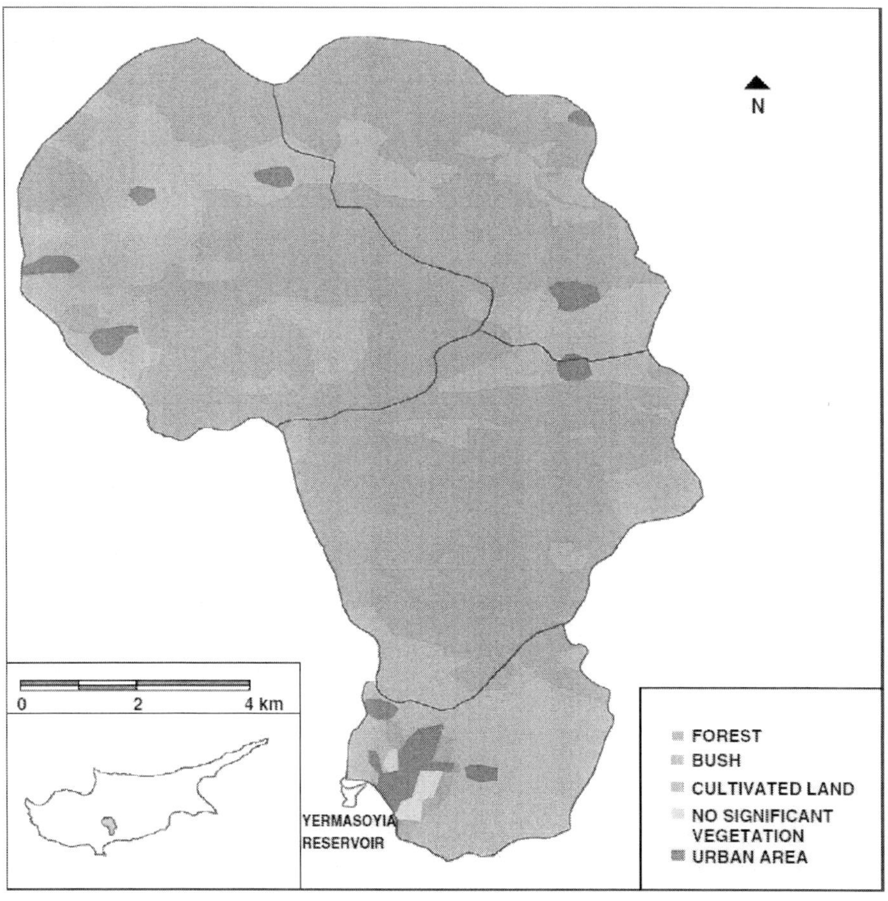

Fig. 3. Soil cover map of the basin of Yermasoyia Reservoir

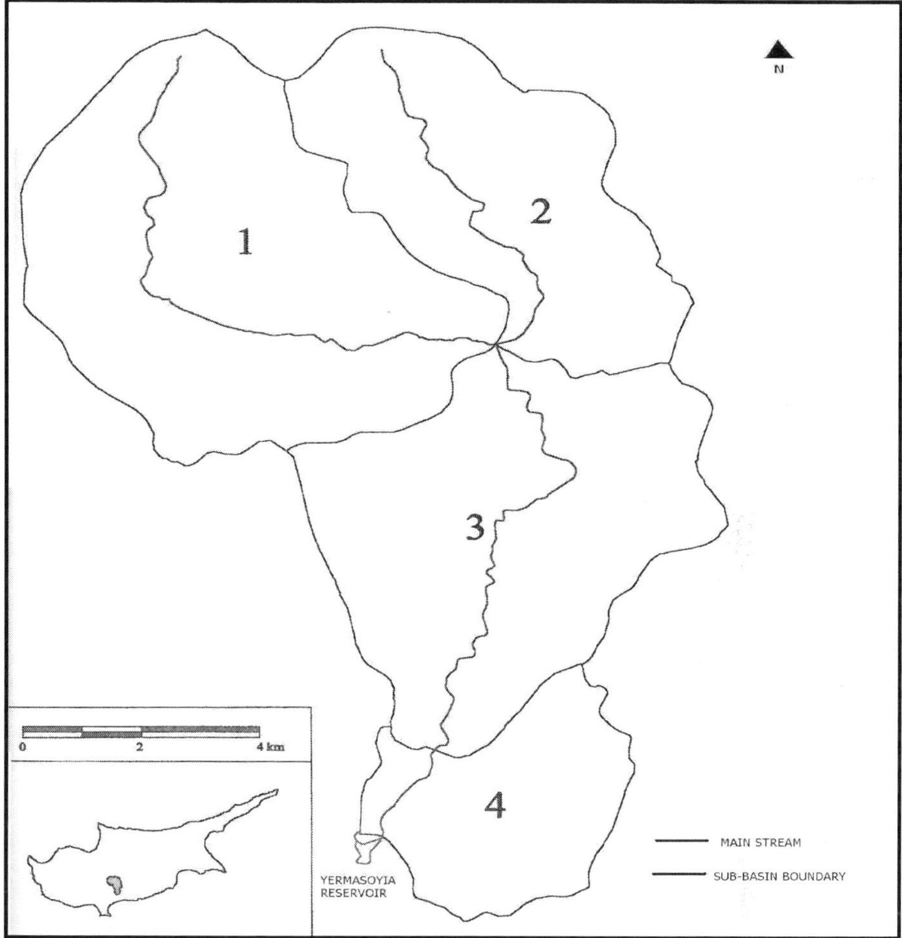

Fig. 4. Main streams of the four sub-basins of Yermasoyia Reservoir basin

5.2.1 Discussion - conclusions

The measured mean annual runoff volume at the outlet of sub-basin 3 is slightly overestimated by the model. Generally, for models of the same structure as the model used in this example, the rainfall volume and the parameter curve number influence the monthly sediment yield at the basin outlet more strongly than the other input data (Hrissanthou, 2002).

The soil erosion submodel also overestimates the mean annual rate of soil erosion for the whole basin. At this point, it has to be noted that runoff resulting from the hydrologic submodel serves as input to the soil erosion submodel. Apart from that, the following factors also contribute to the deviation between the computed and the measured values of soil erosion:

- The equations for soil erosion described above are applied, in this study, to relatively small sub-basins, whereas they were developed initially for small experimental fields.
- Snowmelt runoff, gully and bank erosion are neglected in this study.
- The erosion measurements are indirect, as explained previously.

The sediment inflow into the reservoir resulting from the stream sediment transport submodel is also overestimated, because the overestimated soil erosion quantity serves as input to the stream sediment transport submodel. Apart from that, it must be added that bank erosion was neglected in the stream sediment transport submodel. Moreover, in the present case study, the bed of the main streams of the sub-basins consists of sand and gravel, and the slope of the main streams of two sub-basins exceeds the application limit of the Yang formula, which is valid for sandy beds.

The overestimation of the sediment inflow into the reservoir implies the underestimation of the useful life of the reservoir. In any case, there is no immediate and sharp danger of reservoir filling with sediments.

The mean annual sediment inflow would be more reliable if the computations could be carried out for as many years as possible. Unfortunately, meteorological data were available only for four years.

In the computations performed through the mathematical model, the sub-basin was the space unit and the month was the time unit. However, it has to be stressed that performing the calculations on an event basis is a reasonable way for the quantification of runoff, erosion and sediment transport processes, provided that pertinent detailed data are available.

In the case of large basins, for which mean annual values of soil erosion and sediment yield are required, the monthly time basis of the computations constitutes a temporally detailed approach. Additionally, the computational experience has shown that performing the calculations on a daily time basis, especially in large basins, leads to a disagreement of high order between computed and measured daily values of sediment yield (Hrissanthou, 1990). In contrast, performing the calculations on a monthly time basis results in a relatively good agreement between the computed and measured monthly values of sediment yield, which is due to the integrating effect obtained through use of a sub-basin (as a space unit) and a long simulation period. However, the performance of the calculations on a monthly time basis has as the consequence that some variables of the model equations, e.g. the rainfall intensity, characterizing single storm events lose their physical meaning.

Most parameters of the mathematical model used were estimated by means of tables and topographic, vegetation and soil maps. To take into account the diversity of the sub-basins with respect to soil, topography and vegetation, mean weighted values of the parameters for each sub-basin were computed.

The proportionality coefficient k of the rainfall-runoff submodel was determined on the basis of a value of 35.5% of the runoff coefficient (ratio of rainfall excess to rainfall). This value was estimated from the available rainfall and runoff volume data.

5.3 Application to the basin of Kastoria Lake

The mathematical model, consisting of the rainfall-runoff submodel of water balance, the soil erosion submodel of Schmidt and the stream sediment transport submodel of Yang and Stall, was applied to the basin of Kastoria Lake. Kastoria Lake is located in northwestern Greece, near the Albanian border. The mean water surface area of the lake is about 28 km^2, while the whole basin of the lake is about 253 km^2. The soil cover of the basin consists of forest (29%), pasture (44%), cultivated area (24%) and urban area (3%). The highest altitude of the basin is about 1900 m. The rocks were divided into permeable (34%), impermeable (50%) and semi-permeable (16%).

For a more precise computation of runoff, soil erosion and stream sediment transport, the whole basin was divided into ten natural sub-basins (Fig. 5). The area of the sub-basins varies between 2 and 64 km^2. The mean soil slope of the sub-basins varies between 10 and 49%, while the mean slope of the main streams of the sub-basins varies between 0.5 and 13%.

The main streams of the sub-basins located around Kastoria Lake transport both water and sediment into the lake. The inflowing sediment causes the reduction of the water volume capacity of the lake with time, endangering the existence of this environmental resource. The water volume inflowing into the lake through the streams originates mainly from the rainfall-runoff process in the sub-basins. Groundwater recharge contributes significantly to the volumetric budget of the lake, as well as to the budget of the whole basin. The sediment load reaching the lake arises from the soil erosion, due to rainfall and runoff, of the sub-basins and from the stream bed erosion.

The following data were available:

- Monthly rainfall data from six rainfall stations for 33 hydrologic years (1961/62-1993/94).
- Monthly air temperature data from four meteorological stations for 33 hydrologic years (1961/62-1993/94).
- Individual baseflow measurements in some streams discharging into the lake, for the years 1998 and 1999.

The mean annual value of the rainfall amount from the six stations varies between 563 and 876 mm. The air temperature data were used for the estimation of the potential evapotranspiration according to the method of Thornthwaite. The baseflow measurements belong to the input data of the stream sediment transport submodel.

The mathematical model was applied to each sub-basin separately and for every month of a certain hydrologic year. The monthly values of total flow volume and sediment yield, due to soil and stream bed erosion, at the outlet of each sub-basin, resulting from the mathematical model for a certain hydrologic year, were added to produce the annual value of total flow volume vog and sediment yield ya, respectively. As is well-known, the total flow volume is the sum of direct runoff volume and baseflow volume. The annual soil erosion amount for each sub-basin is symbolized with yd. The ratio of ya to yd is called the sediment delivery ratio (dr).

In Table 3, the mean annual values of vog, ya and yd for 33 hydrologic years (1961/62-1993/94), for the sub-basins, are given. In the same table, the ratio dr of the mean annual values ya / yd is contained (Hrissanthou et al., 2003).

The mean annual value of soil erosion for the whole basin, YD, amounts to 881 000 t, while the mean annual value of sediment yield at the outlets of the sub-basins, YA, amounts to 289 000 t. This means that the sediment delivery ratio for the whole basin, DR, is 33%.

The general remarks of Section 5.2.1 concerning the application of the mathematical model to a relatively large basin, are also valid for the application of the model to the basin of Kastoria Lake.

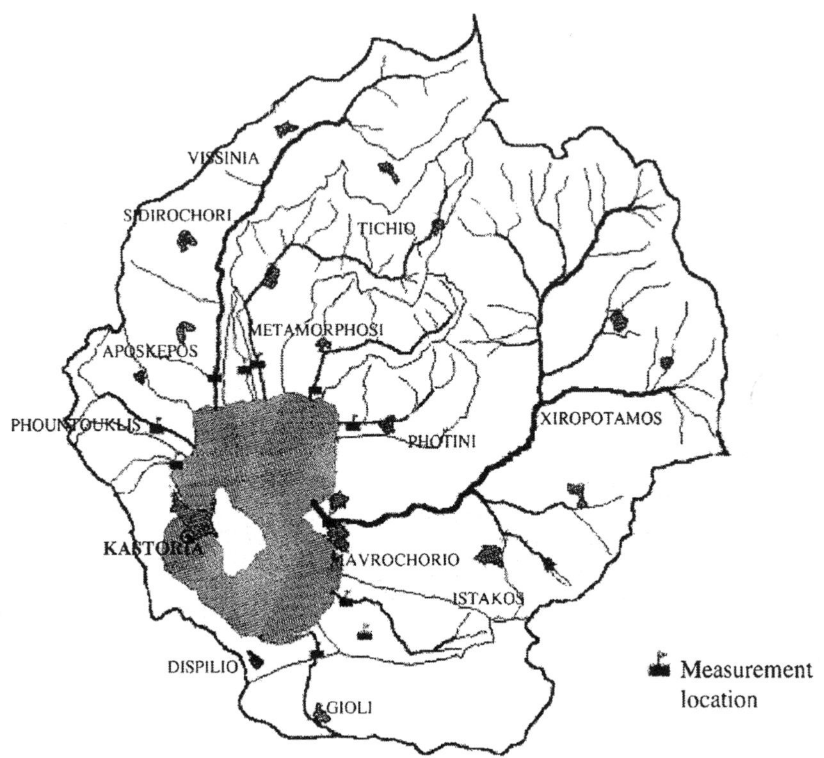

Fig. 5. Sub-basins of Kastoria Lake

Sub-basin	vog (10^6 m^3)	ya (t)	yd (t)	dr (%)
Xiropotamos	23.39	109 200	327 000	33
Vissinia	6.77	45 300	163 100	28
Tichio	7.01	41 000	123 000	33
Kastoria-Dispilio	0.78	27 000	27 000	100
Metamorphosi	3.34	23 500	48 600	48
Aposkepos	2.96	15 400	72 400	21
Photini	1.06	10 200	39 600	26
Istakos	5.16	8 600	48 700	18
Phountouklis	1.04	6 000	17 300	35
Agios Athanasios	1.99	2 400	14 000	17

Table 3. Mean annual values of vog , ya and yd - Ratio dr

5.3.1 Practical use of the model

The simulation model described above can be used for the identification of those sub-basins where sediment control measures have to be implemented. The sediment control measures, which aim at the reduction of sediment inflowing into Kastoria Lake from the sub-basins, are classified into three groups:

- Soil erosion control measures in the sub-basins through the establishment of vegetative cover (e.g. afforestation).
- Check dams in the torrents (mountain part of the sub-basins) to trap bed load and to prevent bed degradation.
- Detention basins for bed load in the alluvial fans (cone-shaped depositions) of the streams.

According to Table 3, the sub-basins, which deliver most sediment load to the lake, are those of Xiropotamos, Vissinia and Tichio. Therefore, the sediment control measures must be implemented , in order of priority, in the sub-basins of Xiropotamos, Vissinia and Tichio. In the other sub-basins, of course, sediment control measures can also be performed.

5.4 Application to the basin of Vistonis Lake

The mathematical model, consisting of the rainfall-runoff submodel of water balance, the soil erosion submodel of Schmidt and the stream sediment transport submodel of Yang and Stall, was also applied to the basin of Vistonis Lake (Thrace, northeastern Greece), which is one of the most important wetlands in Greece. The water surface area of the lake is about 45 km^2. In concrete terms, the model was applied separately to the mountainous part of the three basins of Kompsatos, Kossynthos and Travos (Aspropotamos) Rivers, respectively, which discharge their water into Vistonis Lake (Fig. 6).

The application of the mathematical model to the basin of Vistonis Lake is given, in this section, in more detail compared to the three foregoing case studies. For each of the three basins, five maps were drawn and digitized: a stream system map, a contour map, a vegetation map, a geologic map, and a map of Thiessen polygons. On the basis of the first four maps, the physiographic characteristics for each sub-basin of the three basins (e.g. main stream length, main stream bottom slope, soil slope gradient, sub-basin area, soil cover, rock classification), as well as parameter values depending on the physiographic characteristics (e.g. curve number, soil cover factor, soil roughness coefficient, critical erosion velocity of the soil surface) were calculated. At this point, it must be noted that both soil permeability and rock permeability are of importance for the quantification of the runoff process, because soil moisture variation is influenced by both soil permeability and rock permeability. Therefore, the soils and rocks were divided into permeable, impermeable and semi-permeable on the basis of soil texture and deep percolation percentage, respectively. By means of the fifth map (Thiessen polygons), a mean weighted value of rainfall depth for each sub-basin of the three basins was computed. The parameter values serve as input data to the mathematical model.

The input data for the rainfall-runoff submodel are: monthly rainfall depth, mean daily air temperature, sunlight hours per day, mean daily relative humidity of the atmosphere, mean daily wind velocity, altitude, latitude, soil cover - land use, and hydrologic soil group. The additional input data for the soil erosion submodel, with reference to the rainfall-runoff submodel, are: mean slope angle of soil surface, sub-basin area, soil cover factor (C-factor of USLE), length of the main stream of the sub-basins, roughness coefficient of soil surface, critical erosion velocity, water and sediment density. The additional input data for the stream sediment transport submodel, with reference to the foregoing submodels, concern the main stream of the sub-basins: baseflow, bottom slope, bottom width, bed roughness, diameter of suspended particles, grain diameter of bed material, and kinematic viscosity of water.

Finally, a sediment routing plan is necessary in order to specify the sediment motion from sub-basin to sub-basin.

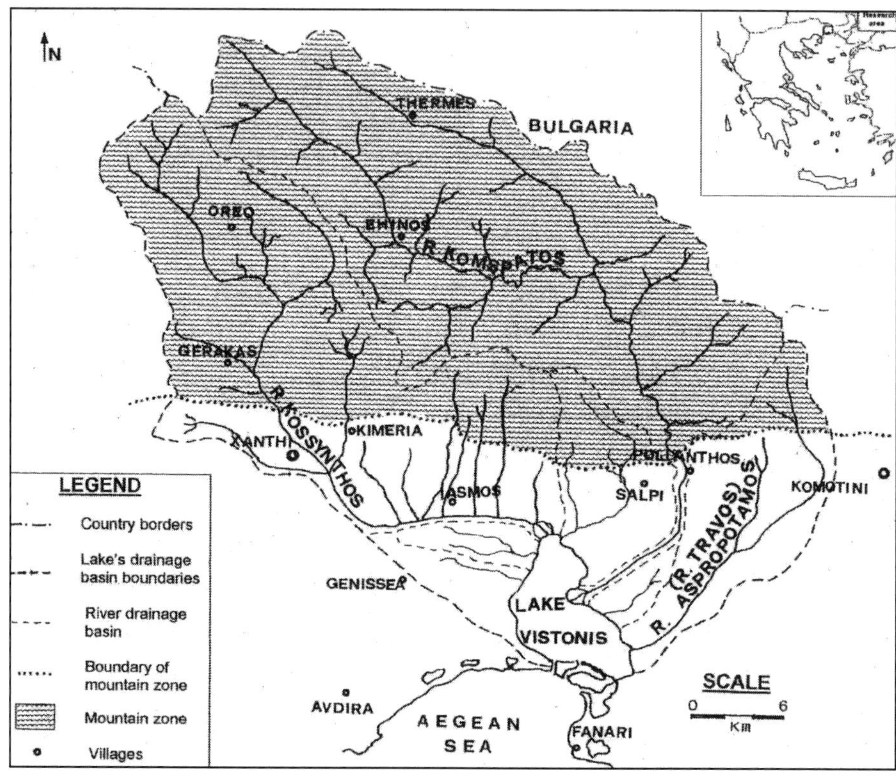

Fig. 6. Rivers discharging into Vistonis Lake

5.4.1 Application of the model to Kompsatos River basin

The basin of Kompsatos River has an area of about 567 km^2 consisting of forest (37.3%), bush (36.1%), cultivated land (26.1%) and urban area (0.5%). The dominant rocks are marble, amphibolite, rhyolite and gneiss-granite. The highest altitude of the basin is about 1200 m. The length of the main stream of the basin is about 57 km. For more precise calculations, the basin was divided into 18 natural sub-basins (area: between 13 and 50 km^2, Fig. 7). The mean soil slope gradient of the sub-basins is about 26%.

Monthly rainfall data for 27 years (1966-1992) from four rainfall stations (Dimario, Thermes, Ehinos and Trikorpho, Fig. 6) were available. The mean annual rainfall at these stations amounts to 1038 mm. For every month of the 27 years, mean daily values of air temperature from three meteorological stations (Thermes, Ehinos and Xanthi, Fig. 6) and mean daily values of sunlight hours from a meteorological station (Egiros) were available. Additionally, mean daily values of relative humidity of the atmosphere for every month of the year 1985 were obtained from the meteorological station of Xanthi, and mean daily values of wind velocity for every month of the year 1995 were obtained from the meteorological station of Genissea (Fig. 6). Finally, daily baseflow data were available from measurements for the years 1991 and 1992.

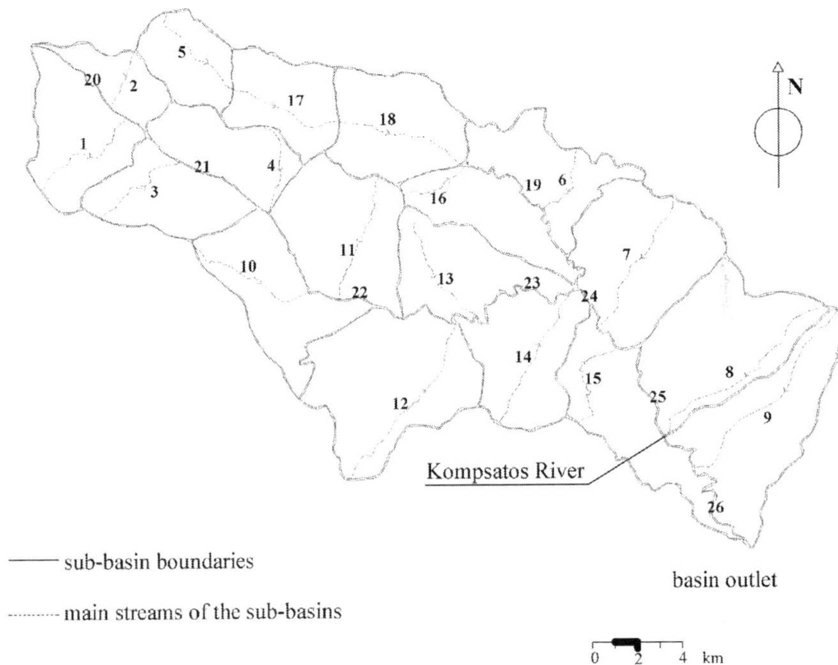

Fig. 7. Kompsatos River basin - Sub-basins with main streams

5.4.2 Application of the model to Kossynthos River basin

The basin of Kossynthos River has an area of about 237 km^2 consisting of forest (74%), bush (4.5%), urban area (1.5%) and an area with no significant vegetation (20%). The dominant rocks are granite-diorite, marble, gneiss-granite and migmatite. The highest altitude of the basin is about 1700 m. The length of the main stream of the basin is about 35 km. For more precise calculations, the basin was divided into ten natural sub-basins (area: between 16 and 35 km^2, Fig. 8). The mean soil slope gradient of the sub-basins is about 37%.

Monthly rainfall data from six rainfall stations (Livaditis, Ehinos, Xanthi, Gerakas, Oreo and Lykodromio, Fig. 6) for eleven years (1980-1990) were available. The mean annual rainfall at these stations amounts to 883 mm. Apart from the rainfall data, mean daily values of air temperature, sunlight hours and relative humidity of the atmosphere from the meteorological station of Xanthi (Fig. 6) for every month of the years 1980-1990 were available. As far as the wind velocity is concerned, the same data were used as for the basin of Kompsatos River. Daily baseflow data were available from measurements for the years 1991 and 1992. Finally, median values of grain diameters of bed material were available from samples taken in the lower and upper parts of Kossynthos River basin, in June 1997.

5.4.3 Application of the model to Travos River basin

The basin of Travos (Aspropotamos) River has an area of about 40 km^2 consisting of forest (15.7%), bush (82.1%), urban area (0.8%) and an area with no significant vegetation (1.4%). The dominant rocks are granite-diorite, gneiss, marble and amphibolite. The highest altitude of the basin is about 1400 m. The length of the main stream of the basin is about 8.5 km. The mean soil slope gradient of the basin is about 53%.
Monthly rainfall data for 27 years (1966-1992), as well as mean daily values of air temperature, sunlight hours and relative humidity of the atmosphere for every month of the above years were available from the meteorological station of Egiros, which is located near the basin, but in the flat part of the region. The mean annual rainfall at this station amounts to 665 mm. Moreover, mean daily values of wind velocity for certain months of the years 1997 and 1998 were obtained from the meteorological station of Xanthi (Fig. 6).

5.4.4 Computational results

The mathematical simulation model was applied separately to each sub-basin of the three basins. The computations were performed on a monthly time basis, because monthly rainfall data were available. The discussion on the monthly time basis is found in Section 5.2.1.
The monthly values of sediment yield at the outlet of each of the three basins for a certain year were added to produce the annual value of sediment yield. The mean annual value of sediment yield, YA, for the three basins has as follows (Hrissanthou et al., 2010):

- Kompsatos River basin, $YA = 447\ 000$ t
- Kossynthos River basin, $YA = 192\ 000$ t
- Travos River basin, $YA = 28\ 000$ t

The monthly values of erosion amount in each of the three basins for a certain year were added to produce the annual value of erosion amount. The mean annual value of erosion amount, YD, for the three basins has as follows (Hrissanthou et al., 2010):

- Kompsatos River basin, $YD = 1\ 026\ 000$ t
- Kossynthos River basin, $YD = 409\ 000$ t
- Travos River basin, $YD = 112\ 000$ t

The ratio of the annual sediment at the outlet of a basin to the annual erosion amount in the basin is called the sediment delivery ratio, as mentioned in the other case studies. The mean annual value of the sediment delivery ratio, DR, for the three basins has as follows (Hrissanthou et al., 2010):

- Kompsatos River basin, $DR = 38\%$
- Kossynthos River basin, $DR = 49\%$
- Travos River basin, $DR = 25\%$

On the basis of the above arithmetic results, the calculated total mean annual sediment yield, which flows into Vistonis Lake, is 667 000 t. For the most unfavourable case, that the whole sediment inflowing into the lake is trapped by the lake, the mean annual volume of the deposited sediment will be 445 000 m^3. The convertion of mass to volume was made on the basis of the assumption that sediment bulk density is 1.5 t/m^3. This is a typical value for submerged reservoir deposits consisting of a sand-silt mixture.

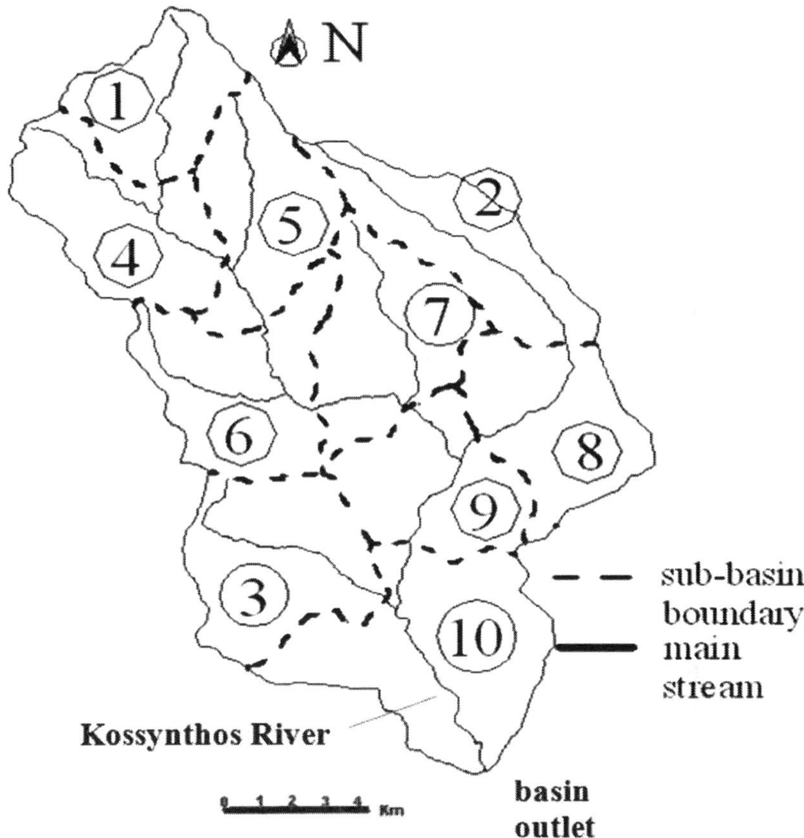

Fig. 8. Kossynthos River basin – Sub-basins with main streams

5.4.5 Estimate of the accumulated sediment in Vistonis Lake

From topographic maps of Vistonis Lake for the years 1949 and 1970, obtained from the Hydrographic Service of the Greek Navy, in which the lake boundaries are also shown (Fig. 9), resulted that the decrease rate of the lake water volume because of sediment deposition is 347 000 m^3/year.

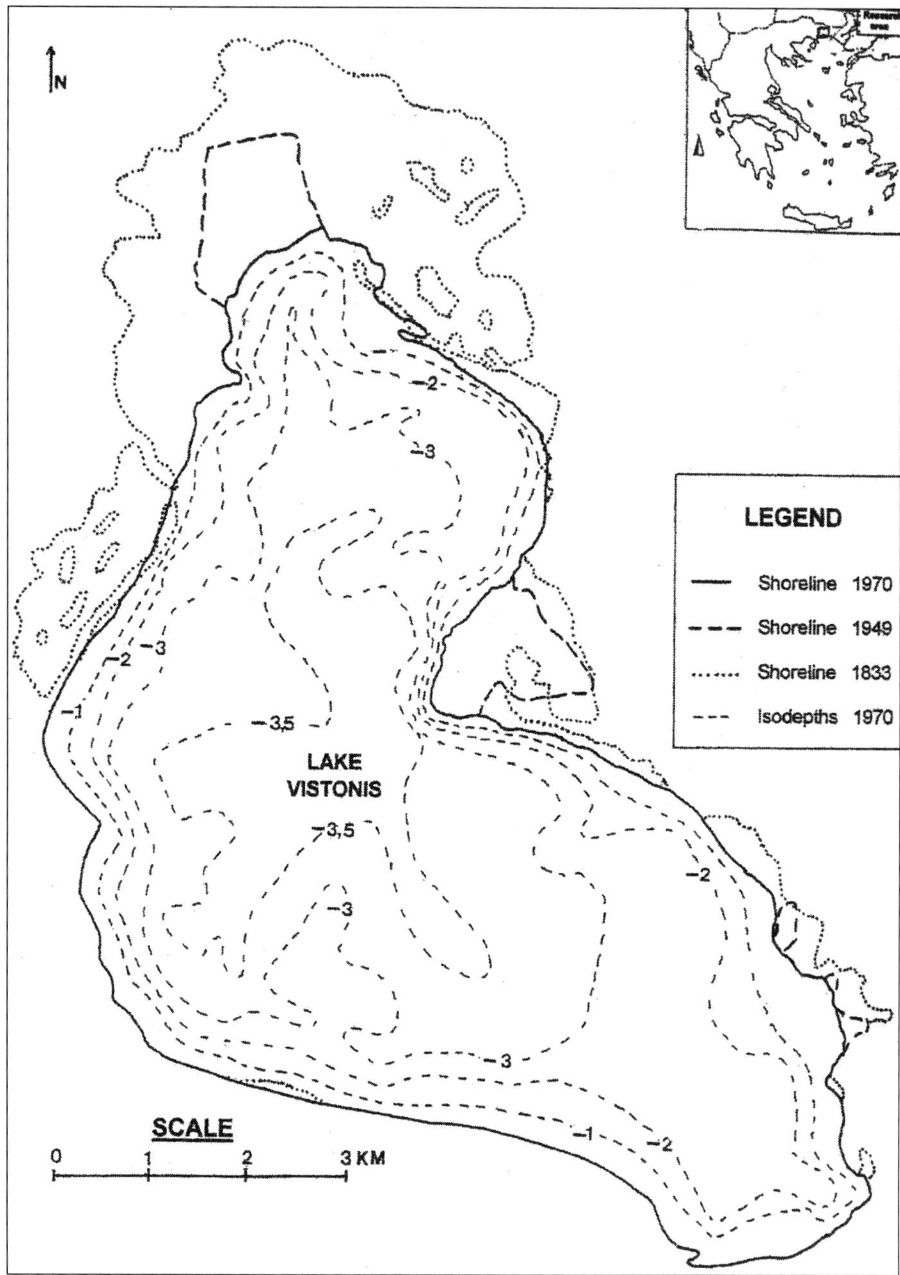

Fig. 9. Shoreline of Vistinis Lake – Isobath contours

5.4.6 Discussion on the comparison between computations and estimates

A comparison between computations by means of the mathematical model and estimates by means of the topographic maps indicates that the mean annual sediment volume accumulated in the lake was overestimated about 28%, according to the computations. The deviations between computations by means of the mathematical model and estimates by means of the maps is attributed, among others, to the following reasons:

- It was not taken into account that a small part of the deposited sediment may be transported to the sea, because the lake communicates with the sea. In other words, the trap efficiency of the lake was assumed to be as equal to 100%.
- The mean annual sediment yield at the basin outlet of Kompsatos and Travos Rivers was computed for the same time period (1966-1992), while the mean annual sediment yield at the basin outlet of Kossynthos River was computed for a different time period (1980-1990). Moreover, the mean annual sediment accumulation in Vistonis Lake was estimated approximatively by means of topographic maps also for a different time period (1949-1970).
- The mean annual sediment yield was computed at the outlet of the mountainous part of the basins considered. It means that the erosion and sediment transport in the flat part of the basins was not taken into account. Anyway, it is believed that a considerable part of sediment load reaching the outlets of the mountainous part of the basins will be deposited in the flat part of the basins. However, in the case of flood events, the sediment deposited in the plains will be transported to the lake.

Finally, the assumptions and simplifications of the mathematical model mentioned in Section 5.2.1, contribute to the inaccurate computation of sediment yield. Additionally, the following remarks are given for the case study of Vistonis Lake basin:

- The proportionality factor k of the hydrological submodel was determined on the basis of the empirical assumption that the runoff coefficient (ratio of rainfall excess to rainfall) amounts to 40% on a monthly time basis, especially for winter months or months with a considerable rainfall amount. The above assumption cannot be avoided, because the factor k is the only parameter of the whole mathematical model that cannot be estimated directly, by means of tables or maps. The final value of factor k used in the hydrologic submodel is a mean value resulting from the k - values determined for different months.
- Some input data (e.g. bottom channel width) of the model used were determined in an empirical way, namely by optical estimation, while the values of other input data (e.g. diameter of suspended particles, sediment density) were determined by assumption. However, these parameters do not influence strongly the sediment yield value at the basin outlet.

6. General conclusion

From the preceding discussions on the computational results of the mathematical simulation models described in this chapter, it is concluded that these models are applicable to lake or reservoir basins for which both hydrometeorological data and topographic, vegetation and soil maps are available, in order to predict roughly lake or reservoir sedimentation in terms of soil erosion.

However, it has to be emphasized that some of the model imperfections given above lead to an overestimation and some others to an underestimation of the sediment yield at the basin outlets, which has as a favourable consequence the compensation of the deviations between computations and estimates or measurements.

7. References

Bathurst, J.C. (2002). Physically-based erosion and sediment yield modeling: the SHETRAN concept, In: *Modelling erosion, sediment transport and sediment yield*, edited by W. Summer & D.E. Walling, IHP, UNESCO, Paris, France

Beasley, D.B., Huggins, L.F. & Monke, E.J. (1980). ANSWERS: a model for watershed planning, *Transactions of the American Society of Agricultural Engineers*, Vol. 23, No. 4, pp. 938-944

Brune, G.M. (1953). Trap efficiency of reservoirs, *Transactions, American Geophysical Union*, Vol. 34, No. 3, pp. 407-418

Das, G. & Agarwal, A. (1990). Development of a conceptual sediment graph model, *Transactions of the American Society of Agricultural Engineers*, Vol. 33, No. 1, pp. 100-104

De Aragão, R., Srinivasan, V.S., Suzuki, K., Kadota, A., Oguro, M. & Sakata, Y. (2005). Evaluation of a physically-based model to simulate the runoff and erosion processes in a semiarid region of Brazil, In: *Sediment Budgets 2*, edited by A.J. Horowitz and D.E. Walling, Proceedings Foz do Iguaço Symposium, April 2005, pp. 85-93, IAHS Publication No. 292, ISBN 1-901502-92-9

De Roo, A.P.J., Wesseling, C.G. & Ritsema, C.J. (1996). LISEM: A single-event physically based hydrological and soil erosion model for drainage basins. I: Theory, input and output, *Hydrological Processes*, Vol. 10, pp. 1107-1117

De Vente, J., Poesen, J. & Verstraeten, G. (2005). The application of semi-quantitative methods and reservoir sedimentation rates for the prediction of basin sediment yield in Spain, *Journal of Hydrology*, Vol. 305, pp. 63-86

Doorenbos, J. & Pruitt, W.O. (1977). Crop water requirements, *FAO, Irrigation and Drainage Paper 24 (revised)*, FAO, Rome, Italy

Einstein, H.A. (1950). The bed-load function for sediment transportation in open channel flows, *Technical Bulletin No. 1026, US Department of Agriculture, Soil Conservation Service*, Washington

Engelund, F. & Hansen, E. (1967). *A monograph on sediment transport in alluvial streams.* Teknisk Forlag, Copenhagen, Denmark

Flanagan, D.C. & Nearing, M.A. (1995). USDA-Water Erosion Prediction Project hillslope profile and watershed model documentation, *NSERL Report No. 10, USDA-ARS National Soil Erosion Research Laboratory*, West Lafayette, Indiana

Foster, G.R. (1982). Modelling the erosion process, In: *Hydrologic Modeling of Small Watersheds*, edited by C.T. Haan, H.P. Johnson & D.L. Brakensiek, Chapter 5, pp. 297-380, American Society of Agricultural Engineers Monograph No. 5, ISBN 0-916150-44-5

Foster, G.R., Meyer, L.D. & Onstad, C.A. (1977). A runoff erosivity factor and variable slope length exponents for soil loss estimates, *Transactions of the American Society of Agricultural Engineers*, Vol. 20, No. 4, pp. 683-687

Giakoumakis, S., Tsakiris, G. & Efremides, D. (1991). On the rainfall-runoff modeling in a Mediterranean island environment, In: *Advances in Water Resources Technology*, Balkema, Rotterdam, pp. 137-148.

Hairsine, P.B. & Rose, C.W. (1992a). Modeling water erosion due to overland flow using physical principles, 1. Sheet flow, *Water Resources Research*, Vol. 28, No. 1, pp. 237-243

Hairsine, P.B. & Rose, C.W. (1992b). Modeling water erosion due to overland flow using physical principles, 2. Rill flow, *Water Resources Research*, Vol. 28, No. 1, pp. 245-250

Hrissanthou, V. (1986). Mathematical model for the computation of sediment yield from a basin, In: *Drainage Basin Sediment Delivery*, edited by R.F. Hadley, Proceedings of the Albuquerque Symposium, USA, August 1986, pp. 393-401, IAHS Publication No. 159, ISBN 0-947571-80-9

Hrissanthou, V. (1988). Simulation model for the computation of sediment yield due to upland and channel erosion from a large basin, In: *Sediment Budgets*, edited by M.P. Bordas & D.E. Walling, Proceedings of the Porto Alegre Symposium, Brazil, December 1988, IAHS Publication No. 174, ISBN 0-947571-56-6

Hrissanthou, V. (1990). Application of a sediment routing model to a Middle European watershed, *Water Resources Bulletin*, Vol. 26, No. 5, pp. 801-810

Hrissanthou, V. (2002). Comparative application of two erosion models to a basin, *Hydrological Sciences Journal*, Vol. 47, No. 2, pp. 279-292

Hrissanthou, V. (2006). Comparative application of two mathematical models to predict sedimentation in Yermasoyia Reservoir, Cyprus, *Hydrological Processes*, Vol. 20, No. 18, pp. 3939-3952

Hrissanthou, V., Mylopoulos, N., Tolikas, D. & Mylopoulos, Y. (2003). Simulation modeling of runoff, groundwater flow and sediment transport into Kastoria Lake, Greece, *Water Resources Management*, Vol. 17, pp. 223-242

Hrissanthou, V., Delimani, P. & Xeidakis, G. (2010). Estimate of sediment inflow into Vistonis Lake, Greece, *International Journal of Sediment Research*, Vol. 25, No. 2, pp. 161-174

Huggins, L.F. & Burney, J.R. (1982). Surface runoff, storage and routing, In: *Hydrologic Modeling of Small Watersheds*, edited by C.T. Haan, H.P. Johnson & D.L. Brakensiek, Chapter 5, pp. 209-214, American Society of Agricultural Engineers Monograph No. 5, ISBN 0-916150-44-5

Johnson, B.E. & Julien, P.Y. (2000). The two-dimensional upland erosion model CASC2D-SED, In: *The Hydrology-Geomorphology Interface: Rainfall, Floods, Sedimentation, Land Use*, edited by M. Hassan, O. Slaymaker & S. Berkowicz, Proceedings Jerusalem Conference, May 1999, pp. 107-125, IAHS Publication No. 261, ISBN 1-901502-16-3

Karim, M.F. & Kennedy, J.F. (1983). Computer-based predictors for sediment discharges and friction factor of alluvial streams, *Proceedings Second International Symposium on River Sedimentation*, Nanjing, China, pp. 219-233

Lukey, B.T., Sheffield, J., Bathurst, J.C., Hiley, R.A. & Mathys, N. (2000). Test of the SHETRAN technology for modeling the impact of reforestation on badlands runoff and sediment yield at Draix, France, *Journal of Hydrology*, Vol. 235, pp. 44-62

Lutz, W. (1984). Berechnung von Hochwasserabflüssen unter Anwendung von Gebietskenngrößen, *Mitteilungen des Instituts für Hydrologie und Wasserwirtschaft, Universität Karlsruhe*, Germany, Heft 24

Renard, K.G., Foster, G.R., Weesies, G.A., McCool, D.K. & Yoder, D.C. (1996). Predicting soil erosion by water: A guide to conservation planning with the Revised Universal Soil Loss Equation (RUSLE), *US Department of Agriculture, Agricultural Research Service, Agriculture Handbook No. 703*

Santos, C.A.G., Suzuki, K., Watanabe, M. & Srinivasan, V.S. (1997) Developing a sheet erosion equation for a semiarid region, In: *Human Impact on Erosion and Sedimentation*, edited by D.E. Walling & J.-L. Probst, Proceedings of an International Symposium (Symposium S6) held during the 5th Scientific Assembly of the IAHS at Rabat, Morocco, 23 April-3 May 1997, IAHS Publication No. 245, ISBN 0-901502-30-9

Schmidt, J. (1992). Predicting the sediment yield from agricultural land using a new soil erosion model, *Proceedings 5th International Symposium on River Sedimentation*, edited by P. Larsen & N. Eisenhauer, Karlsruhe, Germany, pp. 1045-1051

Schröder, W. & Theune, C. (1984). Feststoffabtrag und Stauraumverlandung in Mitteleuropa, *Wasserwirtschaft*, Vol. 74, No. 7/8, pp. 374-379

Schwertmann, U. (1981). Die Vorausschätzung des Bodenabtrages durch Wasser in Bayern, *Institut für Bodenkunde, TU München, Weihenstephan*, Germany

Sharma, K.D., Dhir, R.P. & Murthy, J.S.R. (1993). Modelling soil erosion in arid zone drainage basins, In: *Sediment Problems: Strategies for Monitoring, Prediction and Control*, edited by R.F. Hadley & T. Mizuyama, Proceedings Yokohama Symposium, July 1993, pp. 269-276, IAHS Publication No. 217, ISBN 0-947571-78-7

Smith, J.H., Davis, J.R. & Fogel, M. (1977). Determination of sediment yield by transferring rainfall data, *Water Resources Bulletin*, Vol. 13, No. 3, pp. 529-541

SCS (Soil Conservation Service) (1972). *National Engineering Handbook*, Section 4: Hydrology, USDA, SCS, Washington DC

Van Rijn, L.C. (1984a). Sediment transport, Part I: Bed load transport, *Journal of Hydraulic Engineering, ASCE*, Vol. 110, No. 11, pp. 1431-1456

Van Rijn, L.C. (1984b). Sediment transport, Part II: Suspended load transport, *Journal of Hydraulic Engineering, ASCE*, Vol. 110, No. 11, pp. 1613-1641

Vetter, M. (1992). Ein Beitrag zur Berechnung des Feststofftransports in offenen Gerinnen, *Dissertation, Mitteilungen des Instituts für Wasserwesen, Universität der Bundeswehr München*, Germany, Nr. 42

Wischmeier, W.H. & Smith, D.D. (1978). Predicting rainfall erosion losses, A guide to conservation planning, *US Department of Agriculture, Agriculture Handbook No. 537*

Yalin, S. (1963). An expression of bed-load transportation, *Journal of the Hydraulics Division, ASCE*, Vol. 89, No. 3, pp. 221-250

Yang, C.T. (1973). Incipient motion and sediment transport, *Journal of the Hydraulics Division, ASCE*, Vol. 99, No. 10, pp. 1679-1704

Yang, C.T. & Stall, J.B. (1976). Applicability of unit stream power equation, *Journal of the Hydraulics Division, ASCE*, Vol. 102, No. 5, pp. 559-568

Zanke, U. (1982). *Grundlagen der Sedimentbewegung*, Springer Verlag, ISBN 3-540-11672-9, Berlin, Heidelberg, New York

12

Hydrodynamic Influences on Fluid Mud Distribution in the Amazon Subaqueous Delta

Roberto Fioravanti Carelli Fontes[1], Áurea Maria Ciotti[2]
and Belmiro Mendes de Castro[3]
[1]*Universidade Estadual Paulista (Campus do Litoral Paulista)*
[2]*Universidade de São Paulo (Centro de Biologia Marinha)*
[3]*Universidade de São Paulo (Instituto Oceanográfico)*
Brazil

1. Introduction

Transport and distribution of sediments in the Amazon Subaqueous Delta and the Amazon Shelf (ASD; AS) depend upon of the loads in the Amazon River Basin and on the hydrodynamics aspects. The latter, on the other hand, reacts to the distribution of sediment patches due to decreasing of the bottom stress parameter on finer sediments and fluid mud regions, mostly located on the inner shelf. The Amazon River discharge, tides and stratification are the dominant forcing for currents and related phenomena on the inner AS.
In order to study the physical aspects related to sediment transport in the ASD, we applied the Estuarine and Coastal Ocean Model and Sediment Transport (ECOMSED). This model is capable of simulating both hydrodynamics and sediment dynamics processes in coastal regions. In this chapter, we present results from hydrodynamic modeling experiments, taking into account the complexity of the AS dynamics, including the river plume fate and shape, as well as the frontal zone positioning. The North Brazil Current and other oceanic processes have not been considered in this study, since their main influences occur in the outer shelf, far beyond the river mouth.
The Amazon River Estuary (ARE) does not fit into a classical definition of an estuary, once the mixing zone is not constrained by its margins, appearing in the open shelf. The haline front develops further ahead from the river mouth, preserving most of its characteristics, without being in an estuarine channel. The ASD consists of reworked sediment deposits located seward of the river mouth, on the inner continental shelf. Kineke et al. (1996) defined as fluid mud, the extensive regions of dense nearbed suspensions of sediments where concentrations are above 10 g L^{-1} . Thickly patches of fluid mud layers affect circulation by decreasing the bottom stress coefficient and enhancing tidal currents and the sea level oscillations. Model calibration considered a variable bottom stress distributed according to the ASD, accommodating the reworked sediments and fluid mud parameterizations. Values for these parameters ranged from 2.0 10^{-5}, in fluid mud regions, to 3.2 10^{-3}, in the reworked sediments background. The patches of fluid mud and reworked sediments define, on their vicinities, regions of strong bottom stress gradients capable of promoting residual vorticity and residual circulation.
Residual flows in marine environments can be generated by wind stress variability, by horizontal density gradients, by barotropic effects due to remote processes or by nonlinear tide

current interactions, when energy cascades from dominant frequencies to its harmonics. Tidal currents flowing over coastal areas, subjected to irregular bathymetry, produces residual flow due to nonlinear interactions. Results from numerical experiments focused on the generation of residual vorticity, due to nonlinear interactions and anisotropy of sediment distribution, suggest that residual flow may be enhanced on the ASD region. We evaluated and discussed the role and magnitude of various terms related to residual vorticity as tendency; advection; roughness; dissipation; velocity; bathymetry and Coriolis. The roughness term is the most relevant on vicinity of transitional regions of distinguished sediment patches. Although residual currents are about one order of magnitude lesser than tidal currents, they can be relevant in long-term component of suspended sediment transport and transport of living matter, as algae and larvae in the AS.
The Amazon River Plume (ARP) defines the front position on the continental shelf, as well as the region of maximum sediment deposition rates, or maximum turbidity zone. According to the hydraulic control theory, we could define an internal, or composite Froud Number, which aims to describe the region where hydraulic control occurs. Seaward of this control region a hydraulic jump defines the location where the ARP disconnects from the bottom and acquires negative vorticity, by turning southeastward. Afterwards, the trade winds and the North Brazil Current drive the ARP northwestward, along the coast of Amapá.
Finally, we discuss the leading mechanisms on generation and maintenance of the salinity and turbidity fronts, which are the keys on fluid mud layer formation. Tides promote vertical shear homogenization due to interactions of currents with topography, via hydraulic control, acting as a maintainer of the haline front position and defining the maximum turbidity zone in the ASD.

1.1 Outline & rationale

Some charactericstics of the AS are described in Section 2, related to the hydrodynamics and morphology of the ASD and its environmental description. The numerical model ECOMSED and data are desccribed in Section 3, considering the hydrodynamic core of the model and its sediment transport module. In Section 3.2 there is a detailed consideration on influences of sediment patches (including the fluid mud layer), and parameterization of the bottom stress, a crutial step on modeling the dynamics of currents and tides in coastal environments.
The ARE is an unique estuarine environment, mostly related to the positioning of its salinity front at the continental shelf. In Section 3.3, an approximation of the hydraulic control theory aims to explain how tides, stratification and bathymetry act on positioning of the salinity front. These are also fundamental on defining the position of the maximum turbidity region. In Section 4 we discuss the role of tides in fine sediment patch distribitions. Conversely, patches of fine sediments reduces the bottom drag coeficent and promotes tidal sea level and current amplification.
Another consequence of sediment distribution in marine environments may lead to generation of residual vorticity, which arises from non-linear interactions of currents on vicinity of different sediment patches regions. Also in Section 4 we discuss the role of anisotropy of sediment distribution on generating residual circulation in the ASD.

2. Characteristics of the Amazon Shelf Region

The Amazon Subaqueous Delta is part of the Amazon Continental Shelf (North Brazil), which still is a relatively well preserved region and almost free of anthropogenic influences, although significant human influences in the Amazon Basin was already present long time ago in oscillations of the river discharge (Richey et al., 1989). The Amazon River discharge varies seasonally and may be modulated by teleconnection anomalies, as that imposed by the El Niño-Southern Oscillation (ENSO) phenomenon. The mean annual discharge at the river mouth is about 2.0 10^5 m^3s^{-1}, which roughly represents 10% of overall freshwater input into the global ocean system.

The ARE does not fit into a classical definition of estuary due to its characteristics of broad extension and huge discharge (Miranda et al., 2002). Also, the estuarine mixing zone is not constrained by the estuary margins. The salinity front and the maximum turbidity region develop further ahead from the river mouth, which does not occur in a regular estuary.

As the AS is located right at the the equator, there is geostrophy degeneration on the momentum equations and equilibrium is achieved trought others terms. This promotes a very energetic environment where the barotropic tides are fundamental on circulation and mixing processes. Winds and waves are moderated in this region and do not have relative importance on the local dynamics, although it may be relevant on ressuspension and sediment transport during extreme events.

Former studies as (Beardsley et al., 1995; Fontes et al., 2008; Gabioux et al., 2005; Geyer et al., 1996), among others, have measured or modeled the tidal amplification due to supression of the bottom stress, imposed by fluid mud layers. The bathymetry and site locations in the AS are in Fig. 1.

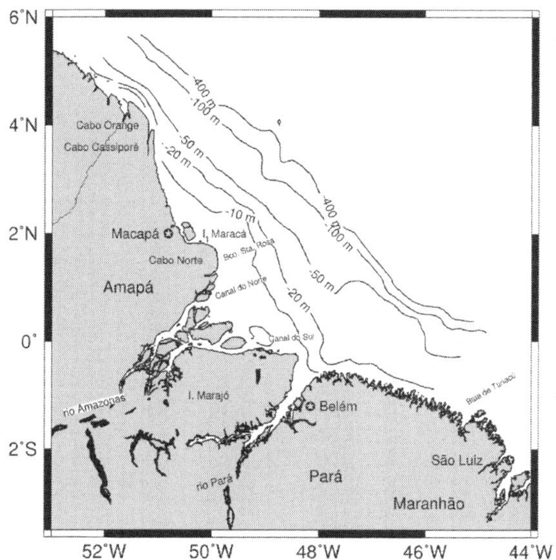

Fig. 1. Site locations in the Amazon Shelf and bathymetry representation.

3. Data & methods

We used the dataset and information provided from the former 1990's AMASSEDS Project, a multidisciplinary effort for compreension of sediment transport on the continental shelf near the Amazon River mouth (AMASSedS, 1990). Measurements of currents, salinity and temperature, tides and sediment concentration were available for proper modeling calibration and validation, mostly derived from the compilation in Alessi et al. (1992).

3.1 The ECOMSED numerical model

The Estuarine and Coastal Ocean Model and Sediment Trasport (ECOMSED) is an integrated suite of fortran routines for solving hydrodynamics and sediment transport in estuaries and coastal seas (Blumberg, 1996). The development of this suite has its origins with the Princeton Ocean Model (POM), a pioneer and consagrated model applied in ocean research (Blumberg & Mellor, 1980; 1983). The model is based on a three-dimensional set of equations that describes the geophysical fluid from the Naavier-Stokes formulation under the slallow water and Boussinesq approximations. The primitive equations solve the horizontal mode terms, including geostrophy and the baroclinic effect. The parameterization of Reynolds stress and flux terms account for the turbulent diffusion of salt, heat and momentum. The vertical mixing coefficients are obtained by solution of a second-order turbulence scheeme described in Mellor & Yamada (1982).

We applied a lattice with 81×181 horizontal grid cells and 17 vertical levels. The vertical $z-$coordinate fomulation is redefined in terms of σ-coordinates, in order to better represent the geometry of bottom and subsurface boundaries layers, in both shallow and deeper portions of the AS.

The sediment transport module (SED) employs the hydrodynamic results from the hydrodynamic core, in the same numerical grid. This SED module simulates the transport of suspended sediments for cohesive and non-cohesive sediment classes, as well as deposition, ressuspension and bed armoring. The same dynamic features from the hydrodynamic core are employed in the SED module: temperature, salinity, viscosity and turbulence diffusivity. For ordinary estuarine ocean models, density is primarly a function of temperature and salinity. As the ASD is also characterized for having substantial large nepheloid layers of high concentration of sediments, these fluid mud layer affects sea water density, amongst salinity and temperature Felix et al. (2006).

We included the contribution of cohesive sediment concentration in density calculation as in Wang (2002),

$$\rho = \rho_w + (1 - \frac{\rho_w}{\rho_s})C \tag{1}$$

where ρ_w is the clear water density, ρ_s is the cohesive sediment density and C is the suspended sediment concentration.

The suspended transport of fine sands is calculated using the van Rijn's method (van Rijn, 1993). The bed load transport is not considered in this module because most of the sediment transport in marine system is in suspension as the rate of moviment of coarse materials is limited by the transport capacity of the environmental flow Haan et al. (1994).

3.2 Effect of sediments on hydrodynamics

Wang (2002) studied the dynamics of nepheloid layer in an idealized estuary, considering the coulpling effect of seawater and resuspended sediment concentration. The author found a two layer sediment distribution structure formed as a lutocline is developed above a nepheloid layer. A vertical sediment concentration gradient is of maximum at the former, and this vertical structure is found in regions as the ASD.

Bottom shear stress and the dynamics of the bottom boundary layer (bbl) in shallow marine environments are highly influenced by winds and currents, as well as distribution of sediment classes and its nature. The presence of submersed hills and valleys, mud deposits and bottom roughness are also relevant on the bbl dynamics. Near the bottom, the bottom stress and the turbulent kinetic energy are due to the combined effect of wave and currents (Grant &

Madsen, 1986). The parameterization of the drag coefficient (C_d) depends on structure of the water-bed interface,

$$C_d = \frac{\kappa^2}{\ln^2(z/z_o)} \tag{2}$$

where κ is the von Kármán's constant; z is the height in the bbl and z_o is the roughness length scale. Adams & Weatherly (1981) defined a vertical profile in the bbl based on extended consideration of combined physical, biological and morphological effects,

$$u(z) = \frac{u_*}{\kappa}[\log(\frac{z}{z_{oc}}) + \beta \int_{z_{oc}}^{z} \frac{Ri_H}{z}\, dz] \tag{3}$$

where Ri_H is the Richardson's Number as defined by Heathershaw (1979),

$$Ri_H = \frac{w_s\, \kappa\, z\, g\, c}{\rho\, u_*^3}(1 - \rho/\rho_s) \tag{4}$$

and,

w_s: sediment settling velocity;
c : concentration of sediments;
ρ_s: density of sediments.

The bottom stress ($C_d = u_*^2/u^2$) can be obtained from Equation 3,

$$C_d \approx \kappa^2/[(1 + 4.7 < Ri_H >)\, log(\frac{z}{z_{oc}})]^2 \tag{5}$$

where $< Ri_H >$ represents the vertical integrated Richardson's Number.
We prescribed the drag coefficient in the bbl directly from sediment distribution, instead applying the original model formulation. From Dyer (1986) we were able to relate a wide range of sediment classes with bottom roughness, considering a low stratified fluid into the bbl. We set this bottom stress formulation by compilation of sediment distribution over the AS (Fig. 2) and definition of fluid mud distributions obtained during the AMASSEDS Project. The the C_d modeling parameterization considered the sediment class and fluid mud distributions over the AS. Fluid mud regions have lowest C_d values while mud- and sand-mixtures assume intermediate values.

3.3 Approximation of the hydraulic control theory

The hydraulic control theory (HTC) must define the mechanisms for maintenance and positioning of the salinity front in the ASD and, therefore, relate it with the maximum turbidity zone definition.
(Chao & Paluszkiewicz, 1991) applyed the HCT on channels as in presence of lateral or bottom constrictions with two vertical density layers. Besides some restrictions on the aplication of HCT in marine environments, we used the approach developed by Cudaback & Jay (1996) in the Columbia River (OR, EUA). Fontes et al. (2008) followed the same method on the inner Amazon Shelf, near the river mouth.
According to the HCT, the river discharge may be subject to changes on its dynamical state, where the dynamic condition state turns from super-critical to sub-critical when it passes trough an internal hydraulic jump. A control point between the river mouth and the coastal zone defines the bottom salinity front and the maximum turbidity zone.

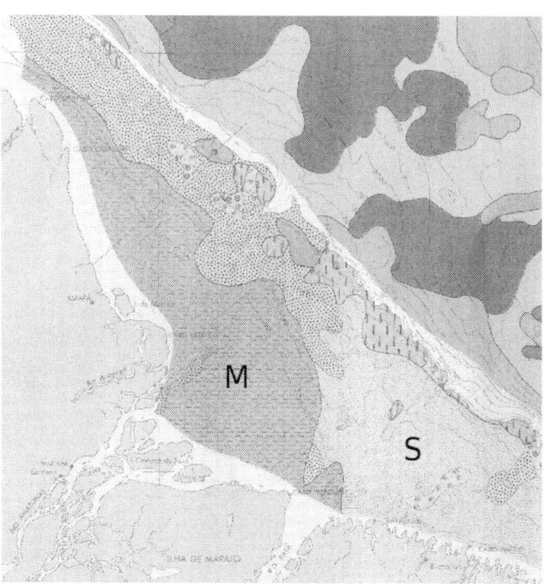

Fig. 2. Faciologic map of the Amazon Shelf from BRASIL (1979). Mud (M) and sand (S) sediment patches are annotated in the figure, representing regions of extreme C_d parametrization.

Armi & Farmer (1986) extended the application of HCT by defining an internal, or composite Froud number,

$$G^2 = F_1^2 + F_2^2 = \frac{u_1^2}{g'.h_1} + \frac{u_2^2}{g'.h_2} \tag{6}$$

where, $g' = g(\rho_1 - \rho_2)/\rho_1$ is the reduced gravity; ρ_i and h_i are density and layer thickness respectively [i=1(top),2(bottom)]; and F_i is the Froud number defined in each layer. This composite Froud number was calculated for an outflow cross shelf section at the Canal do Norte (North Channel).

The hydraulic control point occurs where the bottom slope is strongest, starting from -0.125 m km^{-1} and reaching up to -0.385 m km^{-1}. The region of maximum gradient in bathymetry occurs at 15 m deep, around 100 km from the coast. Figure 3 shows the line section along the ASD where the Froud number was evaluated.

3.4 Small scale vorticity generation mechanisms

Residual flow is an important effect in coastal oceanography and is commonly related to the subtidal flow, where tides and wind driving circulation are filtered out as well as other ambient influences. The residual influences are only due to rectification processes related to non-linear interactions from oscillatory flows. They are taken into account in the local circulation in order to affect advection. Regardless of that common assumption, it is convenient to redefine residual flow of a generic property a throughout a whole cycle of the

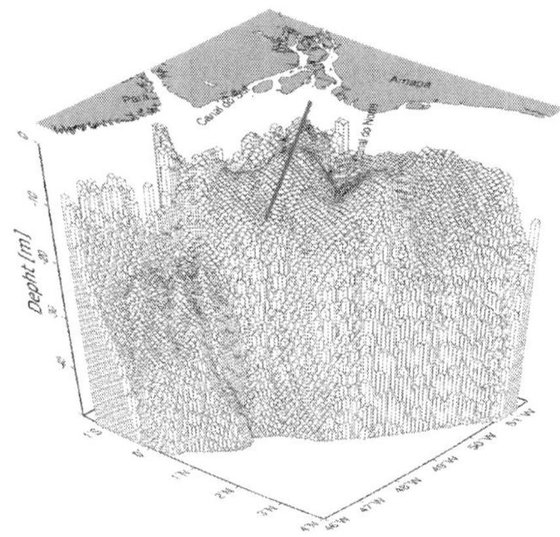

Fig. 3. ASD bathymetry representation and the section (red line) along with the Froud number was evaluated.

major tidal constituent M_2, as defined by Robinson (1983),

$$< a >= \frac{1}{T} \int_{t_0}^{t_0+T} a \, dt \tag{7}$$

where T is the period of the M_2 tidal component. In this case we are interested in the residual flow, so that depth integrated velocity at position x_0 is,

$$\vec{v} = \frac{1}{\bar{H} + \eta} \int_{-\bar{H}}^{\eta} \vec{v}_h \, dz \tag{8}$$

where $\bar{H}$ is the mean local depth. The residual term is named the Eulerian ($\vec{v}_E$), as properly justified, and is represented by integration over time,

$$\vec{v}_E = \frac{1}{T} \int_0^{t_0+T} \left(\frac{1}{\bar{H} + \eta} \int_{-\bar{H}}^{\eta} \vec{v}_h \, dz \right) dt \tag{9}$$

To retain the aspects of residual estimation we only considered the effects of tides (no winds, nor river discharge). Although tidal currents on the AS are energetic, their essentially oscillatory behavior does not result in significant net transport. The sediment load transport, for instance, is subject of long-term processes developed over the continental shelf, where residual flows can play a special feature on advection and consequently in the net transport. Residual flows can be generated by wind stress anisotropy over the shelf, by horizontal density gradients, horizontal gradients of sea surface due to remote processes or by non-linear tide interactions, when energy migrates from dominant frequencies to their harmonics and mean. It is known that tidal currents flowing over coastal areas subjected to irregular bathymetry produces residual flow due to non-linear interactions (Tee, 1994).

Anisotropy in the fields of properties like sediment distribution (bottom stress), Coriolis, bathymetry, velocity and dissipation are important on marine environments as the ASD. Those terms are defined from the vorticity Equation (Gross & Werner, 1994),

$$\frac{d\vec{\omega}}{dt} = -\frac{C_d|\vec{v}_H|}{H+\eta}\vec{\omega} + \frac{\vec{f}+\vec{\omega}}{H+\eta}\frac{d(H+\eta)}{dt} + \frac{C_D\vec{v}_H|\vec{v}_H|}{H+\eta} \times \left[\frac{\nabla_h|\vec{v}_H|}{|\vec{v}_H|} - \frac{\nabla_h(H+\eta)}{H+\eta} + \frac{\nabla_h C_D}{C_D}\right] \tag{10}$$

where $\omega = \frac{\partial v}{\partial x} - \frac{\partial u}{\partial y}$ is the vertical component of relative vorticity, $C_D = f(x,y)$ is the bottom stress horizontal distribution and $\vec{v}_H$ is the barotropic velocity. The η, H and f are sea surface displacement, depth and Coriolis' parameter, respectively.

4. Results

4.1 The Influence of tides on sediment dynamics

Using the same modeling dynamics described in Fontes et al. (2008), we considered the evaluation of cohesive sediments from the river discharge into the AS and ASD. The environmental dynamic conditions and charges of sediments represent climatological conditions of the AS, which are in Table 1,

Dynamical Mode	*Prognostic*
run time	600 h
River outflow discharge	2.0 $10^5 m^3 s^{-1}$
Salinity discharge	0.0
Temperature discharge	25.0°C
Cohesive sed. conc. in discharge	200 mg L^{-1}
Ambient initial salinity	35.0
Ambient initial temperature	25.0°C
Ambient coh. sed. conc.	5.00 mg L^{-1}
Spatially variable C_D	2.0 $10^{-4} \rightarrow 3.2$ 10^{-3}
Tidal components	semidiurnals Luni-solar and Solar M_2 and S_2
Winds	climatology (5.0 m s^{-1} - NE)

Table 1. Modeling conditions and parameterization for cohesive sediment transport in the ASD.

The evolution of cohesive sediment concentrations over the ASD is in Fig. 4 for both, bottom and surface distributions. When leaving the river mouth, they extend hundreds of kilometers Norhtwestward along the coast of Amapá. The higher concentrations at the bottom most layers (> 10 mg L^{-1}) are better defined than the plume of sediments at the surface, where concentrations are at least one order of magnitude lesser than those near the bottom.

The formation and positioning of the sediment and salinity fronts (not shown) have similar dynamic aspects. Tides, bathymetry and the dynamical state represented by the Froud Number are capable of define them. The sediment dynamics differs by intrinsic phenomena as floculation and deposition, which are relevant in the formation of the nepheloid layers (concentrations above 10 gL^{-1}) as in Fig. 5.

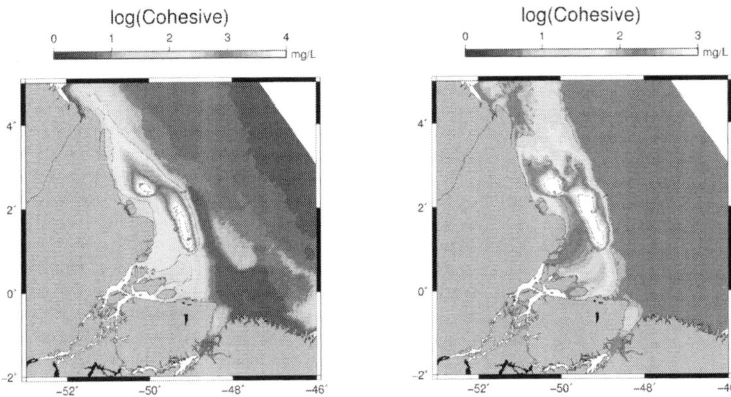

Fig. 4. Distribution of cohesive sediments in the bottom most (*left panel*) and near surface layer (*right panel*). Concentrations are in log_{10} scale and they differ in one order of magnitude.

4.2 Residual vorticity estimation

The tidal excursion along the ASD can promote residual vorticity when integrated between one tidal cicle, as previously described. For the ASD application we found the roughness gradient term the most important amongst the terms in Equation 10, regarding residual flux generation,

$$\frac{C_D \vec{v}_H |\vec{v}_H|}{H+\eta} \times \frac{\nabla_h C_D}{C_D}$$

The vorticity terms evaluated in the ASD are in Fig. 6, where the most relevant derives from integration and not from graphical correlation.

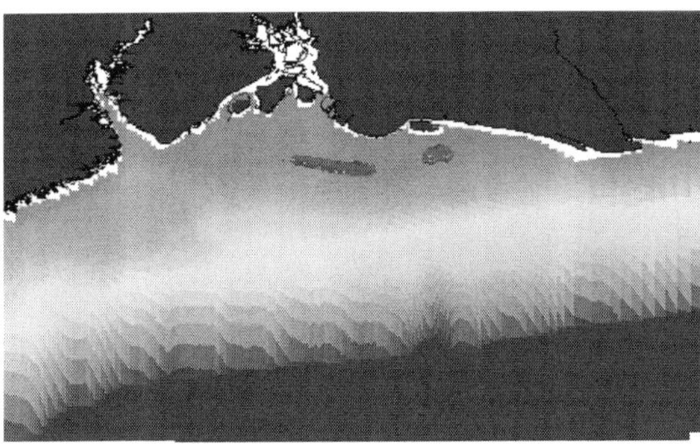

Fig. 5. Nepheloid layers located off Cabo Norte and Maraca Island. The red isosurface defines a 10 g L^{-1} concentration value for cohesive sediments.

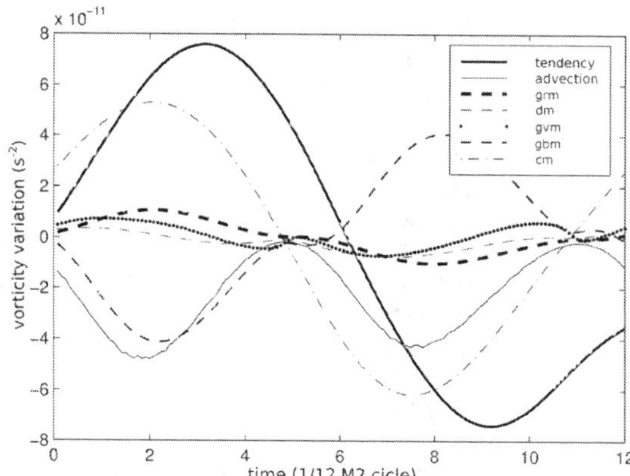

Fig. 6. Vorticity evaluated through 1 cycle of M_2 component, near Maracá Island (frontier of fluid mud region). The results are obtained according to Equation 10.

Estimation of residual flux was investigated through analysis of vorticity described on Section 3.4. For this residual flow estimation we choose a point located at the edge of the fluid mud frontier, nearby Maracá Island (Fig. 7). In coastal regions of the AS, the tidal ellipses are degenerated and highly polarized so they can be approximated by its rectilinear form,

$$\vec{\mathcal{q}}_h = \mathcal{V} \cos \sigma t \, \vec{s} \tag{11}$$

according to a natural system of coordinates $(\vec{s}, \vec{n})$, where $\vec{s}$ is tangent to the stream current and $\vec{n}$ is the normal, left-oriented from the displacement. In this way, the vorticity Equation 10 in its scalar form can be expressed by:

$$\begin{gathered}\frac{\partial \omega}{\partial t} + \mathcal{V} \cos \sigma t \frac{\partial \omega}{\partial s} = A(s) \cos \sigma t + \\ +[B(s) + B'(s)] \cos \sigma t \cos |\sigma t| - C(s) |\cos \sigma t| \omega \end{gathered} \tag{12}$$

which is the Eulerian form as defined in Robinson (1983) and simplified by removing the lower order terms and making some other assumptions. σ is the M_2 tidal frequency, $\mathcal{V}$ is the tidal current amplitude in $\vec{s}$ direction,

$$A(s) = \frac{f\mathcal{V}}{H} \frac{\partial H}{\partial s}$$

$$B(s) = C_D \, \mathcal{V} \frac{\partial}{\partial n} (\mathcal{V}/H)$$

$$B'(s) = \frac{\mathcal{V}|\mathcal{V}|}{H} \frac{\partial C_D}{\partial n}$$

$$C(s) = \frac{C_D \, \mathcal{V}}{H}$$

are the terms of vorticity generation due to specific interactions: Coriolis mechanism **CM**; bathymetry and velocity gradients mechanism **GM**; roughness gradient mechanism **RM** and dissipative mechanism **DM**.The last term was held constant over the domain. A

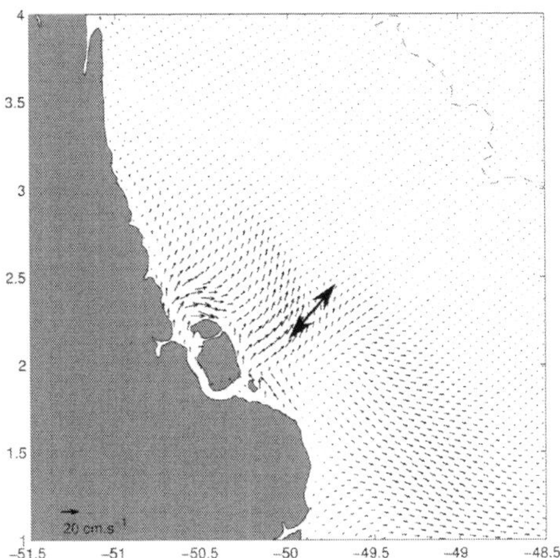

Fig. 7. Magnitude of the residual currents evaluated through one cycle of the M_2 tidal component, near Maracá Island. Higher values match transitional region of fluid-mud/bulk sediments.

detailed discussion considering different approaches for solving this problem can be found in Robinson (1983). At this point it is necessary to define a spatial scale related to residual vorticity,

$$E_M = [\frac{2}{T}\int_0^{T/2} \mathcal{V} \sin{(\sigma t)}\, dt]\frac{T}{2} = \frac{\mathcal{V}}{\pi}T \tag{13}$$

Miranda et al. (2002) called this the "tidal excursion"; $T \approx 12.42$ h is the period of M_2. Estimation of the residual vorticity at the chosen point was made by computing the contributions of individual mechanisms listed above. Results were obtained by the model, considering that point and its neighborhood,

$$\begin{aligned}
\mathcal{V}(22,96) &= 1.5\, ms^{-1} \\
C_D(22,96) &= 3.27 \times 10^{-4} \\
H(22,96) &= 6.39\, m \\
\sigma_{M_2} &= 2.8 \times 10^{-4}\, s^{-1} \\
\partial n &\approx 8.15 \times 10^{3} \\
\mathcal{V}(21,95) &= 1.3\, ms^{-1} \\
C_D(21,95) &= 2.0 \times 10^{-4} \\
H(21,95) &= 5.30\, m
\end{aligned}$$

The total residual vorticity computed at the point was $\omega_{res} = 2.11 \times 10^{-5}\, s^{-1}$. An estimation of residual velocity was obtained by applying the circulation theorem. Let the vorticity be distributed over an area scaled by the tidal excursion, E_M. In this way, by applying Equation 13 it comes,

$$E_M = \frac{1.5}{\pi} 12.42(3600) = 2.1 \times 10^4 \; m$$

$$\mathcal{V}_{res} = \frac{\omega\pi(E_M/2)^2}{\pi E_M} \approx 0.1126 \; ms^{-1}$$

close to the value obtained by the model at that considered point, $\mathcal{V}'_{res} = 0.1193$.

5. Discussion and conclusions

Tides and the river discharge are the most energetic features on the inner Amazon Shelf dynamic system and promote, thought the hydraulic control theory, a reasonable explanation for positioning the salinity front and the maximum turbidity zone. Tides act as a stirring mechanism for the front generation and a control point located at 15 m depth, at the threshold, denotes where a hydraulic jump occurs.

We find the density field strongly affected by concentration of cohesive sediments, so this defines the formation of nepheloid layers in the ASD. The model reproduced the shape and position of fluid mud patches nearby Cabo Norte and MaracÃą Island, where concentrations were higher than 10 g L^{-1}.

Ocean color satellite imagery allows the retrieval of products such as particulate inorganic carbon (D. Clark personal communication, 2003). Fig. 8 illustrates an estimative of "climatological" sediment dispersion evaluated for the period of July 2002 through December 2007, with concentration value 2.0 10^{-2} mol m^{-3} denoting the higher values. Although the compilation of sattelite data has low resolution near the coast, it suffices to contour the influence of sediments in the ASD. Concentration values of the 4.0 10^{-2} mol m^{-3} isosurface defines a front that roughly matches the 100 mg L^{-1} isoline for cohesive sediment in the ASD (Fig. 4).

Although the large inertial flow imposed by the Amazon River accounts for most of the advection throughout the estuary, residual flow can locally contribute with long-term advective processes. This is mostly due to a rectification process that results from net transport integration throughout a semidiurnal tidal cycle. The tidally driven residual flow can contribute with advective processes such as sediment transport and pollutant advection and biological ones, as organic and larvae dispersion. As the Amazon River Estuary does not fit in the classical definition of the estuary, the high load sediment concentration flow occur in the ASD favoring the formation of mud deposits that extend for kilometers. Modeling the transport of cohesive sediments in marine environments, as the AS, requires parameterization of substantial oceanographic, meteorologic and sedimentological data. Others, like bathymetry and hydrology are equally fundamental. Also, parameterization of natural environments like estuaries and coastal seas is a hard task, most of the time, once those environments have distinct behavior from the test fluid in laboratories.

A broad scale of temporal and spatial phenomena (from turbulence to tides and mesoscale variations) and the unpredictable occurrence of extreme events as storms and oceanic rings from current systems. Models are not always capable to deal with phenomena like these. Nevertheless, they are usable since the problem definition and efforts on its implementation are focused on a narrower set of physical phenomena. Under the hydraulic control theory, the estuarine dynamics and the nature of sediments in the AS we were able to understand how physical aspects can act in the deposition and transport related phenomena.

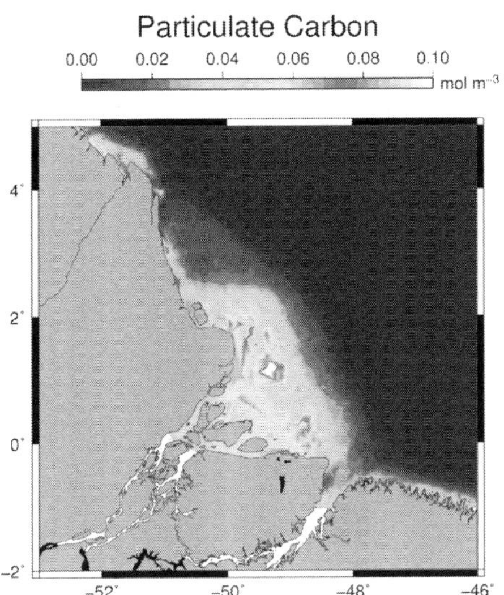

Fig. 8. Particulate inorganic carbon derived from sattelite imagery for the period of July 2002 through Dec. 2007 (D. Clark personal communication, 2003).

6. References

Adams, C. E. & Weatherly, G. L. (1981). Some effects of suspended sediment stratification on an oceanic bottom boundary layer, *J. geophys. Res.* (86): 4161–4172.

Alessi, C., Lentz, S. J., Beardsley, R. C., Castro, B. M. & Geyer, W. R. (1992). A multidisciplinary amazon shelf sediment study (amasseds): Physical oceanography moored array component, *Technical Report WHOI-92-36*, Woods Hole Oceanography Institution.

AMASSedS (1990). A multi-disciplinary amazon shelf sediment study (amasseds). Eos, Trans. AGU, 71, 1771.

Armi, L. & Farmer, D. M. (1986). Maximal two-layer exchange through a contraction with barotropic net flow, *J. Fluid Mech.* 164: 27–51.

Beardsley, R. C., Candela, J., Limeburner, R., Geyer, W. R., Lentz, S. J., Castro, B. M., Cacchione, D. & Carneiro, N. (1995). The m2 tide in the amazon shelf, *J. geophys. Res.* 100(C2): 2283–2319.

Blumberg, A. F. (1996). An estuarine and coastal ocean version of pom, *Proceedings of the Princeton Ocean Model Users Meeting (POM96).*

Blumberg, A. F. & Mellor, G. L. (1980). A coastal ocean numerical model, *in* J. Sundermann & K. P. Holz (eds), *Mathematical Modelling of Estuarine Physics, Proceedings of an International Symposium*, Spring-Verlag, Berlin.

Blumberg, A. F. & Mellor, G. L. (1983). Diagnostic and prognostic numerical circulation studies of the south atlantic bight, *J. geophys. Res.* (88): 4579–4592.

BRASIL (1979). Margem continental norte: Mapa faciológico dos sedimentos superficiais da plataforma e da sedimentação quaternária no oceano profundo, cartographic chart, scale=1:3.500.000. Projeto REMAC - Reconhecimento Global da Margem Continental Brasileira.

Chao, S.-Y. & Paluszkiewicz, T. (1991). The hydraulics of density currents over estuarine sills, *J. Geophys. Res.* 96(C4): 7065–7076.

Cudaback, C. N. & Jay, D. A. (1996). Formation of columbia river plume: Hydraulic control in action?, *in* D. G. Aubrey & C. T. Friedrichs (eds), *Buoyance Effects on Coastal and Estuarine Dynamics*, Vol. 53 of *Coastal and Estuarine Studies*, American Geophysical Union, Washington, DC, pp. 139–174.

Dyer, K. R. (1986). *Coastal and estuarine sediment dynamics*, John Wiley and Sons Ltd., Great Britain.

Felix, M., Peakall, J. & McCaffrey, W. D. (2006). Relative importance of processes that govern the generation of particulate hyperpycnal flows, *Journal of Sedimentary Research* 76: 382–387.

Fontes, R. F. C., Castro, B. M. & Beardsley, R. C. (2008). Numerical study of circulation on the inner amazon shelf, *Ocean Dynamics* .

Gabioux, M., Vinzon, S. B. & Paiva, A. M. (2005). Tidal propagation over fluid mud layers on the amazon shelf, *Continental Shelf Research* (25): 113–125.

Geyer, W. R., Beardsley, R. C., Lentz, S. J., Candela, J., Limeburner, R., Johns, W. E., Castro, B. M. & Soares, I. D. (1996). Physical oceanography of the amazon shelf, *Continent. Shelf Res.* 5/6(16): 575–616.

Grant, W. D. & Madsen, O. S. (1986). The continental-shelf bottom boundary layer, *Ann. Rev. Fluid Mech.* 18: 265–305.

Gross, T. F. & Werner, F. E. (1994). Residual circulations due to bottom roughness variability under tidal flows, *J. phys. Oceanogr.* 24: 1494–1502.

Haan, C. T., Barfield, B. J. & Hayes, J. C. (1994). *Design Hydrology and Sedimentology for Small Catchments*, Academic Press, Inc.

Heathershaw, A. D. (1979). The turbulent structure of the bottom boundary layer in a tidal current, *Geophys. J. Astron. Soc.* 58: 395–430.

Kineke, G. C., Sternberg, R. W., Trowbridge, J. H. & Geyer, W. R. (1996). Fluid-mud processes on the amazon continental shelf, *Continent. Shelf Res.* 16(5/6): 667–696.

Mellor, G. & Yamada, T. (1982). Development of a turbulence closure model for geophysical fluid problems, *Revs. Geophys. Space Phys.* (20): 851–875.

Miranda, L. B., Castro, B. M. & Kjefve, B. (2002). *Principios de Oceanografia Fisica de Estuarios*, Edusp.

Richey, J. E., Nobre, C. & Deser, C. (1989). Amazon river discharge and climate variability: 1903 to 1985, *Science, New Series* 246(4926): 101–103.

Robinson, I. (1983). *Tidally induced residual flows*, Elsevier Oceanography Series, 35, Elsevier, Whiteknights, England. Physical oceanography of coastal and shelf seas.

Tee, K. T. (1994). Dynamics of a two-dimensional topographic rectification process, *J. phys. Oceanogr.* 24.

van Rijn, L. C. (1993). *Principles of sediment transport in rivers, estuaries and coastal seas*, Aqua Publications, Amsterdam.

Wang, X. H. (2002). Tide-induced sediment resuspension and the bottom boundary layer in an idealized estuary with a muddy bed, *Journal of Physical Oceanography* 32: 3113–3131.

13

Hydrodynamic Effects of Sedimentation on Mass Transport Properties in Dead Water Zone of Natural Rivers

Michio Sanjou
Department of Civil and Earth Resources Engineering, Kyoto University
Japan

1. Introduction

We can often see dead water zones composed of consecutive groynes in natural rivers. The groynes are generally constructed in the bank of actual rivers in order to navigate stream direction and to prevent bank erosion. Dead water zones such as side cavities are also observed in harbors of rivers, and it is well known that there are significant differences between streamwise velocities of the mainstream and the cavity zone. Of particular significance is that shear instability related to the velocity difference induces coherent horizontal vortex along the boundary of the mainstream and the cavity. Further, large-scale gyres are formed in the dead water zone which conveys suspended sediment from the main-channel, and local sedimentations are promoted in the cavity as shown in photo 1. So, it is necessary to reveal the hydrodynamic properties included with turbulence phenomena in order to control sedimentation reasonably. Akkerman et al.(2004) conducted the sensitivity analysis with a 1-dimensional morphodynamic model, and they discussed the several effects on sedimentation and flood water depth after the occurrence of groyne damage. Recently, permeable groynes are proposed to realize stable bed condition. For example, Kadota & Suzuki (2010) discussed experimentally effects of the permeability and the scales of the groynes and stone gabion in submerged and emerged flow conditions. Tominaga & Sakaki (2010) conducted ADV measurements around the permeable groynes in a natural river, and they evaluated distributions of bed shear stress accurately.

In these cavities, not only sedimentation but also congestion of pollutants is often highlighted. It is thus very important to investigate mass & momentum exchanges between the main-channel and the side-cavity in river environment and hydraulic engineering. The above-mentioned horizontal gyres and coherent turbulent structures play significant roles on mass and momentum exchanges. Uijttewaal et al.(2001) measured distribution of dye concentration and pointed out that aspect ratio of side cavity has significant effects on exchange rate of mass between the mainstream and the dead water zone. Weitbrecht et al.(2007) have also conducted laboratory measurements, in which distribution of velocity components and dye were obtained. They examined the relation between the exchange rate and the bed configuration of the cavity.

Photo 1. Sedimentation in cavity zone of natural river (Yada river in Nagoya, Japan)

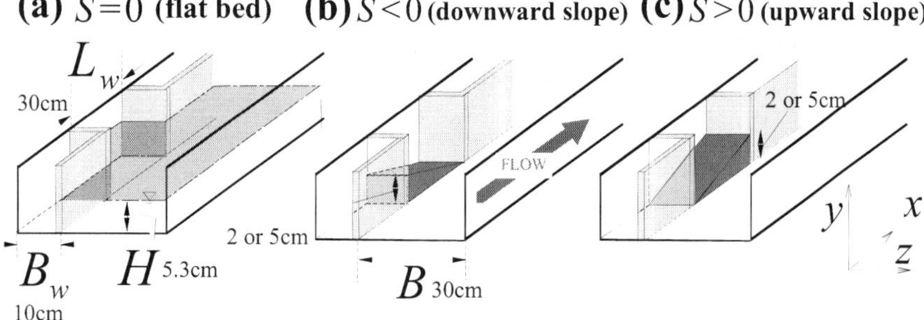

Fig. 1. Bed configurations in the present experiments

U_m (cm/s)	H (cm)	B (cm)	B_w (cm)	L_w (cm/s)	Fr	Re	S
12	5.3	30	10	30	0.17	6360	-1/6
							-1/15
							0
							1/15
							1/6

Table 1. Hydraulic condition

However, there remain uncertainties about accurate evaluations of exchange velocity, effects of the bed formation on mass transport and physical modeling of mass exchange in the side-cavity flows. Therefore, in this study, PIV and LIF were conducted in a laboratory flume using a laser light sheet and high-speed CMOS camera in order to reveal hydrodynamic characteristics and turbulence structure and evaluate exchange properties of dye concentration across the junction.

2. Experimental procedure

2.1 Hydrodynamic condition and bed configuration

The experiments were conducted in a 10m long, 40cm width glass-made tilting flume. The coordinate was chosen that (x, y, z) correspond to the streamwise, vertical and spanwise directions, respectively. The channel bed was chosen as the vertical origin (y=0). In the measurement section, 7m downstream from the channel entrance, a side-cavity is placed by acrylic plates. The widths of the main-channel and the cavity are B=30cm and B_w=10cm , respectively. The streamwise lenght of the cavity is L_w=30cm. Table 1 shows hydraulic condition, in which the water depth is H=5.3cm, the bulk-mean velocity is U_m=12cm/s, Froude number is Fr=0.166 and Reynolds number is Re=6360. We chosen several kinds of bed configurations in the cavity as shown in Fig.1. S is a parameter to indicate the bed configuration. Negative sign of S means downward incline as shown in Fig.1 (b), and positive one means upward incline as shown in Fig.1 (c). S =0 means the flat bed condition. The absolute value of S means the inclination of the bed. In the present study, the elevation gaps are 2cm and 5cm for the mild and steep conditions, respectively.

2.2 Measurement method

The 2mm thick laser light sheet (LLS) was generated by 3W Ar-ion laser using a cylindrical lens. The illuminated flow images were taken by a CMOS camera (1000×1000pixels) with 100Hz frame-rate and 30Hz sampling-rate. The sampling duration is 60 seconds for each case. The instantaneous velocity components $(\tilde{u}, \tilde{w})$ on the horizontal plane were calculated by the PIV algorithm. Further, in the non-flat conditions, LLS was also projected along the bed slope. This is called as "inclined LLS measurements". In the "horizontal-LLS measurements", the elevations of the LLS were y=1cm, 2cm, 2.5cm, 5cm for the all cases. The "inclined LLS measurements" were carried out for the steep slope conditions, in which the gap of the LLS and the bed surface is 1cm.
The distribution of dye concentration is measured by the LIF method, in which the sharp-cut filter was put on the lens of the CMOS camera in order to obtain clear concentration image illuminated by the LLS. The instantaneous distribution of the dye concentration C was calculated by using brightness values of these LIF images. The Dye (Rhodamine-B), the concentration of which is 0.2mg/l, was dissolved in the cavity. At the initial stage of the measurements, the cavity and the main-channel are separated by the plate. After remove of the plate, the dye exchange and transfer motions were captured by the CMOS camera. Borg et al. (2001) pointed out that there exists a linear relation between image brightness and dye concentration under the small concentration conditon($C<0.2$mg/l). This fact was surely recognized in this study by using the present data, and a linear calibration curve was obtained.

3. Currents and turbulence structure

In the present section, the results of velocity measurements using PIV techniques are introduced. Time-averaged velocity distribution, turbulence properties and the coherent structure are examined. Further, the influences of the bed configurations on the hydrodynamic characteristics are also considered.

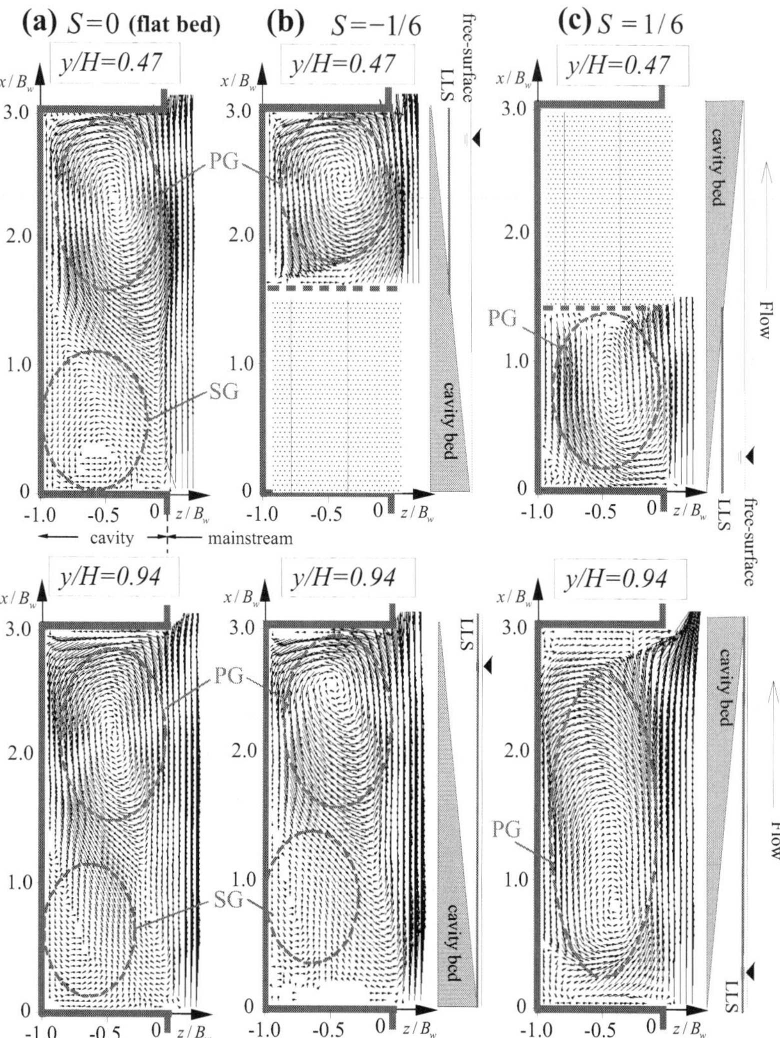

Fig. 2. Time-averaged horizontal velocity vectors

3.1 Time-averaged horizontal velocity components

Fig.2 shows time-averaged horizontal velocity vectors (U,W) at y/H=0.47 and 0.94. At the near free-surface, y/H=0.94, two kinds of large scale gyres with counter rotations are observed in the S =0 and -1/6. Whereas, single gyre structure appears and it covers a whole region of the canopy in the S =1/6. The PG is a horizontal circulation induced by the main-stream, and it was observed, irrespective of the elevation and the geometry of the canopy bed. The secondary gyre (SG) seems to be generated by the MG. In the S =1/6, the contact area between the main stream and the cavity zone decreases downstream, and the inflow from the mainstream toward the side canopy becomes smaller compared with those observed in the S =0 and -1/6. Thus, the position of PG is shifted toward upstream-side of the side canopy, and the sufficient space could be not kept to generate SG in the S =1/6. At the mid-depth layer, y/H=0.47, the PG and the SG are observed in the same manner as those of y/H=0.94 in the flat condition. In contrast the only PG is formed in the mid depth layer in the non-flat conditions.

3.2 Scale evaluations of horizontal gyres

Figs. 3 and 4 show the distributions of vorticity with vertical axis, $\Omega \equiv \partial U / \partial z - \partial W / \partial x$ and delta value Δ at y/H=0.94, for S =-1/6, 0 and 1/6, respectively.

Chong and Perry (1990) considered the Eigenvalue σ of the velocity shear tensor $(\partial \tilde{u}_i/\partial x_i)$ in shear layers. The Eigenvalue equation of 2D shear flow is given by

$$\sigma^2 - P\sigma + Q = 0 \tag{1}$$

in which

$$P \equiv \frac{\partial \tilde{u}_i}{\partial x_i} = 0 \quad \text{and } Q \equiv \frac{1}{2}\left(\left(\frac{\partial \tilde{u}_i}{\partial x_i}\right)^2 - \frac{\partial \tilde{u}_i}{\partial x_i}\frac{\partial \tilde{u}_j}{\partial x_j}\right) \tag{2}$$

$$\Delta \equiv P^2 - 4Q \tag{3}$$

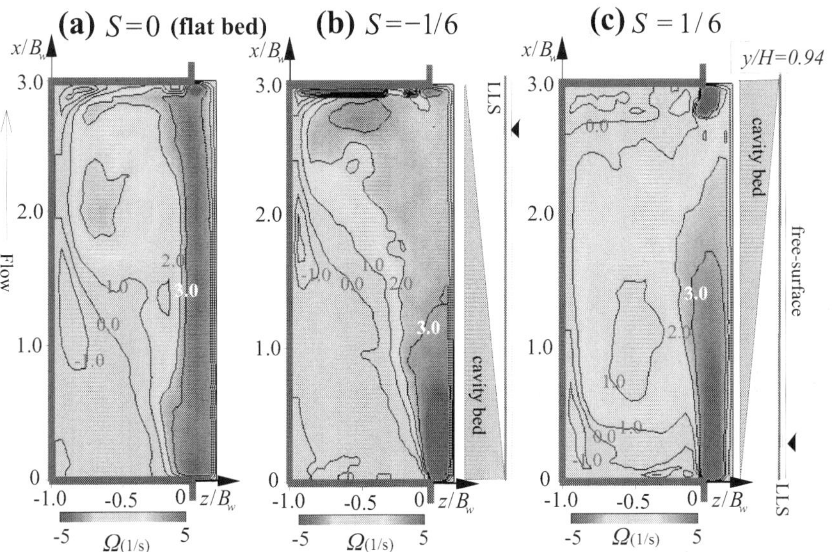

Fig. 3. Distributions of time-averaged vorticity with vertical axis

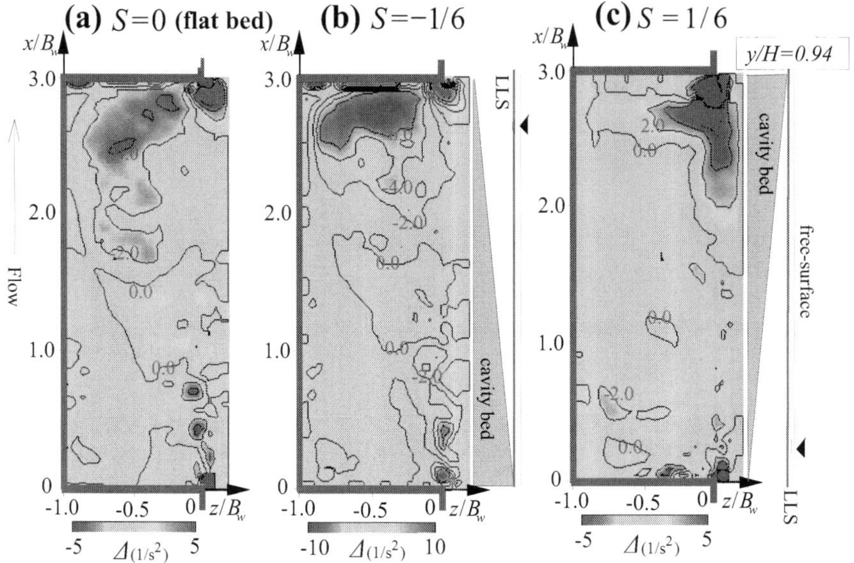

Fig. 4. Horizontal distributions of delta-value

where $\tilde{u}_i$=instantaneous velocity components with i=1 and 2 herein. Complex Eigenvalues occur if the discriminant Δ is negative. Chong and Perry assumed that Δ<0 corresponds to the existence of a vortex core, and verified the validity of such a vortex detection, referred to as the *delta method* in shear layers.

Positive and negative values of the vorticity mean anti-clockwise and clockwise rotations, respectively. It should be noticed that even if the fluid does not rotate, large vorticity appears in the shear layer, where spatial gradient of the velocity components is dominant. Therefore, in the present study, not only the vorticity but also the delta value is used to detect gyre regions and to measure the corresponding areas.

In the flat bed condition (S =0), positive distribution of Ω is observed along the junction, in which streamwise velocity shear $\partial U / \partial z$ is very large and furthermore small-scale shedding vortices are generated periodically by the shear instability. It is noted that the positive and negative regions appears corresponding to the PG and SG as shown in Fig.2. It is found that the delta value becomes negative in these regions.

In S =-1/6, the distributions of the vorticity and delta seem to be similar to those of the flat condition. The positive voriticity region corresponding to the PG shifts toward the downstream-side compared with the flat condition. This property is also observed in the distribution of the delta. This is because the PG is transported toward the downstream-side in the cavity due to the effect of the bed configuration.

In contrast, in S =1/6, the positive vorticity and the negative delta are dominant in the cavity. These properties correspond well with the horizontal distributions of velocity vectors as shown in Fig.2. In the upstream side of the cavity, the negative vorticity and the negative delta appear corresponding to the SG which becomes very small due to the bed configuration effect.

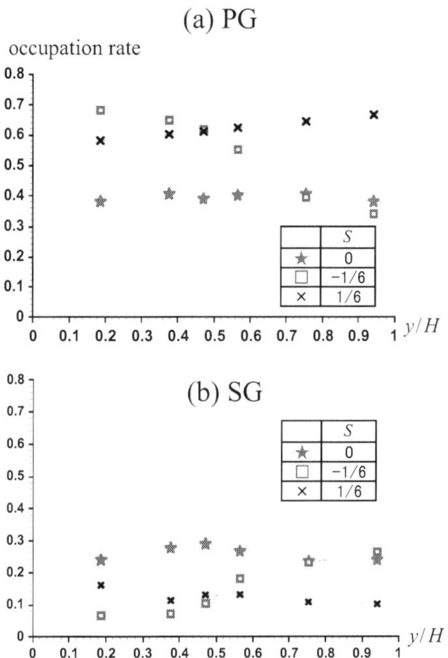

Fig. 5. Area occupation rates of primary gyre and secondary gyre in the cavity

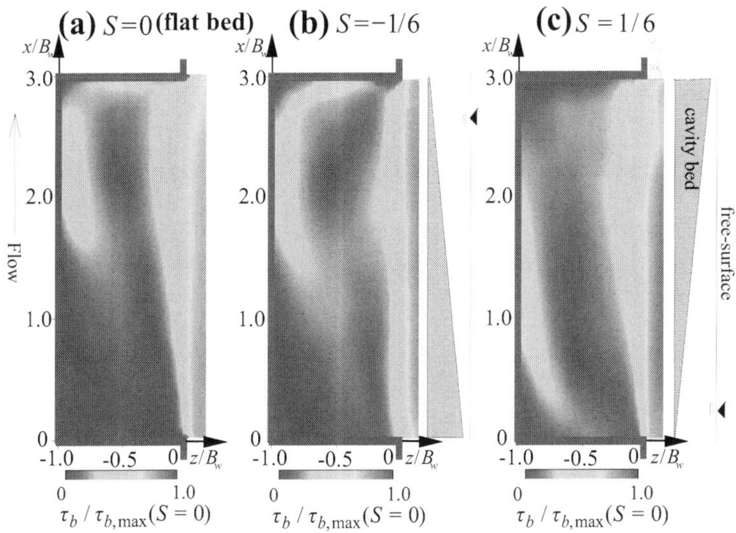

Fig. 6. Distributions of bed shear stress in the cavity

The total area of the PG was calculated by summing the local square, in which the positive vorticity and negative delta appear. In the same way, the total area of SG was calculated by the local square, where negative vorticity and negative delta exist. Fig.5 (a) and (b) show areas of PG and SG normalized by a total cavity area. In the S =0 and -1/6, the area of PG is comparable with that of SG near the free surface. In contrast, in the S =1/6, the area of PG is larger and that of SG is smaller than those observed in S =0, -1/6.

3.3 Distribution of bed shear stress

Figs. 6 shows the distributions of bed shear stress τ_b in the cavity. τ_b was calculated by using the viscosity μ and the velocity gradient, i.e., $\mu\partial\sqrt{U^2+W^2}/\partial y$. For the slope conditions, The LLS was projected pararell to the cavity bed with the 1cm gap between them, and thus, the near-bed velocities parallel to the bed surface (U, W) could be obtained by the PIV. It is found that small bed shear stress zones are observed in the core of the PG. They also appear in the SG region. In S =1/6, the small bed shear stress zone relevant to the PG is shifted toward the upstream-side compared with S =0 and 5. It should be noticed that local sedimentations are promoted significantly in these low-speed regions. In contrast, it is expected that a scour process is observed in the outer side of PG. The present results allow us to understand that the formations of PG and SG play significant roles on the sedimentation in the cavity.

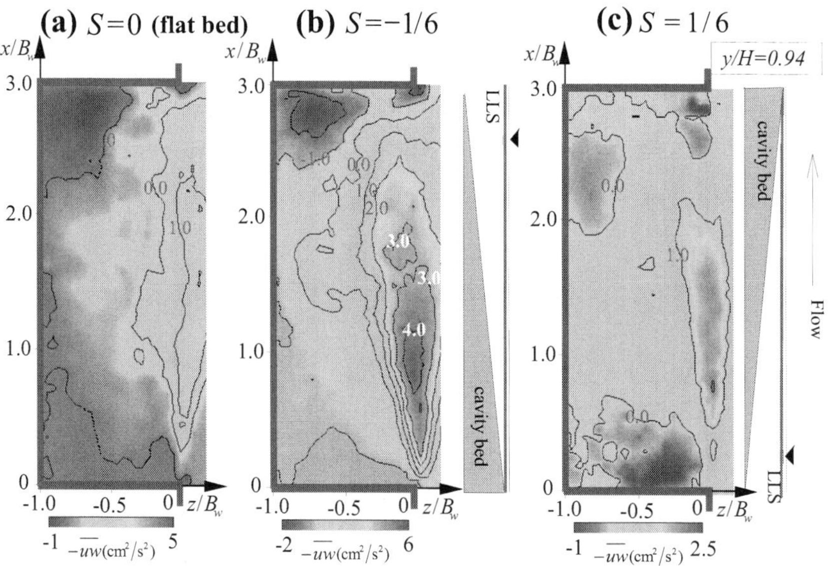

Fig. 7. Horizontal distributions of Reynolds stress

3.4 Distribution of Reynolds stress

Fig.7 shows horizontal distributions of Reynolds stress $-\overline{uw}$ at y/H=0.94. The large and positive Reynolds stress zone is formed along the junction between the mainstream and the cavity. The shear instability induces such a strong turbulence layer irrespectively of the bed configuration. It is found that Reynolds stress becomes largest in the downward slope condition, S =-1/6. The spanwise currents from the cavity toward the mainstream appear the upstream-side of the PG in Fig.2. They are accelerated more nearer the upstream-side of the cavity, where the water depth decreases due to the bed configuration. Thus, larger Reynolds stress may be generated much more in the down slope condition than other bed configurations.

Fig.7 shows spatial variation of Reynolds stress in the junction. A peak Reynolds stress appears at x / B_w =1.0 in the down slope condition, S =-1/6. In contrast, all most constant value is observed except the upstream and downstream walls in the cavity. The Reynolds stress of the downward slope condition (S =-1/6) is twice as large as those of the flat (S =0) and the upward slope conditions (S =1/6), in which same order Reynolds stress is observed.

3.5 Instantaneous velocity properties

Fig. 9 shows time-series of the distributions of instantaneous Reynolds stress $-uw$ and discriptions of horizontal velocity components $(\tilde{u},\tilde{w})$ every 1s for S =0. The elevation is y/H=0.94. At t=0s, the circle A indicates the coherent structure of fluid percel which is transported strongly toward the mainstream. This is called as sweep motion in the present study. The shedding vortex follows the sweep as indicated by the circle B, in which the fluid percel intrudes into the cavity accompanied with the strong positive Reynolds stress. This is called as ejection. These coherent motions are convected downstream at t=1s, and in contrast, the sedding vortex in the circle B is disappearing. A new sweep motion appear as indicated by the circle C. At t= 2s, it is found that a large amont of fluid is transported toward the cavity zone by the sweep (B). Furhter, a shedding vortex (D) is generated together with positve Reynolds stress. This result suggetst that mass transfer process between the mainstream and the cavity is promoted significantly by such coherent strucures.

Figs. 10 (a) and (b) show examples of the instantanesous velocity fields for S =-1/6 and 1/6. In S =-1/6, the sedding vortex is observed as indicated by the circle A, and the sweep motion (B) is formed. The sweep seem to be larger than that obserbed in S =0. It is found that the PG is also visualized even in the instantaneous velocity field. In S =1/6, the ejection is observed as indicated by circle A. However, the sedding vortex is not be regonnized. This result correspond well to the finding that Reynolds stress is smaller in the S =1/6 than the S =0 and -1/6 as shown in Fig.7. It is therefore found that the bed configuration has striking impacts on turbulence production in the cavity.

Therefore, horizontal and vertical LLSs are projected simultaneously, and these illuminated planes are taken by dual CMOS cameras. Fig.11 shows examples of time series of instantaneous velocity vectors in the horizontal and vertical planes every 0.6 seconds. The positions of these planes are y/H=0.94. (near free-surface) and z/B=-0.2 (near the junction between the main-channel and the cavity). The white broken line of horizontal views and the red one of vertical views indicate the positions of vertical and horizontal illuminated planes, respectively. The contours mean distributions of instantaneous Reynolds stress $-uw$ and vertical velcotiy $\tilde{v}$ for the horizontal and vertical views, respectively.

In the S =0, red circle indicates large Reynolds stress observed at the junction near the upstream side of the cavity. It is inferred that shedding vortex produces locally the large Reynolds stress distribution. Further, the Reynolds stress seems to intrude toward the cavity. It is found by comparison of the horizontal and vertical views that upward currents appear in the intruding region of large Reynolds stress. The downward current follows the upward one, and it is thus suggested that strong 3-D structure is formed near the junction, when shedding vortex enter to the cavity. In the S =-1/6, a similar 3-D structure is observed in the same manner as one of S =0. Particularly, after the transfer of Reynolds stress toward the cavity, it is accompanied with downward current in the downstream side of the cavity. In the S =1/6, upward currents are dominant in the whole area of the vertical plane. Of particular significance is that larger upward flows are generated than the surrounding positions at the inrushing time of sweep motion.

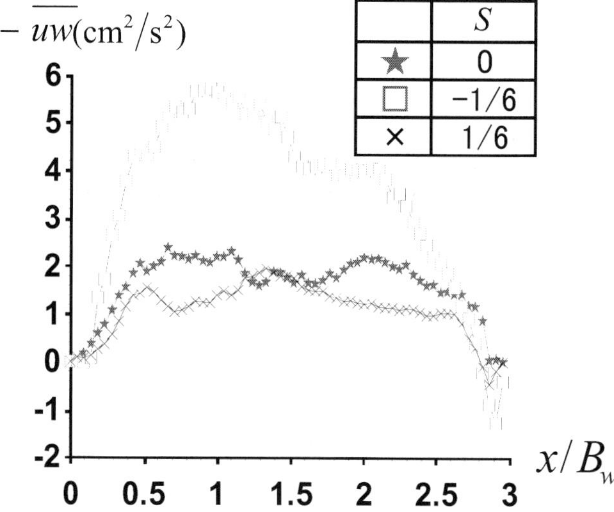

Fig. 8. Comparison of Reynolds stress profiles among different bed configurations

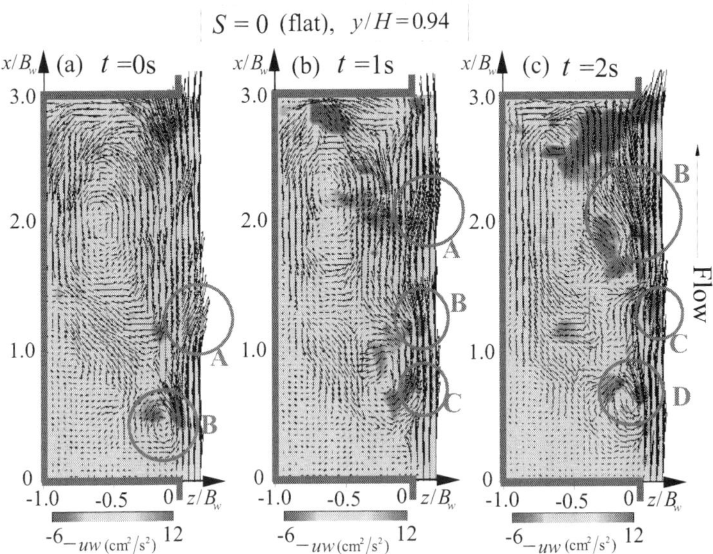

Fig. 9. Time-series of the distributions of instantaneous Reynolds stress and horizontal velocity components for the flat bed configuration

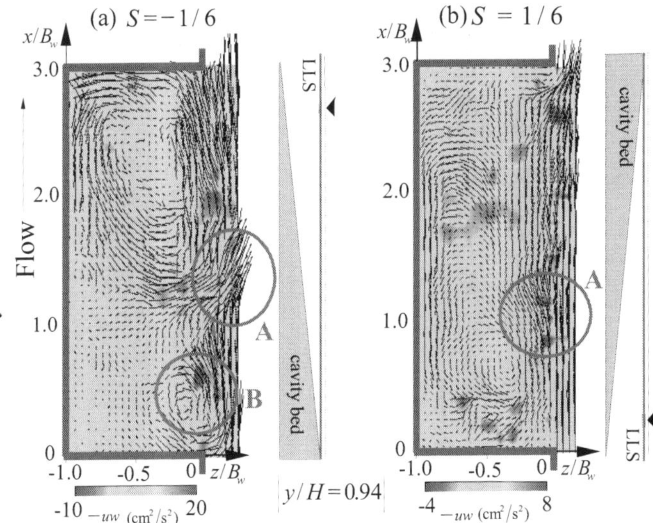

Fig. 10. Distributions of instantaneous Reynolds stress and horizontal velocity components for the slope bed configurations

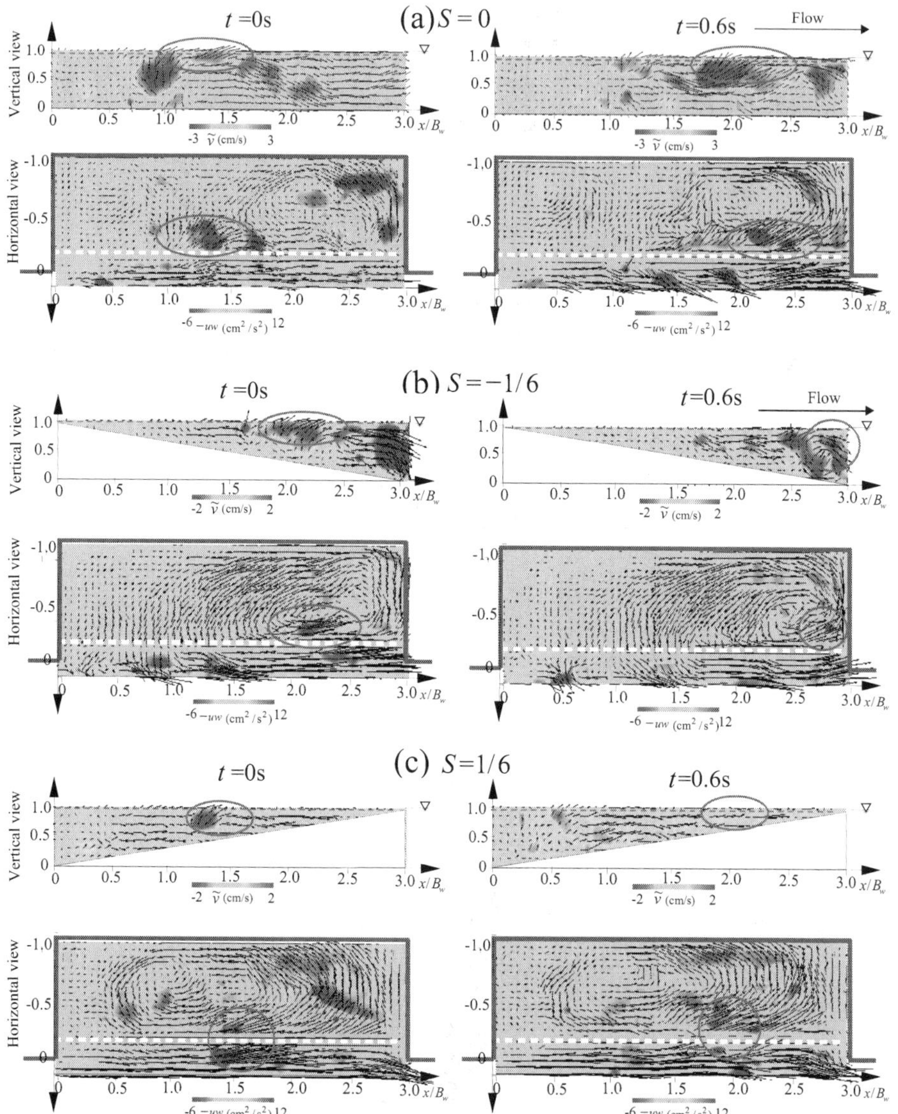

Fig. 11. Instantaneous velocity vectors in the horizontal and vertical planes

In Figs.9 and 10, only horizontal currents are focused, and thus, there remain many uncertainties about vertical component of velocity. Particularly, it is very significant to reveal how horizontal vortices and gyres induces upward and downward flows.

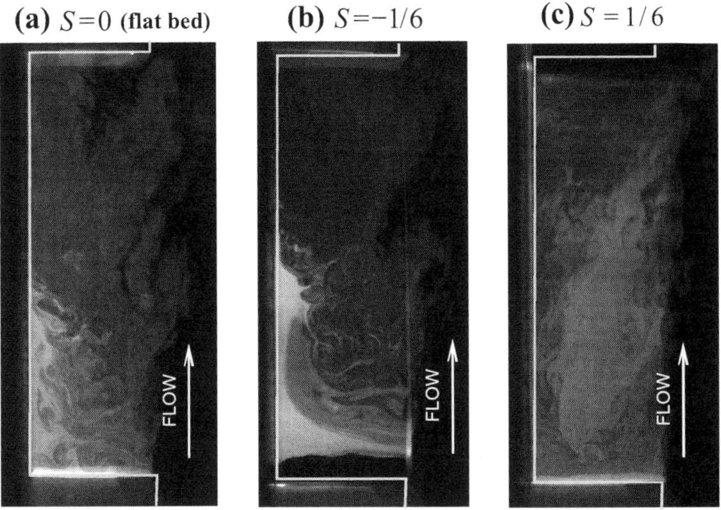

Fig. 12. Dye transport in different types of bed conditions

4. Mass transfer properties between mainstream and cavity

The coherent turbulent structures such as shedding vortex, sweep and ejection play significant roles on mass and sediment transfers. However, there are many uncertainties about relation between the turbulence and mass transfers. In this section, exchange properties of dye concentration and effects of sedimentation on them are considered.

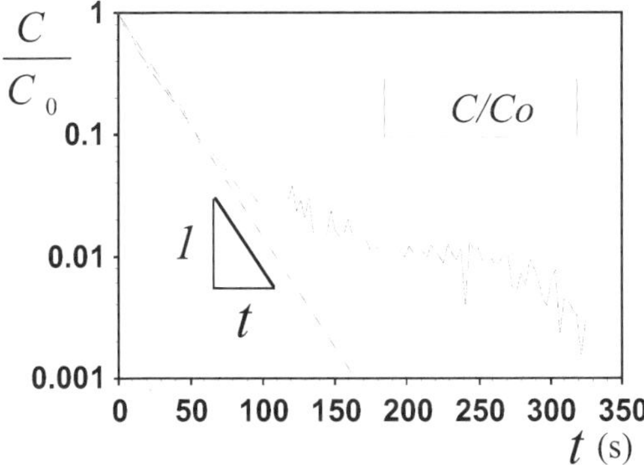

Fig. 13. Time-variation of dye concentration for the flat condition

4.1 Dye visualization of mass transfer

Fig.12 shows instantaneous distribution of dye concentration at y/H=0.94. The mass exchange is promoted significantly in the PG region, and, in contrast, the dye is accumulated in the SG region for S =0 and -1/6. These results coincide well with those of previous experiments conducted by Uijttewaal et al and Weitbrecht et al. Whereas, for the S =1/6, although horizontal area of PG covers the almost whole cavity zone, circulation velocities may comparatively weak, and dye remains in the PG region. Of particular significance is that mass transport in the side cavity depends strongly on the formation properties of the horizontal gyres.

4.2 Evaluation of transfer coefficient

Transport equation for concentration exchange between the mainstream and the dead water zone such as side cavity, is follows as;

$$DC/Dt=(kU_m/B_w)(C-C_0) \tag{4}$$

C is instantaneous concentration, C_0 is initial concentration value and k is transfer coefficient given by following equation.

$$k=B_m/(UT) \tag{5}$$

T is the specific time when C/C_0 becomes 1/10. It is calculated by using time-variation of C during linearly decrease stage.

Fig.13 shows time-series of concentration at y/H=0.94. for the flat-bed condition. The concentaration decreases linearly from t=0s to 50s, and thereafter, time-variation becomes smaller. As pointed out by Uijttewaal et al, the linear-decrease corresponds to the mass exchange promoted by the PG, and incontrast, after the contribution of the PG, the time-variation becomes small due to stagnating induced by the SG. This suggests that the PG plays significant rolls on the mass transfer beneath interface between the main-channel and the dead water zone. Therefore, the transfer coefficient k was evaluated by applying the time-gradient of concentration to Eqs.(4) and (5).

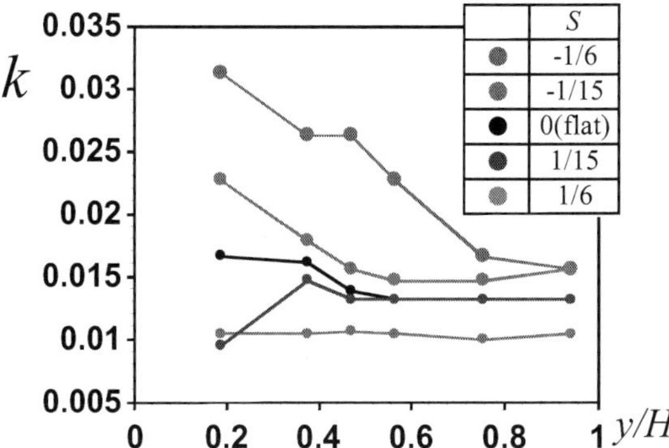

Fig. 14. Vertical profile of transfer coefficient

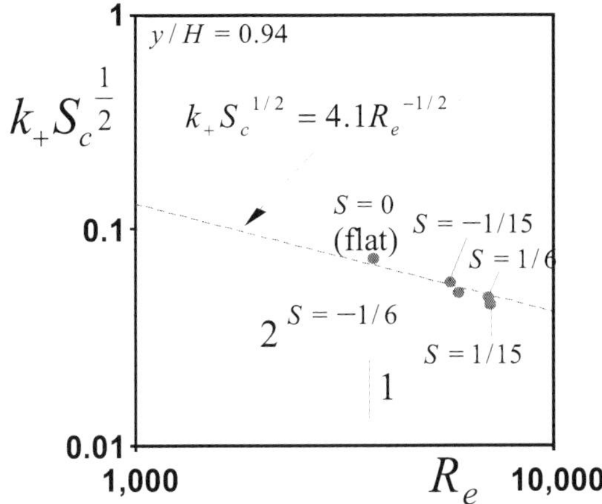

Fig. 15. Relation between transfer velocity and Reynolds number

Fig.14 shows vertical profile of transfer coefficient. It decreases toward the free-surface in the S <0 series, and this is because area of the SG, in which mass is accumulated as mentioned above, becomes larger nearer the free-surface. In the flat bed case, the structures of the PG and SG are almost constant in the vertical direction, and vertical variation of k is comparatively small. In the S >0 series, the single gyre is formed in the whole depth region, and dye concentration is captured in the PG region without transfer to the main stream. It is thus found that variations of k are smaller than those observed in the S =0 and S <0 .

4.3 Parameterization

In the studies on gas transfer phenomena, reasonable parameterizations are conducted using various physical models. In the present study, a following large-eddy model, in which effects of large-scale turbulence can be considered, is used for our measured data.

$$k_+S_c^{1/2} \propto R_e^{-1/2} \tag{6}$$

in which, $k_+ \equiv kU_m / u'_c$ is transfer velocity normalized by velocity-scales. u'_c is space-averaged turbulence intensity along the boundary between the main-channel and the side-cavity. $\mathrm{Re} \equiv u'_c L_p / \nu$ is Reynolds number. L_p is the length scale of PG. Sc is Schmidt number and Sc is chozen as 1. Fig.15 shows relation between the normalized transfer velocity and Reynolds number. It is found that the present data coincide well with the large-eddy model. So, it is suggested that when the length scale of PG and turbulent intensity are accuraltely given, transfer velocity could be predicted reasonably, irrespective of bed configrations.

5. Conclusion

In this chapter, the open-channel flows with dead water zone were highlighted, and the effects of sedimentation on horizontal circulation structure and related mass-exchanges were considered on the basis of the measured data obtained by the laboratory experiments.

The both of MG and SG are observed in the flat-bed and downward slope conditions, and in contrast, the only MG appears in the upward slope condition. It was found that these formations of the horizontal circulations play significant roles on the mass transfer properties between the mainstream and the cavity. So, we should consider the effects of the bed configuration, when the turbulence structure and the related mass transfer phenomena are evaluated in detail.

6. References

Akkerman, G.J., van Heereveld, M.A., van der Wal, M. & Stam, J.M.T.(2004). Groyne optimization and river hydrodynamics, *Proc. Riverflow* 2004, Napoli, on CD-ROM.

Borg,A., Bolinger,J. & Fuchs,L.(2001). Simultaneous velocity and concentration measurements in the near field of a turbulent low-pressure jet by digital particle image velocimetry-planar laser-induced fluorescence, *Exp fluids* 31, 140-152.

Chong, M.S. & Perry, A.E. (1990). A general classification of three-dimensional flow fields. *Phys. Fluids* A, 2, 765-777.

Kadota, A. & Suzuki, K. (2010). Local scour and development of sand wave around a permeable groyne of stone gabion, *Environmental Hydraulics, Christodoulou & Stamou (eds.)*, 807-812.

Tominaga, A. & Sakaki, T. (2010). Evaluation of bed shear stress from velocity measurements in gravel-bed river with local non-uniformity, *Riverflow 2010, Dittrich, Koll, Aberle & Geisenhainer (eds)*, 187-194.

Uijttewaal, W.S.J., D. Lehmann & van Mazijk, A. (2001). Exchange processes between a river and its groyne fields: Model experiments, *Journal of Hydraulic Engineering*, ASCE, 1273(11), 928-936.

Weitbrecht, V., Socolofsky,S.A. & Jirka, G.H. (2007). Experiments on mass exchage between groin fields and main stream in rivers, *Journal of Hydraulic Engineering*, ASCE, 134(2), 173-183.

14

Sediment Transport and River Channel Dynamics in Romania – Variability and Control Factors

Liliana Zaharia[1], Florina Grecu[1],
Gabriela Ioana-Toroimac[1] and Gianina Neculau[2]
[1]*University of Bucharest, Faculty of Geography*
[2]*National Institute of Hydrology and Water Management*
Romania

1. Introduction

Sediment transport and river channel dynamics are the result of the complex interaction between natural and human factors, at both local and regional scale. The study of sediment transport and river channel dynamics may be an important way to better know and understand the mechanisms that rule the functioning of fluvial system, allowing forecasts of its future evolution to be made and appropriate adaptation measures to be taken by society in front of the risks related to the fluvial dynamics and sediment transfer.

The purpose of this chapter is to present specific aspects concerning the sediment transport and river channel dynamics in Romania and to reveal the role of various control factors. Our contribution consists mainly in regional analyses that highlight the following aspects: i) the spatial and temporal variability of suspended sediment transport; ii) the relationships between suspended sediment yield/load and some control variables (precipitation, water discharge, catchment's characteristics, and human activities), and iii) the vertical and lateral dynamics of selected river channels.

The chapter summarizes a part of the significant results of our previous studies and some more recent researches in the field of sediment transport and channel dynamics in Romania. It also includes new issues and approaches, based especially on case studies, which further develop the proposed topic. The analyses focus preeminently on two areas: the Carpathian's Curvature region, and the central part of the Romanian Plain (Fig. 1A). We have chosen the first area because it has the highest sediment yield in Romania: over 20 - 25 t ha^{-1} yr^{-1}, meaning more than 10 times the average sediment yield on a national scale, which is about 2 t ha^{-1} yr^{-1} (Mociorniță & Birtu, 1997). The central part of the Romanian Plain is characterized by an intense lateral dynamics of the river channels, making it vulnerable to the risks induced by fluvial dynamics, because of the economic importance of the area and the significant density of population and settlements. The data used and the methodologies employed are mentioned in the text.

This chapter is structured in four main parts concerning: i) previous researches on sediment transport and fluvial dynamics in Romania; ii) sediment transport and control variables; iii) river channel dynamics and iii) human impact on sediment transfer.

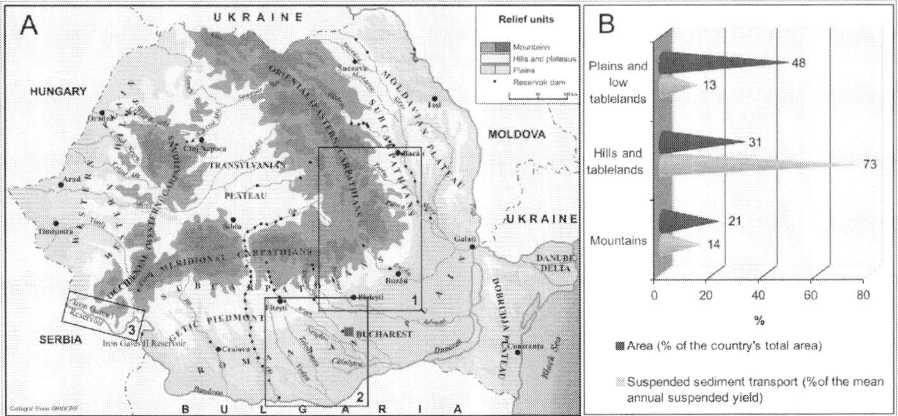

Fig. 1. A The main morphological units in Romania and the location of the study areas: 1: Carpathian's Curvature region; 2: Central part of the Romanian Plain; 3: Iron Gates Region. B. The distribution of suspended sediment yield over the Romanian territory at the scale of the major landform units.

2. Overview on previous researches on sediment transport and river bed dynamics in Romania

In Romania, although researches in the fields of hydrology and fluvial geomorphology date back to the early 20th century, the systematic gauging of sediment load within the national hydrometric network started after 1950. Initially, the recordings were only concerned with suspended sediment concentration (turbidity) and sediment load. In the 1960s, however, the efforts were also aimed at the systematic investigation of sediment granulometry and bed load transport. Yet, over the time, most of the recordings have dealt with suspended sediment load, while the measurements of bed load amounts have been fewer. An important role in the scientific research in this field was played by the specialized national institutes (of hydrology, pedology, geography, etc.), the universities and the "Stejarul" research station in Piatra Neamţ.

Significant researches on sediment transport. One of the first significant synthesis studies on suspended sediment load of the Romanian rivers belongs to Diaconu (1964), who highlighted the vertical zonation of suspended sediment yield. Seven years later, in 1971, Diaconu published a more ample study, which became a landmark for all subsequent investigations concerning river sediment transport in the country. In it, he presented the results given by the processing of suspended load data recorded between 1952 and 1967 on 202 catchments. For the first time in Romania, the author included maps showing the spatial variability of suspended sediment yield and suspended sediment concentration (turbidity) on a national scale. In 1972, Ujvári identified the relationships between the rivers turbidity and the suspended sediment load on the one hand, and between the former and the elevation, on the other hand, under different geological and morphological conditions. Another reference work dealing with the investigation of suspended sediment load in Romania was accomplished by Moţoc in 1984. The study emphasized the contribution of erosion processes and land use to suspended load transport along the Romanian streams. In

1987, Mociorniță and Birtu updated the previous information regarding the spatial and temporal variability of suspended load in Romania. The results and discussions relied on the processing of the data recorded between 1950 and 1984 by the national gauging network. These datasets also included information about the exceptional flood events, which occurred in 1970, 1972, 1975 and 1979, and about the alterations of the flow and sediment load regimes induced by the increased number of reservoirs.

A more recent synthesis study that highlighted the relationship between suspended sediment yield and the catchment area for different morpholithological conditions of Romania was the one conducted by Rădoane (2005). Many studies (methodological and at a regional scale) dealing with sediment origin, transport, alluvial budget and granulometry, as well as with the factors controlling the river sediment load have been published in recent decades. The most relevant ones are authored by Dumitriu (2007), Ichim (1992), Ichim & Rădoane (1990), Ichim et al. (1998), Olariu et al. (1999), Rădoane & Rădoane (2001, 2005), Rădoane et al. (1998, 2003a, 2006), Roşca & Teodor (1990), Teodor (1992, 1999), Zaharia (1998), Zaharia & Ioana-Toroimac (2009), Zăvoianu & Mustățea (1992). Important papers on sediments and fluvial processes are included in the four volumes of the Symposium "Proveniența şi efluența aluviunilor" ("Sediments Origin and Outflow") held in Piatra Neamț (1986, 1988, 1990 and 1992).

Researches on river bed dynamics. In Romania, the studies on riverbed dynamics started in the 1960s. Of the first papers, those accomplished by Diaconu et al. (1962) and Urziceanu (1967), clearly stand out from the others. For instance, the study authored by Urziceanu is important, inasmuch as it identifies on a national scale some zonal relations among the channel dynamics, the sediments diameter and the water velocity based on an index expressing the channels mobility. In his turn, Diaconu (1971) tackles the channels variations by analyzing the influence of water flow on the hydraulic elements of the channel's cross section (area, mean and maximum depth). Other important contributions to the study of the channel vertical dynamics and the river long profiles are those of Ichim et al., (1989), Pandi & Sorocovschi (2009), Rădoane et al. (1991, 2003b, 2008a, 2010), Rădoane & Rădoane (2007), etc. Aspects concerning fluvial dynamics are presented in the works authored by Cocoş & Cocoş (2005), Feier & Rădoane (2007), Ioana-Toroimac et al. (2010), Rădoane et al. (2008b), and Grecu et al. (2010 a, b, c).

Given its practical significance in the context of sustainable development, the hydrogeomorphological dynamics has either turned into a major issue or has only been tangentially tackled in many doctoral theses, of which are worth mentioning the papers accomplished by Popa-Burdulea (2007), Ioana-Toroimac (2009), Ghiță (2009), Cârciumaru (2010) and Văcaru (2010).

3. Suspended sediment transport – the role of control variables

The spatial and temporal variability of sediment transport is controlled by a combination of different variables, including the natural features of the catchments and human activities. Because these variables generally act in a synergistic manner, sometimes it may be very difficult to isolate the effect of a particular one (Charlton, 2008).

3.1 Spatial variability of suspended sediment transport in Romania

The Romanian rivers, except for the Danube, carry, on the average, 1550 kg s^{-1} of suspended sediments every year, which means a sediment yield of 48.9 million t yr^{-1} and a specific sediment yield of 2.06 t ha^{-1} yr^{-1} (Mociorniță & Birtu, 1987). These values account for a relative high erosion rate in Romania, which makes it necessary that adequate measures to reduce it should be taken in order to mitigate the negative consequences generated by this phenomenon.

The analysis of the distribution of liquid flow and suspended sediment yield at the scale of the major landform units in Romania shows that 73% of the mean annual suspended sediment yield come from the hilly and piedmontane areas (with elevations of 300 to 800 m), which account for 31% of the Romanian territory and 24% of the mean annual water volume of the rivers (Fig. 1B). These areas are extremly favourable for sediment production. In the mountain realm, as well as in the plains and the low tableland regions the percentage of suspended sediment yield is only 14% and 13% respectively of the mean annual sediment yield, while these areas hold 21% and 48% respectively of the entire Romanian territory. One should also notice that the mountain realm is responsible for two-thirds (66%) of the mean annual water volume of the rivers, while the plain areas contribute by only 10% (Mociorniţă & Birtu, 1987; Pişota & Zaharia, 2002).

The largest amounts of suspended load are carried by the rivers draining the Subcarpathian region, where the values of the mean specific suspended sediment yield exceed 10 t ha^{-1} yr^{-1}. As far as the maximum values are concerned, these are found in the Curvature Subcarpathians (more than 20 - 25 t ha^{-1} yr^{-1}) (Fig. 2A). The high erosion rate in this area, which is controlled by natural (geological, morphological, climatic and hydrological) and anthropogenic causes (massive deforestations), is mirrored by the extremely high values of water turbidity, which exceeds 25,000 g m^{-3} (Fig. 2B). In the Carpathian realm and the plain areas, the values of specific suspended sediment yield are less than 0.5 t ha^{-1} yr^{-1}, while water turbidity keeps below 250 g m^{-3}.

The values of specific sediment yield increase with the elevation from the sea level to 200 - 600 m a.s.l., where they reach the maximum. From here upwards, the values generally decrease as the elevation grows. At low altitudes, although the rocks are crumbly, the faint slopes and low water velocities hinder the erosion. At high elevations, even though the steep grades encourage erosion, the protective vegetal cover and especially the hard rocks are responsible for the low suspended load values. The most favorable erosion conditions, which explain the high values of suspended sediment transport, are encountered between 200 and 600 m a.s.l. (Diaconu & Şerban, 1994).

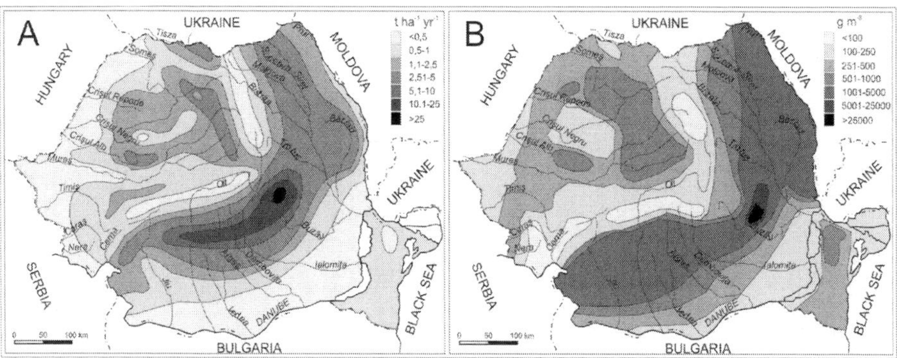

Fig. 2. The map of suspended sediment yield (A) and river turbidity in Romania (B) (according to Mociorniţă & Birtu, 1987, for A, and to Diaconu, 1971, for B)

The year-to-year variability of suspended sediment load is more significant than the water discharge. The study accomplished by Mociorniță & Birtu (1987) at the scale of 10 catchments having different geographical conditions (for the period 1950 - 1985) shows variation coefficients for suspended sediment load ranging from 0.5 to 1.12, which is two to three times higher than those of the mean annual water discharge. As far as the yearly variability of the mean sediment transport is concerned, this is mainly determined by the variability of precipitation regime and water flow under the specific geographical conditions of each drainage basin. For Romania, the highest suspended sediment transport occurs during the interval March - July, when the abundant precipitation combined with the spring snow melting, on the one hand, and the summer rain showers, on the other hand, encourage sediment supply. This happens because the erosion processes along the slopes and within the channels intensify when water discharge increases. During this interval, the percentage of the mean monthly suspended sediment load can exceed 20 - 25% of the mean annual volume. The lowest suspended sediment loads (less than 5% of the mean annual volume) occur in autumn, when both precipitation and water discharges are low (Mociorniță & Birtu, 1987).

3.2 The Carpathians' Curvature region – the area with the most important suspended sediment transport in Romania

The analysis takes into account the external part of the Carpathians' Curvature region, which records the highest values of suspended sediment yield in Romania. The study area includes two distinct morphological units: the Carpathians (lying to the north-northeast) and the Subcarpathians (which are found east-southeast from the Carpathian area) (Fig. 3).
One of the main factors that encourage sediment formation and sediment transport is **geology**. In the Subcarpathian region, the crumbly sedimentary rocks belonging to the Miocene and Pliocene molasse prevail, while in the Carpathian area, Cretaceous and Paleogene flysch formations are common. They include a mosaic of rocks, like sandstones, marls, limestones, conglomerates, marly-sandy schists, etc (Mutihac et al., 2007).
The Carpathian realm is affected by uplift movements of 2 - 3 mm per year, whereas the Subcarpathian area rises by 0.5 to 2 mm per year (Cornea et al., 1979). These movements influence the channel dynamics and consequently the sediment transport. Besides, because the area has a high degree of seismic activity it is often rattled by earthquakes, some of them with magnitudes exceeding 6 - 7 degrees on the Richter scale. Of the most powerful events of this kind, which occurred during the last century, one can mention the earthquakes of 10 November 1940 and 4 March 1977, with magnitudes of 7.4 and 7.2 respectively on the Richter scale. The shock waves generated by earthquakes may trigger various relief shaping processes, such as landslides, mudflows, and collapses, which bring sediments into the river channels (Bălteanu, 1983; Zaharia, 1999).
A major control for the erosion processes and sediment transport is represented by **climatic conditions** in general, and by precipitation in particular. The annual amounts of precipitation in the mountain area have values ranging from 70 to 1000 mm in the mountain realm and 600 to 700 mm in the Subcarpathian area (Zaharia, 2005). The high erosion rate, mirrored by the high amounts of suspended sediment load, is explained by the precipitation regime, corroborated with the lithological conditions. In June and July, due to the torrential regime of precipitation the maximum amounts fallen in 24 hours may exceed 100 or even 250 mm (Administrația Națională de Meteorologie, 2008; Bogdan & Mihai, 1981).

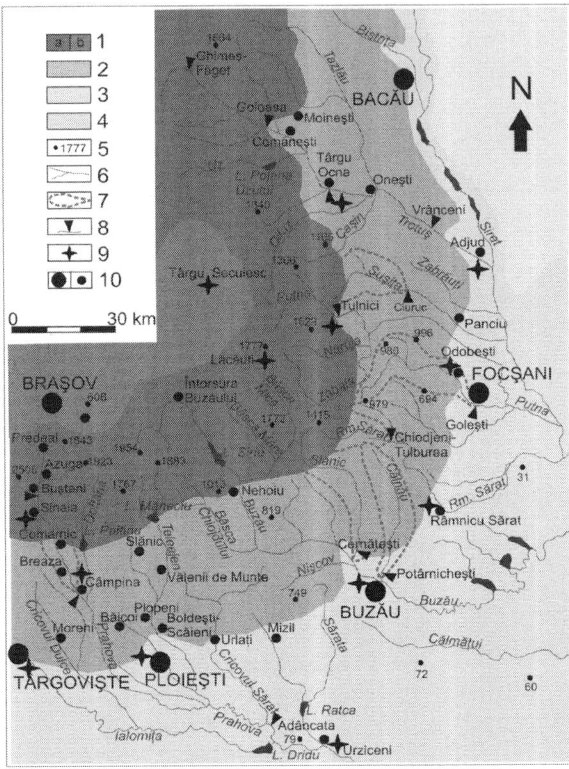

Fig. 3. The Outer Carpathians' Curvature: location of the study catchments, and of hydrometric and meteorological stations. 1: Carpathian area (a: mountainous area; b: depression area), 2: Subcarpathian area; 3: Tableland region; 4: Plain area; 5: Elevation (m a.s.l.); 6: Drainage network; 7: Watershed boundary of the study catchments; 8: Hydrometric station; 9: Weather station; 10: Cities.

Suspended load transport is closely related to **water discharge**. Throughout the outer area of the Carpathians' Curvature, the mean specific discharge varies between 10 and 16 l s^{-1} km^{-2}, in the Carpathian realm and between 3 and 12 l s^{-1} km^{-2}, in the Subcarpathian area. The flow regime is characterized by high discharges in spring (41 - 43 % of the mean annual water volume) and summer (24 - 30%), whilst during the winter and fall the flow is more reduced (less than 15% of the mean annual water volume). The highest sediment amounts are carried during the flood events. These are generated by the abundant rainfalls falling especially in summer and spring (Zaharia, 2005). For instance, during the flood event of July 2005, the Milcov River recorded at Goleşti 94,000 kg s^{-1} of suspended sediments, which is over 5,000 times higher than the mean multiannual sediment load (according to the data provided by the Vrancea Water Management System).

The erosion processes affecting the slopes are controlled to a large extent by **vegetation cover**. In the Carpathian catchments, the degree of afforestation averages 50 - 70%, while in the Subcarpathian area it drops to less than 30 - 40% (according to AQUAPROIECT, 1992).

Massive deforestations took place in the Carpathians' Curvature region in the second half of the 19th century and they continued until the third decade of the 20th century. After 1976, with the implementation of a national program to control soil erosion, significant reforestations associated with anti-erosion measures were accomplished in this area (Zaharia, 1998). After 1989, the restitution of the forest to the former owners was followed by massive and abusive deforestations.

Human activities. The main activities in the area, which control the sediment transport in the study region, are the massive deforestations, pasture expansion, grazing, mining of sand and gravel from river channels, and dam building (in this area there are four important reservoirs and a small impoundment). Because of the specific lithological conditions of the Curvature Subcarpathians, deforestations and overgrazing have induced intense slope erosion processes and landslides, which generated significant amounts of sediments.

The favorable natural features and the human activities have accelerated the erosion processes in the Subcarpathian area and consequently the values of suspended sediment yield are high: over 10 t ha^{-1} yr^{-1} and for some parts even 20 - 25 t ha^{-1} yr^{-1}. In the mountain area, suspended sediment yield is less than 5 t ha^{-1} yr^{-1}. This is explained by the hardness of flysch rocks and the higher degree of aforestation, which mitigate the action of the rather abundant precipitation.

During the year, suspended sediment load varies from month to month as a result of hydrological and meteorological conditions. The rivers carry the largest amounts of suspended sediments in the interval April - July. In the Putna catchment, the percentage of each month within this interval may exceed 20 - 30% of the mean annual suspended sediment yield. The winter months have the lowest share of suspended load (less than 3 - 4% of the total mean annual suspended sediment yield) (according to Zaharia, 1999, for the period 1956 - 1992).

The interannual variability of the mean suspended sediment load is more significant for the river catchments that are mostly developed in the Subcarpathian area, where the values of the variation coefficient (Cv) of the mean annual suspended sediment load are generally higher than 1. As far as the rivers flowing in the Carpathian area are concerned, they have Cv values lower than 1 (Zaharia, 2005).

3.2.1 Relationships between suspended sediment yield and hydrometeorological variables

As mentioned previously, an important part in sediment formation and sediment transport is played by climatic conditions (especially precipitation) and flow discharge features. With this in mind, we will try to reveal the influence of precipitation and runoff on suspended sediment yield.

3.2.1.1 Precipitation role in suspended sediment load

In a previous study (Zaharia & Ioana-Toroimac, 2009) we analyzed the link between the erosion dynamics and precipitation in the Carpathians' Curvature region. Now, we will briefly discuss again the results making at the same time some additional comments. The study was based on the monthly and annual precipitation values recorded at 24 weather stations lying in the Carpathians' Curvature area and in its immediate vicinity during the period 1961 - 2000 (data provided by the National Meteorological Administration - N.M.A.). Likewise, we took into account the water discharge and suspended sediment load (mean monthly and annual values) recorded for seven catchments during the same period (data provided by the National Institute of Hydrology and Water Management - N.I.H.W.M.). Table 1 gives the morphometric and geographical features of the study catchments, whereas Fig. 3 shows their geographical location.

River	Hydrometric station	A (km^2)	H_m (m)	S_m (degree)	FR (%)	CMP (1961 - 2000)*
Şuşiţa	Ciuruc	172	556	2.6	74	634.7
Putna	Tulnici	365	1014	5.6	54	713.5
Milcov	Goleşti	406	422	3.1	55	630
Rm. Sărat	Chiojdeni-Tulburea	177	790	4.5	58	571.3
Câlnău	Potârnicheşti	194	338	2.2	27	544.2
Slănic	Cernăteşti	422	524	3.8	47	569.4
Prahova	Câmpina	484	1124	5.5	58	926.4

* Based on the data provided by the N.M.A.

Table 1. Morphometric and geographical features of the study catchments (according to Zaharia & Ioana-Toroimac, 2009, with additions). A: catchment area; H_m: catchment mean elevation; S_m: catchment mean gradient; FR: forest ratio; CMP: catchment mean precipitation.

By interpolating the mean annual values of precipitation recorded at the 24 weather stations (employing the Natural Neighbor method in Vertical Mapper module from MapInfo soft, version 6.5.) and by overlaying the results on the maps of the study catchments, the mean annual precipitation falling on each of them was estimated. The data regarding the suspended sediment load allowed us to determine the mean specific sediment yield, computed as a ratio of suspended sediment load, expressed in t yr^{-1}, and the catchment area (in hectares). Table 2 shows data on water discharge and suspended sediment load for the study catchments.

River	Hydrometric station	Q_m* (1961-2000)		Qmax*/year (m^3 s^{-1})	R (1961-2000)		CVQ_m	CVR
		(m^3 s^{-1})	(l s^{-1} km^{-2})		(kg s^{-1})*	(t ha^{-1} yr^{-1})		
Şuşiţa	Ciuruc	1.39	8.08	491/1991	3.33	6.1	0.51	1.17
Putna	Tulnici	4.75	13.0	333/1971	3.92	3.38	0.36	0.93
Milcov	Goleşti	1.50	3.69	560/1970	18.4	14.3	0.63	1.01
Rm. Sărat	Chiojdeni-Tulburea	1.51	8.53	335/1991	9.19	16.4	0.47	0.94
Câlnău	Potârnicheşti	0.392	2.02	146/1972	8.27	13.4	0.77	1.19
Slănic	Cernăteşti	1.38	3.27	410/1975	15.4	11.5	0.47	1.23
Prahova	Câmpina	8.27	17.1	369/1988	10.9	7.1	0.25	0.95

* Based on the data provided by the N.I.H.W.M.

Table 2. Data on water discharge and suspended sediment load for the study catchments (according to Zaharia & Ioana-Toroimac, 2009, with additions). Q_m: mean multiannual discharge; R: mean multiannual suspended sediment load; CVQ_m: variation coefficient of the mean annual discharge; CVR: variation coefficient of the mean annual suspended sediment load.

The analysis of the linear correlations between the mean annual precipitation fallen over the catchment and the mean suspended sediment yield highlights significant statistical links (for an error risk α = 0.02, according to the Bravais-Pearson statistical test), despite the fact that correlation coefficients are relatively low (between 0.42 and 0.64) (Table 3). The correlation between the mean multiannual suspended sediment yield and the mean multiannual precipitation fallen over the catchment accomplished for the seven study catchments shows a correlation coefficient of 0.57, which points at a relatively low connection between the two parameters.

In order to identify the precipitation regional trends and the suspended sediment yield variability (at annual scale) the statistical method of Principal Component Analysis (PCA), applied with turnover, was employed (with XLSTAT 5.1. soft). This method allowed the determination of the regional synthetic indexes for the catchment mean precipitation (PC1P), as well as for the mean specific suspended sediment yield (PC1r). These indexes correspond to the information retained by the first principal component. The Mann-Kendall test was applied to establish the statistical significance of the temporal variability trends of the two regional synthetic indexes (PC1r and PC1P). In order to identify the intensity of the relationship between the mean specific suspended sediment yield and the precipitation variability on regional scale, the linear correlation between the two indexes was accomplished, and correlation and determination coefficients were analyzed. The statistical relevance of the correlations was established by using the Bravais-Pearson test (Chadule, 1997).

River	Hydrometric station	Correlation coefficients between suspended sediment yield and:		
		CMP	Q_{max}	Q_m
Şuşiţa	Ciuruc	0.64	0.78	0.69
Putna	Tulnici	0.62	0.72	0.62
Milcov	Goleşti	0.42	0.81	0.8
Rm. Sărat	Chiojdeni-Tulburea	0.46	0.56	0.78
Câlnău	Potârnicheşti	0.54	0.54	0.83
Slănic	Cernăteşti	0.56	0.54	0.74
Prahova	Câmpina	0.55	0.62	0.64

Table 3. Correlation coefficients of the relationships between suspended sediment yield and the hydrometeorological parameters of the study catchments (1961 – 2000). CMP: catchment mean annual precipitation (mm); Q_{max}: maximum annual discharge ($m^3 s^{-1}$); Q_m: mean annual discharge ($m^3 s^{-1}$). (according toZaharia & Ioana-Toroimac, 2009, with alterations)

On regional scale, the linear correlation between the annual synthetic index of specific suspended sediment yield (PC1r) and the annual synthetic index of precipitation (PC1P) has a correlation coefficient (r) of 0.66 (Fig. 4). According to Bravais-Pearson statistical test, for $\alpha = 0.02$ level of significance, the correlation between the analyzed parameters is statistically significant. However, the value of the determination coefficient ($R^2 = 0.45$) shows a low quality of the correlation, which reveals the importance of other local factors in controlling the sediment transport.

Based on the mean monthly and annual amounts of precipitation recorded at the 24 weather stations taken into account, we computed the climatic erosive index for each of them as the ratio between the squared precipitation of the month with the maximum pluviometric value and the total annual amount of precipitation (Pătroescu, 1996). The values of these indexes range from 25 (at Râmnicu Sărat, 152 m a.s.l.) to 39.2 (at Lăcăuţi, 1776 m a.s.l.) (Fig. 6A). The highest values (over 35) are specific for altitudes between 400 and 1700 m a.s.l. (Fig. 6B). The high values of climatic erosive index, in conjunction with the specific lithological conditions and the low degree of afforestation, reflect a great potential for sediment generation.

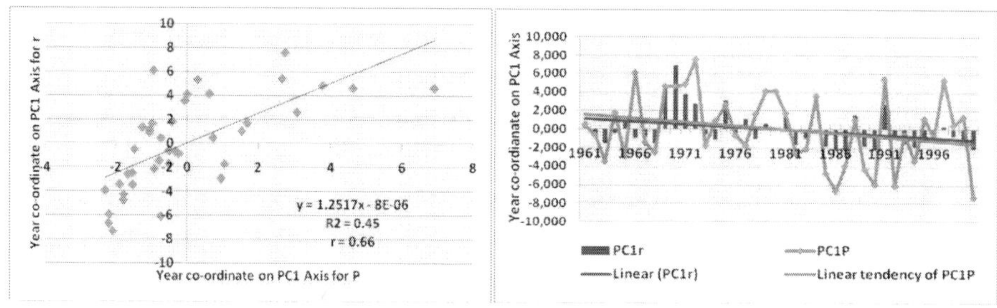

Fig. 4. (left). Linear correlation between the regional index for the annual precipitation (year co-ordinate on PC1 Axis for P) and the regional index for the annual specific suspended sediment yield (year co-ordinate on PC1 Axis for r) in the Carpathians' Curvature region (according to Zaharia & Ioana-Toroimac, 2009, modified)

Fig. 5. (right). Interannual variability of the regional index for the annual precipitation (PC1P) and of the regional index for the annual specific suspended sediment yield (PC1r) in the Carpathians' Curvature region (according to Zaharia & Ioana-Toroimac, 2009, modified)

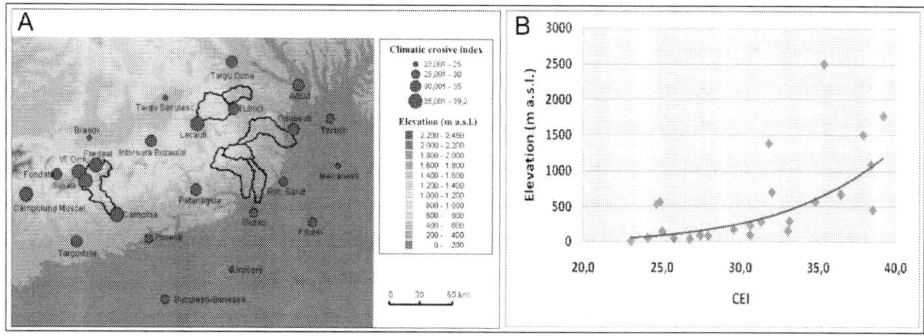

Fig. 6. The variation of climatic erosive indexes in the Carpathians' Curvature region (A) and their correlation with the weather station's elevation (B) (1961 - 2000) (the delimited areas in Fig. A correspond to the studied watershed boundaries).

3.2.1.2 The flow regime and its influence on suspended sediment load

Suspended sediment load is directly influenced by liquid flow. The investigated rivers have mean multiannual water discharges ranging from 0.392 to 8.27 $m^3 s^{-1}$ (corresponding to specific discharges of 2.02 and 8.27 $l s^{-1} km^{-2}$ respectively) (Table 2). During the flood events, water discharges may become 300 times higher than the mean multiannual values and consequently the rivers carry huge amounts of sediments. In the study area, the highest floods within the investigated period occurred in 1970, 1971, 1975, 1981, 1988 and 1991. These were also responsible for the maximum annual discharges (Table 2).

The influence of liquid flow on suspended sediment load is emphasized by the relatively high values (0.54 - 0.83) of the linear correlation coefficients between suspended sediment yield and the flow parameters (mean and maximum water discharge) (Table 3). However, the interannual variability of suspended sediment yield (expressed by the variation coefficient - CVR) is two to three times higher than the mean annual water discharge (Table 2), which mirrors a higher irregularity of sediment supply. Generally, the sediment regime is similar to the flow one. Yet, no simple relationship exists between suspended sediment concentration and water discharge for a given cross-section (Charlton, 2008). Figure 7 shows the monthly and seasonal variability of suspended sediment load of the Slănic River.

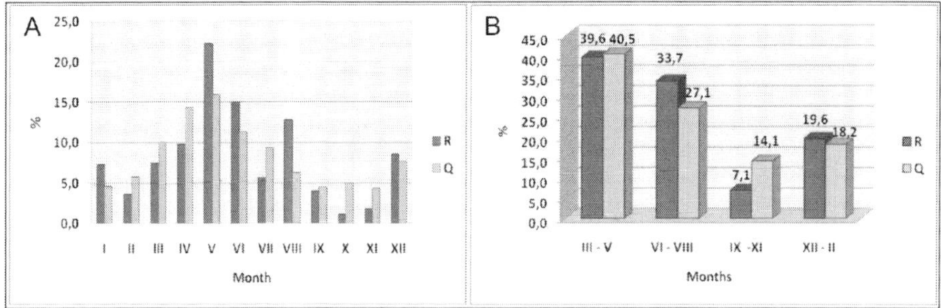

Fig. 7. Monthly (A) and seasonal (B) variations of suspended sediment load (R) and water discharge (Q) for the Slănic River at Cernăteşti gauging station (1971 - 2000) (% from the mean annual total volume)

3.2.2 Relationships between suspended sediment yield and catchment features

In our previous study (Zaharia & Ioana-Toroimac, 2009) we analyzed the outer Carpathian Curvature area from the standpoint of the intensity of the relationships between suspended sediment yield and three features of the catchments: catchment mean elevation, catchment mean gradient and forest ratio. In addition to the present study, we have also taken into account the drainage area.

The accomplished correlations between suspended sediment yield (for the period 1961 - 2000) and the above mentioned characteristics of the catchments, show that the most important role in suspended sediment transport is played by the catchment mean altitude (correlation coefficient = 0.52). Catchment size, catchment mean slope and the degree of afforestation have a lower influence, which is mirrored by the low correlation and determination coefficients (Table 4). The four control variables are inversely related to suspended sediment yield, which is shown by the negative values of the "a" parameter of the regression line equation (Table 4).

Relationship	Correlation coefficient	Coefficient of determination	Regression line equation
r - A	0.25	0.063	y = -6.578x + 386.08
r - H_m	0.52	0.28	y = -31.305x + 1009.2
r - S_m	0.38	0.15	y = -0.1035x + 4.9842
r - Fr	0.37	0.14	y = -0.0106x + 0.6441

Table 4. Correlation and determination coefficients of the relationships between suspended sediment yield and the catchment's characteristics (according to Zaharia & Ioana-Toroimac, 2009, with additions). r: mean multiannual specific suspended sediment yield (t ha^{-1} yr^{-1}; 1961 - 2000); A: catchment area; H_m: catchment mean elevation (m a.s.l.); S_m: catchment mean gradient (degrees); Fr: forest ratio.

Because the number of catchments taken into account in the correlative analysis is relatively low, we consider that the results are only informative. They might be further improved by considering more catchments, but this would require additional information, which for the time being is not available.

4. River channel dynamics

River channel morphology is the result of liquid and solid flow transiting the hydrological system, associated with the catchment features and human influences. These control variables may induce channel adjustments at a certain time scale. For example, lateral and vertical adjustments are noticed at a time scale inferior to 10 years (Knighton, 1984). In this general framework, our analysis focuses on four examples (the Trotuş, Prahova, Câlniştea and Vedea rivers) in order to contribute to a better understanding of the river dynamics and ofthe responsible factors.

4.1 Vertical dynamics of the Trotuş riverbed

The study of the bed processes, conducted by Rădoane et al. (2010), for 60 cross-sections belonging to the gauging stations in the Eastern Carpathians, highlighted the fact that there is no standard model for the river channel evolution. According to this study, there is a large variety of evolutions regarding the vertical change of the bed elevation; incision processes are dominant (in 52% of the cases) in spite of the aggradation processes (29% of the cases). In this context, this section aims to analyze the vertical variations of the riverbed elevation of the Trotuş River (162 km in length, with a 4456 km^2 catchment area, and a 34.6 $m^3 s^{-1}$ annual mean discharge at Vrânceni station), which crosses the Eastern Carpathians and Subcarpathians (Fig. 1A and Fig. 3).
The channel's vertical dynamics is analyzed for a period of 45 - 48 years for four gauging stations (Ghimeş-Făget, in the mountain section; Goioasa, at the border between the mountain and Subcarpathian areas; Târgu Ocna, lying in the Subcarpathian realm, and Vrânceni, placed at the outer extremity of the Subcarpathian section) (Fig. 3). This study extends the analysis period with almost 20 years (unlike the studies of Rădoane & Ichim, 1987, and Rădoane et al., 1991, which analyzed only half of this period), taking into account four gauging stations (unlike the study of Rădoane et al., 2010, which relied on the Goioasa gauging station only). The riverbed dynamics is indicated by the "hp oscillation", which represents the difference between the water stage (measured from the "0" elevation of the gauging station' s staff) and the maximum depth, computed for every water discharge measurement (based on the data provided by the N.I.H.W.M.). The values of this parameter allow comparisons to be made between different years, because the "0" elevation of the gauging station's staff does not change (Chirilă, 2010). The evolution of the "hp oscillation" is analyzed through the linear and polynomial trends.
During the entire investigated period, the Trotuş River revealed a general trend to deepen its channel at Ghimeş Făget (Fig. 8A), where a significant incision was noticed after 1973. Downstream, at Goioasa, despite the general trend of aggradation, several intervals of vertical erosion (1971, 1987, 2004-2005) and aggradation (1967-1971, 1975, 1984-1985, 1990-1994, 2007) were noticed (Fig. 8B). Generally, the aggradation intervals are either synchronous or follow important flood events (like those of 1969, 1970, 1975, 1978, 1979, 1983, 1984, 1989, 1991, 2002, 2004, and 2005). This is due to important degradation processes in the river catchment, which are more intense during the floods. The gap between flood

occurrence and aggradation processes depends on other factors belonging to the river catchment, which influence the time of alluvium evacuation into the main river. At Târgu Ocna (Fig. 8C), where the channel has a downcutting tendency, a significant incision was noticed after the construction of a reservoir dam on its tributary, the Uz, between 1967 and 1973. At Vrânceni (Fig. 8D), one can notice a general aggradation trend, which has been more intense after 1972. In 2005, the intense degradation processes can be put to the account of the strong flood waves that occurred along the Trotuş, of which the most powerful was the one in July.

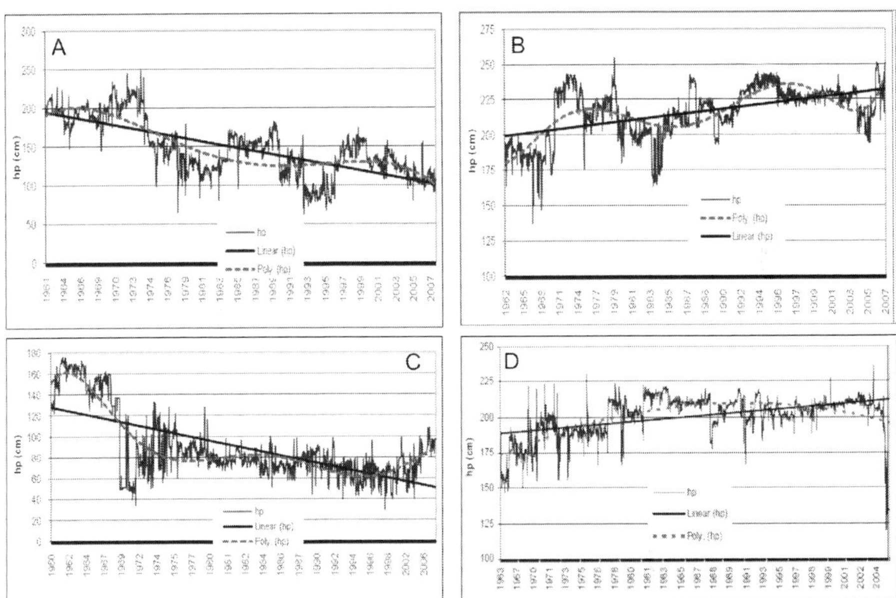

Fig. 8. Temporal variability of the bed elevation (hp) for the Trotuş River and the trends (linear and polynomial) at selected gauging stations: A) Ghimeş-Făget (study period 1961-2007; lying 35 km downstream the river's origin; catchment area, 381 km²; mean elevation of the catchment upstream the gauging station, 1116 m); B) Goioasa (study period 1962-2007; lying 52.4 km downstream the river's origin; catchement, area 765 km²; mean elevation of the catchment upstream the gauging station, 1052 m); C) Târgu Ocna (study period 1960-2007; lying 89 km downstream the river's origin; catchment area, 2091 km²; mean elevation of the catchment upstream the gauging station, 1924 m); D) Vrânceni (study period 1963-2007; lying 126 km downstream the river's origin; catchment area, 407 km²; mean elevation of the catchment upstream the gauging station, 734 m).

4.2 The Prahova's channel lateral adjustments

Several rivers coming down the French and Italian Alps and the Polish Carpathians have recently experienced downward erosion, narrowings and metamorphoses (*sensu* Schumm, 1977) of their braided channels, due to the changes in liquid and solid discharge (Arnaud-Fassetta & Fort, 2004; Bravard et al., 1999; Kondolf at al., 2002; Liébault & Piégay, 2002;

Surian & Rinaldi, 2003; Surian & Cisotto, 2007; Zawiejska & Wyzga, 2009). Within this general framework, this section aims to analyze the lateral dynamics of the braided channels. Because of the lack of information concerning the bedload sediment discharge, a secondary purpose is to provide an overview of the indirect responsible factors that influence this evolution, be they natural or human.

The study watercourse is the Prahova River (193 km long, with a catchment area of 3754 km^2, and a mean annual discharge of 8 $m^3 s^{-1}$ at Câmpina and 27 $m^3 s^{-1}$ at Adâncata), which comes down the Carpathians and crosses the Subcarpathians and the Romanian Plain (respectively the piedmont plain of Ploieşti and the subsidence plain of Gherghiţa-Sărata) (Fig. 9A). This evolution is analyzed starting from a diachronic study, which is based on several documents: Ordnance Survey maps of 1895-1902 (of scale 1:20000), topographic maps of 1953 - 1957 and 1977 - 1980 (of scale 1:25000) and aerial photographs of 2003 - 2005 (of scale 1:5000). These documents are calibrated, georeferenced and corrected in order to have the same geographic reference frame and to eliminate most of the geometric distortions. The analysis focuses on the active braided channels, which represent all the channels filled with water and separated by sandy bars. They indicate an active flooding of this area, which moves the sand grains downstream, and allow us to infer an annual frequency of occurrence (Peiry, 1988). For this discussion, the width of the active braided channel has been measured perpendicularly on its axis, every 250 m along the river.

The reconstruction of channel pattern with this methodological approach indicates that, around the year 1900, the Prahova River formed braided channels both in the Subcarpathians and in the Ploieşti piedmont plain. Between 1900 and 1955, on the first 17 and the last 10 km, the river branches joined into a single channel and the average width was reduced by 33% (Fig. 9B, C). This narrowing is probably the result of a hydrological recovery after the system had suffered the impact of the hydro-climatic hazards that occurred at the end of the 19th century, when several rainy and flood events were recorded in the Prahova's catchment (Mustăţea, 2005). This conclusion is reinforced by the finding that the braided channels of the Doftana River, one of the Prahova's Subcarpathian tributaries, also shrank by 35% (Ioana-Toroimac, 2009). Between 1955 and 1980, the Prahova's active channel narrowed by 15%. Despite the construction of dams, on the Doftana between 1968 and 1971, and on the Prahova in 1968, the relatively small narrowing of the active channel downstream the junction can be explained by the flood of July 1975 (recurrence interval of 20 years at Prahova-Doftana junction according to Ioana-Toroimac, 2009). It probably had a morphogenetic character and neutralized the effects of the human interventions in the hydrological system. Between 1980 and 2005, the last 10 km of the ancient braided stretch turned completely into a single channel, while the average width of the active channel diminished by 48%. This drastic retraction coincides with the creation of some water management engineering works in Romania (Şerban & Gălie, 2006). We can mention here the construction of a small dam on the Carpathian reach of the Prahova River in 1982, the setting up of gravel and sand workings (20 areas in 2005 comparing to 3 in 1980), the reinforcements with gabions and the building of low-head dams. During this period, no significant flood affected the Prahova's catchment. At the same time, a riparian forest covered some sections of the active channel, thus preventing its expansion.

During the entire study period, some shrinkage was noticed upstream, in the Carpathian section, on the Prahova's single channel, but this was less intense (37% between 1900 and 2005 in comparison with 68% for the braided stretch). This difference of intensity is explained by the channel pattern: the braided channels are more sensitive to liquid discharge variations and especially to bedload sediment discharge (Ioana-Toroimac et al., 2010). Nevertheless, it is still difficult to distinguish the role of each factor in the general evolution, because all the factors act simultaneously. At the same time, the feedback loops of the hydrological system are hard to understand and there is a lack of cartographic sources in-between the four analyzed moments.

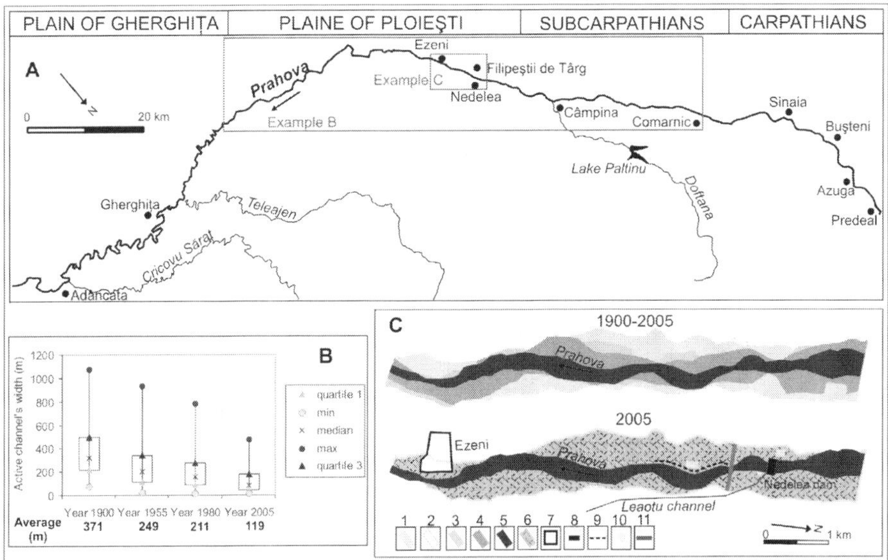

Fig. 9. The Prahova' s active channel lateral adjustments. A. The physiographic units crossed by the Prahova River; B) Active channel' s width variations; C) Retraction of the active channel between 1900 and 2005; in 2005 the ancient active channel, which was covered by pastures, was locally used for settlement expansion and gravel and sand mining; 1: the active channel in 1900; 2: the limits of the active channel in 1900; 3: the active channel in 1955; 4: the active channel in 1980; 5: the active channel in 2005; 6: pasture; 7: town; 8: dam; 9: gabions; 10: gravel and sand extraction area; 11: road.

4.3 Dynamics of the meandering channels: case studies in the Romanian Plain

This section aims to analyze the dynamics of the sinuous channels. The evolution of the meanders mirrors the horizontal dynamics of the channel through lateral cutting, as well as the rivers tendency to acquire the graded condition (Grecu & Palmentola, 2003). River gradient, liquid and solid flow, tectonics and some local conditions may influence the meanders formation and dynamics. The studies accomplished so far on the meander loops of some rivers belonging to the Romanian Plain show that these factors act in a synergistic way (Grecu et al., 2010c).

Our analysis is based on selected rivers in the Romanian Plain, because the winding process represents a specific feature of the evolution of this area during the Holocene (Grecu et al., 2009; Grecu, 2010). Besides, the significant alterations of the river system that took place over the time suggest an evolution similar to that of the Panonic Basin (Popov et al., 2008;

Timar et al., 2005, as cited in Grecu et al., 2010c). The analyzed rivers are the Vedea and Câlniştea, which cross a region affected by tectonic uplifts, more intense in the neighborhood of the Danube, where they exceed 2 mm yr^{-1} (Badea, 1983, 2009). The relationship between tectonics and meandering processes is tackled in a paper authored by Grecu et al. (2010b). Along these rivers, on the winding stretches, morphometric and diachronic analyses were accomplished in order to highlight the channel dynamics in the long and cross profile.

The Vedea River (251 km long, with a drainage area of 5364 km^2 and a mean annual discharge of 11 $m^3 s^{-1}$ according to Ujváry, 1972) has an allochtonous character: it originates in the Getic Piedmont, but more than 2/3 of its length and drainage basin lies in the Romanian Plain area (Fig. 10A). The river has a mean gradient of 2.1‰, which drops in the lower stretch to 0.5‰. This low inclination is one of the responsible factors for the tortuous aspect of the watercourse. In order to highlight the channel dynamics of the Vedea River, we proceeded to a diachronic analysis of the stream stretch lying near the junction of the Teleorman River (Fig. 10B), which is the Vedea's most important tributary (178 km long, with a drainage area of 1408 km2, a mean annual discharge of 3 $m^3 s^{-1}$ and a mean suspended load of 2 kg s^{-1}, according to Ujvári, 1972). This stretch has a high evolution potential due to the water and sediment contribution of the Teleorman, especially during the flood events. For instance, the balance analysis of suspended sediment load shows that between Alexandria and Cervenia (gauging stations placed upstream and downstream the junction of the Vedea and Teleorman rivers), the sediment accumulation amounts to 31,360 t yr^{-1} (value computed based on the data collected by Ujvári, 1972, for the period 1952 - 1967).

The diachronic analysis of this stretch relied on a series of maps developed in 1970 (scale 1:50000) and on aerial photographs taken in 2006 (scale 1:5000), which had been georeferenced and calibrated beforehand. It was ascertained that between 1970 and 2006 the section of the Vedea channel lying within the investigated area grew shorter by 2.5 km (from 12.7 km to 10.2 km). This shortening was due, on the one hand, to the natural adjustments (through rectifications and meanders migration) and on the other hand, to the engineering works. In its turn, the Teleorman River experienced the same pattern of development, and consequently the channel got shorter by 1.1 km, while the mouth changed its position. The Vedea's meanders lying upstream the junction of the Teleorman grew shorter by 600 m and the joint alluvial fan of the Vedea and Teleorman displaced the junction point by 0.83 km (Cârciumaru, 2010). Beside the meanders evolution, the comparison of the channel cross-sections accomplished for Alexandria station for the period 1979 - 2005 allowed us to understand the gradual aggradation of the riverbed.

Meanders formation largely depends on the interactions between the water flow and the materials eroded from the riverbed and the banks. Within the study period the highest discharges of the Vedea River (recorded at Alexandria station) occurred in October 1972 (949 $m^3 s^{-1}$), July 2005 (834 $m^3 s^{-1}$), July 1975 (544 $m^3 s^{-1}$) and July 1970 (463 $m^3 s^{-1}$) (Grecu et al., 2010c). Likewise, during these floods the solid load grew significantly, as it is proven by the flood of July 2005, when the suspended load recorded at Alexandria station amounted to 5,838 kg s^{-1}. Therefore, we may conclude that the natural evolution of the meanders and riverbed changes are influenced by these extreme hydrological events, with high morphogenetic potential.

The **Câlniştea River** (112 km long, with a catchment area of 1748 km^2 and a mean annual discharge of 3 m3 s^{-1} ; according to Ujvári, 1972), which originates in the Romanian Plain,

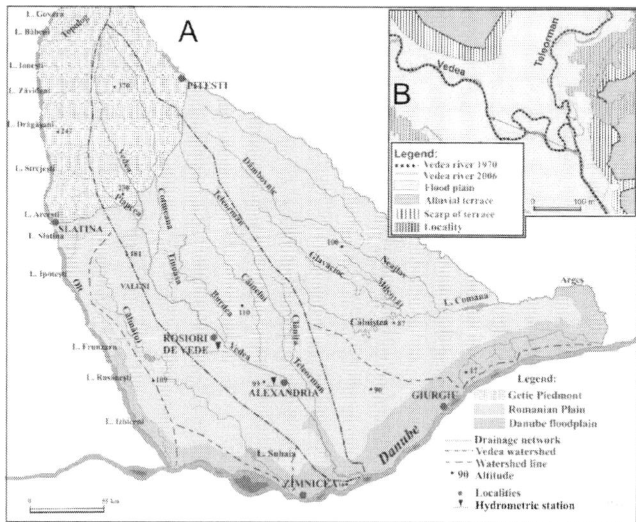

Fig. 10A. The central sector of the Romanian Plain, and the location of the Vedea and Câlniştea rivers. B. The lateral dynamics of the Vedea channel near the junction of the Teleorman between 1970 and 2006 (according to Grecu et al., 2010c, modified). The arrows in B show the most modified sectors.

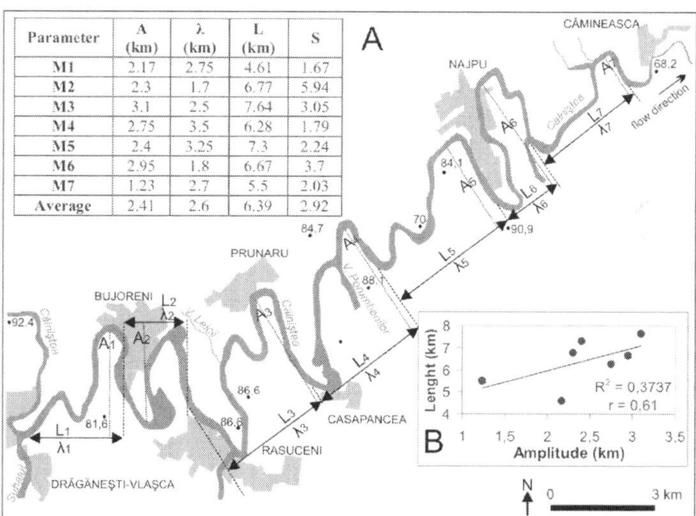

Parameter	A (km)	λ (km)	L (km)	S
M1	2.17	2.75	4.61	1.67
M2	2.3	1.7	6.77	5.94
M3	3.1	2.5	7.64	3.05
M4	2.75	3.5	6.28	1.79
M5	2.4	3.25	7.3	2.24
M6	2.95	1.8	6.67	3.7
M7	1.23	2.7	5.5	2.03
Average	2.41	2.6	6.39	2.92

Fig. 11A. The meander morphometry of the Câlniştea River (according to Grecu et al., 2010b, modified). B. Corellation between the length and the amplitude of the meanders of the Câlniştea River in the study reach.M1...M7: meander number; A: meander amplitude; λ : meander wavelength; L: channel length; S: sinuosity ratio.

has a general west - east flow direction, which might be explained by the presence of a fault in the bedrock (Coteţ,1976). The accomplished analysis focused on a reach of about 45 km long (between Drăgăneşti - Vlaşca and Căminească settlements), and was based on the topographic maps of scale 1:25000 (edited in 1972). This river reach having a sinuosity coefficient of 2.45 (determined as a ratio between channel and valley lengths), was divided into seven sections, each of them encompassing a well-developed meander loop. These river bends have lengths varying from 4.61 to 7.64 km, amplitudes of 2 to 3 km and wavelengths of 1.7 to 3.5 km (Fig. 11A). We have determined a relatively close relationship between the meanders amplitudes and lengths (correlation coefficient r = 0.61) (Fig. 11B).

The meander dynamics was analyzed by overlaying the topographic map of 1972 and the Szathmary map of 1856, both georeferenced and calibrated beforehand. For this period, we noticed a shortening of the meander lengths, a reduction of their amplitudes and a diminution of the channel sinuosity (from 2.89 to 2.45), which on the one hand are the result of a natural rectification process, and on the other hand suggest a possible tendency for the river to reach the graded condition.

5. Human activity and the sediments: the role of dam construction in sediment transport

Human activities are an important control factor both for the processes regarding sediment generation, transfer and accumulation and for the channel dynamics. At the same time, these processes are responsible on short and long run for the alteration of the river channels, which entail negative socio-economic and environmental consequences. If in the previous sections we have mentioned some human activities (deforestation, sand and gravel extraction from river channels, dam construction, etc.), which induce alterations of sediment transport and bring about morphological changes of the channels, now we reveal some significant aspects concerning the impact of reservoirs on sediment yield, based on examples from Romania.

On a national scale, there are at present more than 2100 anthropogenic lakes, serving various functions. Of these, 246 lie behind dams higher than 15 m, for which they are considered reservoirs. From this point of view, Romania holds the 19th position among the 80 member states of the World Commission on Dams, and the 9th position in Europe (Rădoane, 2005). Of the total number of reservoirs, 406 exceed a volume of one million cubic meters of water (according to AQUAPROIECT, 1992) and hold together about 85% of the storage capacity of the anthropogenic lakes in Romania (Zaharia & Pătru, 2009). More than half of the reservoirs (59%) lie in the plain area (Fig. 12), but the highest water volumes (60% of the total capacity) are stored by the reservoirs lying in the hilly and tableland areas, where rocks are crumbly and erosion rates high, which largely explains the intense silting processes.

A study accomplished by Rădoane (2005) shows that on the Romanian territory, during a mean interval of 15 years, the amount of sediments deposited on the bottom of the reservoirs was about 200 million m^3, while the annual deposition rate was 13.4 million m^3, representing more than a quarter (27%) of the mean annual suspended load. Silting processes mostly affect the reservoir chains built on the Olt and Argeş rivers (Fig. 1A), which store nearly half of the total sediment volume deposited in the Romanian reservoirs. Of the 13 impoundments that exist in the upper and middle catchment of the Argeş River, five are more than 70% silted and one is completely filled with sediments (Ogrezeni Reservoir) (Teodor, 1999; Rădoane, 2005).

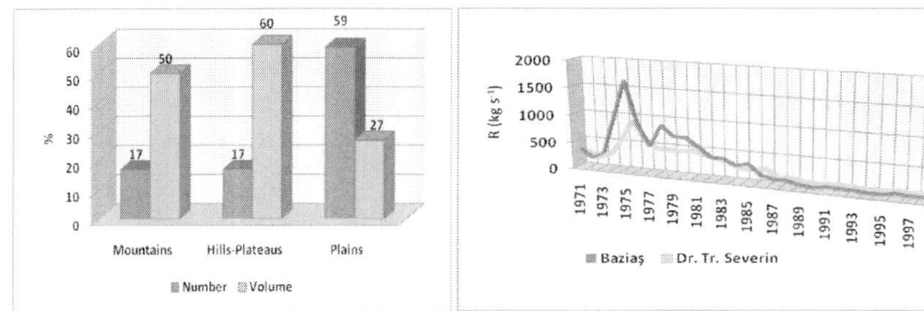

Fig. 12. (left). The percentage of the dam reservoirs of each major relief unit in the total number and in the total volume of the reservoirs exceeding 1 million m^3

Fig. 13. (right). The variability of the Danube mean annual suspended load at Baziaş and Drobeta Tr. Severin (according to the data provided by the "Romanian Waters" National Administration)

The largest reservoir in Romania is the Iron Gates I, placed on the Danube and making up the natural border between Romania and Serbia (Fig. 1A). It came into existence during the period 1964 - 1972, when the Iron Gates I Hydropower and Navigation System (H.N.S) was created, which is one of the most impressive engineering works in Europe and the biggest on the Danube. The reservoir is about 140 km long, covers 100 km^2, and at the maximum retention level (69.5 m a.s.l.) is capable of storing a water volume of 2400 million m^3 (AQUAPROIECT, 1992).

The balance of the sediments deposited in the Iron Gates I reservoir during the period 1971 - 2002, computed based on the cumulated differences between the Baziaş gauging station (lying where the Danube enters the reservoir) and Drobeta Tr. Severin station (lying downstream the dam), shows that the total volume of the deposited sediments amounted to 117.8 million tons (according to the data provided by the "Romanian Waters" National Administration). This volume corresponds to a mean annual rate of 3.7 million tons. The mean multiannual suspended load outflow was 33% lower than the inflow, the values decreasing from 353.2 kg s^{-1} (at Baziaş) to 236.5 kg s^{-1} (at Drobeta Tr. Severin) (data provided by the "Romanian Waters" National Administration). Significant reductions were recorded in the first decade after the reservoir creation and especially in the years with major flood events (1975, 1978). After 1987, however, one can note a relative homogeneousness of the suspended load on the two sections considered, as well as a decrease of sedimentation rate (Fig. 13).

In 1982, the Iron Gates II reservoir (Ostrov) was created 80 km downstream, with an area of 5200 ha and a water volume of 800 million m^3 (AQUAPROIECT, 1992). The creation of the two impoundments along the lower course of the Danube was responsible for the drastic reduction of the sediment load carried downstream. Thus, before entering the Danube Delta (at Ceatal Chilia), the mean annual values of suspended sediment load dropped during the interval 1970 - 2000 by 55% in comparison with the period 1921 - 1960 (from 66 million to 30.4 million t yr^{-1}) .

6. Chapter summary

Sediment transport and river channel dynamics are fluvial processes that may have on short or long run significant societal and environmental consequences. For this reason, they need to be studied thoroughly in order to mitigate their negative effects. These processes are controlled by a combination of variables acting on different spatial and temporal scales. Although it is very difficult to isolate the effect of a particular one, in the present work we have tried to identify the role played by some natural and human control factors on suspended sediment load/yield and on the river channel variability in Romania. In order to do that, we have used several case studies focused on two main physiographic units: the outer Carpathian's Curvature (characterized by the highest erosion rate in Romania) and the central part of the Romanian Plain (which experiences an intense lateral dynamics of the river channels).

The paper is based on three major types of analyses (statistical, correlative, and diachronic), which have helped us emphasize the following aspects: i) the space and time variability of suspended sediment load (at the Romanian scale and in the Carpathian curvature region); ii) the intensity of the relationships between suspended sediment yield and some control variables (precipitation, water discharge, catchment characteristics, and human activities), and iii) the vertical and lateral channel dynamics of some representative rivers in the Carpathian curvature region and the central part of the Romanian Plain.

In the years to come, we intend to improve these results by integrating more control variables (multiple and non-linear correlations), so that to better explain the relationships on regional scale. We wish we could take into account a bigger number of catchments, provided that we get the necessary information. For a more rigorous analysis of the relationship between suspended sediment yield and precipitation, we will take into account not only the amounts of precipitation, but also their intensity and frequency.

7. Acknowledgment

The research on the rivers in the Romanian Plain was accomplished within a PN II IDEI CNCSIS Project (no 994/12.01.2009) - *The hydrogeomorphological system in the concepts of geomorphometry and of modern morphological theories. Applications to hazard and risk diagnosis in areas of the Romanian plain* (Project manager was Dr. Florina Grecu and Dr. Liliana Zaharia was one of the team members).

The authors wish to thank the "Romanian Waters" National Administration, the National Institute of Hydrology and Water Management and the National Meteorological Administration for their kindness to put at our disposal their hydrological and climatic data.

8. References

Administrația Națională de Meteorologie. (2008). *Clima României*, Editura Academiei Române, ISBN 978-973-27-1674-8, Bucharest, Romania

AQUAPROIECT. (1992). *Atlasul Cadastrului apelor din România*, Ministerul Mediului, Bucharest, Romania

Arnaud-Fassetta, G. & Fort, M. (2004). La part respective des facteurs hydroclimatiques et anthropiques dans l'évolution récente (1956-2000) de la bande active du Haut Guil, Queyras, Alpes françaises du Sud, *Méditerranée*, No. 1-2, pp. 143-156, ISSN 0025-8296

Badea, L. (1983). Mişcările neotectonice, In: *Geografia României. I. Geografie fizică*, Badea L., Gâştescu P., Velcea V. (Ed.), pp. 84-88, Editura Academiei Republicii Socialiste România, Bucharest, Romania

Badea, L. (2009). *Relieful României şi neotectonica*, Editura Universitaria, ISBN 978-606-510-682-6, Craiova, Romania

Bălteanu, D. (1983). *Experimentul de teren în geomorfologie. Aplicații în Subcarpații Buzăului*, Editura Academiei, Bucharest, Romania

Bogdan, O. & Mihai, E. (1981). Intensité des pluies dans la région des Subcarpates de Buzău, *Revue roumaine de Géologie, Géophysique, Géographie - Géographie*, Vol. 25, No. 1, pp. 109-120, ISSN 0556-8099

Bravard, J.P.; Landon, N.; Peiry, J.L. & Piégay, H. (1999). Principles of engeneering geomorphology for managing channel erosion and bedload transport, examples from French rivers, *Geomorphology*, No. 31, pp. 291-311, ISSN 0169-555X

Cârciumaru, E. (2010). *Geneza, evoluția şi dinamica actuală a bazinului morfohidrografic Vedea*, PhD thesis, Bucharest University, Romania

Chadule (group) (1997). *Initiation aux pratiques statistiques en géographie*, Editions Masson, ISBN 2-200-01534-8, Paris, France

Charlton, R. (2008). *Fundamentals of fluvial geomorphology*, Routledge, ISBN 978-0-415-33454-9, New York, USA

Chirilă, G. (2010). *Resursele de apă din bazinul râului Trotuş - studiu de hidrologie şi calitatea apelor*, PhD thesis, Bucharest University, Romania

Cocoş, A. & Cocoş, O. (2005). Particularitățile rețelei de văi din bazinul hidrografic al râului Câlniştea, *Comunicări de geografie*, Vol. IX, pp. 311-316, ISSN 1453-5483

Cornea, I.; Drăgoescu, I.; Popescu, M. & Visarion, M. (1979). Harta mişcărilor crustale verticale recente pe teritoriul R.S. România, *Studii şi cercetări de Geologie, Geofizică, Geografie*, Vol. 17, No. 1, pp. 3-30

Coteț, P. (1976). *Câmpia Română - studiu de geomorfologie integrată*, Editura Ceres, Bucharest, Romania

Diaconu, C. (1964). Rezultate noi în studiul scurgerii de aluviuni în suspensie a râurilor R.P.R., *Studii de hidrologie*, No. XI, pp. 81-223, ISSN 0378-9810

Diaconu, C. (1971). Probleme ale scurgerii de aluviuni a râurilor României, *Studii de hidrologie*, No. XXXI, pp. 5-213, ISSN 0378-9810

Diaconu, C.; Avădanei, A.; Ciobanu, S. & Motea, I. (1962). Despre stabilitatea albiilor râurilor R.P.R. în ultimii 30-40 ani, *Studii de hidrologie*, No. III, pp. 53 - 63, ISSN 0378-9810

Diaconu, C. & Şerban, P. (1994). *Sinteze hidrologice*, Editura Tehnică, ISBN 973-31-0638-0, Bucharest, Romania

Dumitriu, D. (2007). *Sistemul aluviunilor din bazinul râului Trotuş*, Editura Universității, ISBN 978-973-666-237-9, Suceava, Romania

Feier, I. & Rădoane, M. (2007). Dinamica în plan orizontal a albiei minore a râului Someşul Mic înainte de lucrările hidrotehnice majore (1870-1968), *Analele Universității Suceava*, No. XVI, pp. 13-26, ISSN 1583-1469

Ghiță, C. (2009). *Geneza, evoluția si dinamica actuală a bazinelor morfohidrografice autohtone din Campia Romană de est cu aplicații la bazinul Mostiştea*, PhD thesis, Bucharest University, Romania

Grecu, F. (2010). *Geografia câmpiilor României*, Editura Universității, ISBN 978-973-737-859-0, Bucharest, Romania

Grecu, F. & Palmentola, G. (2003). *Geomorfologie dinamică*, Editura Tehnică, ISBN 973-31-2132-0, Bucharest, Romania

Grecu, F.; Săcrieru, R.; Ghiță, C. & Văcaru, L. (2009). Geomorphological landmarks of the Eastern Romania plain Holocene evolution, *Zeitschrift für Geomorphologie*, Vol. 53, No. 1, pp. 99-110, ISSN 0372-8854

Grecu, F.; Zaharia, L.; Ghiţă, C.; Ioana-Toroimac, G. & Cârciumaru, E. (2010a). Episodes pluvio-hydrologiques extrêmes et dynamique hydro-géomorphologique dans la Plaine Roumaine: Le cas de la rivière Vedea, *Actes du XXIII-ème Colloque International de Climatologie*, pp. 251-256, ISBN: 978-2-907696-16-6

Grecu, F.; Ghiţă, C. & Săcrieru, R. (2010b). Relation beetwen tectonics and meandering of river channels in the Romanian Plain. Preliminary observation, *Revista de Geomorfologie*, No. 12, pp. 97-104, ISSN 1453-5068

Grecu, F.; Zaharia, L.; Ghiţă, C. & Văcaru, L. (2010c). The dynamic factors of hydrogeomorphic vulnerability in the central sector of the Romanian plain, *Metalurgia International*, Vol . XV, No . 9, pp. 139-148, ISSN 1582-2214

Ichim, I. (1992). Probleme ale cunoaşterii sistemului aluviunilor din România, *Lucr. celui de al IV-lea Simpozion Provenineţa si efluenţa aluviunilor* , No. 4, pp. 7-15

Ichim, I.; Bătucă, D.; Rădoane, M. & Duma, D. (1989). *Morfologia şi dinamica albiilor de râu*, Editura Tehnică, ISBN 9733100056, Bucharest, Romania

Ichim, I. & Rădoane, M. (1990). Channel sediment variability along a river: a case study of the Siret River (Romania), *Earth Surface Processes and Landforms*, Vol. 15, No. 3, pp. 211-225, ISSN 0197-9337

Ichim, I.; Rădoane, N. & Rădoane, M. (1998). *Dinamica sedimentelor: aplicaţii la râul Putna-Vrancea*, Editura Tehnică, ISBN 9733111686, Bucharest, Romania

Ioana-Toroimac, G. (2009). *La dynamique hydrogéomorphologique de la rivière Prahova (Roumanie): fonctionnement actuel, évolution récente et consequences géographiques*, PhD thesis, Université Lille 1, France

Ioana-Toroimac, G. ; Dobre, R.; Grecu, F. & Zaharia, L. (2010). Evolution 2D de la bande active de la Haute Prahova (Roumanie) durant les 150 dernières années, *Géomorphologie : relief, processus, environnement*, No. 3, pp. 275-286, ISBN 2-913282-49-0

Knighton, D. (1984). *Fluvial forms and processes*, Arnold, ISBN 0-7131-6405-0, London, UK

Kondolf, G.M.; Piégay, H. & Landon, N. (2002). Channel response to increased and decreased bedload supply from land-use change: contrasts between two catchments, *Geomorphology*, No. 45, pp. 35-51, ISSN 0169-555X

Liébault, F. & Piégay, H. (2002). Causes of the 20th century channel narrowing in mountain and piedmont rivers of Southeastern France, *Earth Surface Processes and Landforms*, No. 27, pp. 425-444, ISSN 0197-9337

Mociorniţă, C. & Birtu, E. (1987). Unele aspecte privind scurgerea de aluviuni în suspensie în România, *Hidrotehnica*, Vol. 32, No. 7, pp. 241-245, ISSN 0439-0962

Moţoc, M. (1984). Participarea proceselor de eroziune şi a folosinţelor terenului la diferenţierea transportului de aluviuni în suspensie pe râurile din România, *Bul. Inf. ASAS*, No. 13, pp. 16-28

Mustăţea, A. (2005). *Viituri excepţionale pe teritoriul României*, Editura Tipografică S.C. Onesta Com. Prod 94 S.R.L., ISBN 973-0-03611-X, Bucharest, Romania

Mutihac, V.; Stratulat, M.I. & Fechet, R.M. (2007). *Geologia României*, Editura Didactică şi Pedagogică, ISBN 973-30-1858-2, Bucharest, Romania

Pandi, G. & Sorochovschi, V. (2009). Dinamica verticală a albiei râurilor în dealurile Clujului si Dejului, *Riscuri si catastrofe*, Vol. VIII, No. 7, pp. 218-226, ISSN 1584-5273

Pătroescu Nardin, M. (1996). *Subcarpaţii dintre Râmnicu Sarat şi Buzău, Potenţial ecologic şi exploatare biologică*, Editura Carro, ISBN 973-97751-1-X, Bucharest, Romania

Peiry, J.L. (1988). *Approche géographique de la dynamique spatio-temporelle des sédiments d'un cours d'eau intra-montagnard : l'exemple de la plaine alluviale de l'Arve (Haute-Savoie)*, PhD thesis, Université Jean Moulin Lyon 3, France

Pişota, I. & Zaharia, L. (2002). *Hidrologie*, Editura Universităţii din Bucureşti, ISBN 973-575-587-4, Bucharest, Romania

Popa-Burdulea, A. (2007). *Geomorfologia albiei râului Siret*, PhD thesis, „Al.I.Cuza" University Iaşi, Romania

Olariu, P. & Gheroghe, D. (1999). The effects of human activity on land erosion and suspended sediment transport in the Siret hydrographic basin, In: *Vegetation, land use and erosion processes*, Zăvoianu I., Walling D. E., Şerban P. (Ed), pp. 40-50, Institul de Geografie, ISBN 0730009813, Bucharest, Romania

Rădoane, M. (2005). Raport de Cercetare, Grant: A 448, *Revista de Politica Ştiintei şi Scientometrie*, ISSN- 1582-1218

Rădoane, M. & Ichim, I. (1987). Tendinţe în dinamica patului albiilor de râu în timp lung din Carpaţii Orientali, *Lucr. Ses. St. „Curgeri bifazice industriale"*, pp. 2172 - 2173

Rădoane, M.; Ichim, I. & Pandi, G. (1991). Tendinţe actuale în dinamica patului albiilor de râu din Carpaţii Orientali. *St. cerc. geol., geofiz., geogr., ser. geogr.*, Vol. 38, pp. 21 - 31, ISSN 0039-3967

Rădoane, M.; Rădoane, N.; Dumitriu, D. & Miclăuş, C. (1998). Probleme ale transportului de aluviuni în lacuri de interes hidroenergetic din România, *Analele Universităţii Ştefan cel Mare*, Vol. VII, No. 41-57, ISSN 1583-1469

Rădoane, M. & Rădoane, N. (2001). Eroziunea terenurilor şi transportul de aluviuni în sistemele hidrografice Jijia şi Barlad, *Revista de Geomorfologie*, No. 3, pp. 73-86, ISSN 1453-5068

Rădoane, M.; Rădoane, N. & Ichim, I. (2003a). Dinamica sedimentelor în lungul râului Suceava. *Analele Universităţii Ştefan cel Mare*, Vol. X, pp. 37-48, ISSN 1583-1469

Rădoane, M.; Rădoane, N. & Dumitriu, D. (2003b). Geomorphological evolution of river longitudinal profiles, *Geomorphology*, No. 50, pp. 293-306, ISSN 0169-555X

Rădoane, M. & Rădoane, N. (2005). Dams, sediment sources and reservoir silting in Romania, *Geomorphology*, No. 71, pp. 112-125, ISSN 0169-555X

Rădoane, M.; Rădoane, N.; Dumitriu, D. & Miclăuş, C. (2006). Efectele surselor de aluviuni asupra distribuţiei materialului de albie al râurilor est-carpatice, *Studii şi cercetări de Geografie*, No. LI-LIII, pp. 153-168, ISSN 1220-5281

Rădoane, M. & Rădoane, N. (2007). Răspunsul unei albii adâncite în roci coezive la acţiunea factorilor de control naturali şi antropici, *Studii şi cercetări de geografie*, No. LIII-LIV, pp. 117-136, ISSN 1220-5281

Rădoane M., Rădoane N., Cristea I. & Oprea-Gancevici D. (2008a). Evaluarea modificărilor contemporane ale albiei râului Prut pe graniţa românească, *Revista de Geomorfologie*, No. 10, pp. 57-73, ISSN 1453-5068

Rădoane, M.; Rădoane, N.; Cristea, I.; Persoiu, I. & Burdulea, A. (2008b). Quantitative analysis in the fluvial geomorphology, *Geographia technica*, No. 1, pp. 100-111, ISSN 2065-4421

Rădoane, M.; Pandi, G. & Rădoane, N. (2010). Contemporary bed elevation changes from the eastern Carpathians, *Carpathian Journal of Earth and Environnemental Sciences*, Vol. 5, No. 2, pp. 49-60, ISBN 1842-4090

Roşca, D. & Teodor, S. (1990). Influenţa lacurilor de acumulare asupra transportului de aluviuni, *Lucr. celui de al III-lea Simpozion „Provenienţa şi efluenţa aluviunilor"*, pp. 57-65

Schumm, S.A. (1977). *The fluvial system*, Wiley and Sons, ISBN 0-471-01901-1, New York, USA

Surian, N. & Rinaldi, M. (2003). Morphological response to river engineering and management in alluvial channel in Italy, *Geomorphology*, No. 50, pp. 307-326, ISSN 0169-555X

Surian, N. & Cisotto A. (2007). Channel adjustements, bedload transport, and sediment sources in a gravel-bed river, Brenta River, Italy, *Earth Surface Processes and Landforms*, No. 32, pp. 1641-1656, ISSN 0197-9337

Şerban, P. & Gălie, A. (2006). *Managementul apelor. Principii si reglementări europene*, Editura Tipored, ISBN 978-973-86083-8-2, Bucharest, Romania

Teodor, S. (1992). Transportul total de aluviuni (suspensii şi târâte) din bazinul hidrografic Argeş şi unele implicaţii asupra colmatării lacurilor de acumulare, *Lucr. celui de al IV-lea Simpozion „Provenienţa şi efluenţa aluviunilor"*, pp. 165-169

Teodor, S. (1999). *Lacul de baraj şi noua morfodinamică, Studii de caz pentru râul Argeş*, Editura Vergiliu, ISBN 973-98540-4-4, Bucharest, Romania

Ujváry, I. (1972). *Geografia apelor României*, Editura Ştiinţifică, Bucharest, Romania

Urziceanu, D. (1967). Consideraţii privind caractrerizarea stabilităţii albiilor râurilor, *Hidrotehnica, gospodărirea apelor, meteorologie*, Vol. 12, No. 1, pp. 27-31, ISSN 0018-134X

Văcaru, L.C. (2010). *Bazinul hidrografic Neajlov - Studiu de geomorfologie dinamică*, PhD thesis, Bucharest University, Romania

Zaharia, L. (1998). Tendances dans l'évolution des transferts de matières en suspension dans les Subcarpates de Courbure, en relation avec les modifications du milieu naturel, *Géomorphologie: relief, processus, environnement*, No. 1, pp. 3-15, ISSN 1266-5304

Zaharia, L. (1999). *Resursele de apă din bazinul râului Putna. Studiu de hidrologie*, Editura Universităţii din Bucureşti, ISBN 973-575-334-0, Bucharest, Romania

Zaharia, L. (2005). Studiul resurselor de apă din Carpaţii şi Subcarpaţii Curburii, *Lucrări şi rapoarte de cercetare*, Vol. 1, pp. 137-171, ISBN 973-737-012-0

Zaharia, L. & Ioana-Toroimac, G. (2009). Erosion dynamics - precipitation relationship in the Carpathians' curvature region (Romania). *Geografia fisica e dinamica quaternaria*, No. 32, 95-102, ISSN 0391-9838

Zaharia, L. & Pătru, I. (2009). Considerations on the Anthropic Romanian Lakes and their Impact on the Environement, Lakes, reservoirs and ponds, *Romanian Journal of Limnology*, No. 1, pp. 128-132, ISSN: 1844-6477

Zawiejska, J. & Wyzga, B. (2010). Twentieth-century channel changes of the Dunajec River, southern Poland: Patterns, causes and controls. *Geomorphology*, No. 117, pp. 234-246, ISSN 0169-555X

Zăvoianu, I. & Mustăţea, A. (1992). Legătura dintre debitele de apă şi de aluviuni în suspensie pe râurile din România, *Lucr. celui de al IV-lea Simpozion „Provenienţa şi efluenţa aluviunilor"*, pp. 158-164

15

Integrating River Bed Dynamics to Flood Risk Assessment

Clemens Neuhold, Philipp Stanzel
and Hans Peter Nachtnebel
University of Natural Resources and Life Sciences Vienna
Austria

1. Introduction

Risk zonation maps are mostly derived from single design floods which represent a hazard based on a specified return period. The respective delineation of inundated areas and the estimation of flow depths and flow velocities are fundamental inputs for flood risk estimation of exposed objects. For this purpose in most cases 2D hydrodynamic unsteady models are applied (BMFLUW, 2006 a). In the frame of state of the art approaches it is implicitly assumed that the morphology will not change; neither during flood events nor by long term erosion or deposition. However, alluvial river beds are subjected to severe morphological changes during floods which have significant implications for the water level (Nachtnebel & Debene, 2004). It is therefore obvious that the river bed elevation can change quickly and drastically (Neuhold et al., 2009). Observed morphological developments during and after flood events (Fig.1) indicate, to some extent, tremendous changes in river bed elevation due to sediment transport, log jam, rock jam, land slide, etc.

Fig. 1. Sediment accumulations in the Ill river catchment (source: IWHW, BMFLUW, 2005)

The high probability of occurrence of such processes clearly implies the necessity of incorporating calculated and estimated morphological changes to the flood risk assessment procedure (Neuhold et al., 2007). Therefore, the influence of sediment transport on the respective water surface elevation is investigated in this study. As this additional process is considered it is obvious that uncertainty increases (Neuhold et al., 2010 a, b; Neuhold & Nachtnebel, 2011). The identification of potential impacts on the water surface elevation and accordingly the possible inundation depth as well as delineation could, however, lead to an increase of awareness and adaptation of flood risk management strategies.

The study focuses on implications related to hazard assessment covering aspects of hydrology, hydraulics and sediment transport. Further, the study aims at enhancing approaches of vulnerability assessment and therefore, damage estimation by providing a direct link of probability distribution functions of inundation depths with the respective damage functions of flood-prone utilisations (damage-probability relationship).

The concept is tested in the river Ill catchment in the Western Austrian Alps which has suffered major floods during the recent past (1999, 2000 and 2005). The River Ill, with a mean annual discharge of 66 m^3/s and a catchment area of approximately 1300m^2 is the main river catchment in south-eastern Vorarlberg, the most-western federal state of Austria (Fig.2). The catchment area is characterized by torrential tributaries, hydraulic structures, hydropower plants and complex morphological characteristics.

Hydro-meteorological observations of precipitation, air temperature and runoff were gathered. Elevations range from 400 to 3000 m. a. s. l. and the mean annual precipitation averages 1700 mm. A 100-year flood event is estimated at 820 m^3/s. Current, as well as historical surveying data (since 1978), were provided for 60 km of the River Ill and, altogether, 15 km of 8 tributaries comprising cross section measurements (with distances of 100 m on average) and airborne laser scan data. Sediment samples were drawn in 71 locations. Additional information on geographical features of the catchment (elevation, land cover, cadastral information and soil type) and on hydropower influence on the runoff regime is considered.

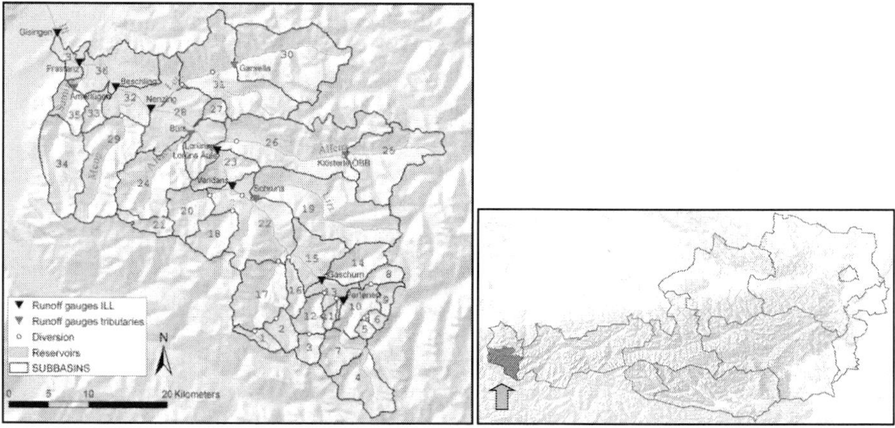

Fig. 2. Study area: Austria and the Ill river catchment in the west

2. Methodology

The applied methodological approach is elaborated to analyse and quantify the variability and uncertainty of single steps of hazard assessment and to enhance vulnerability assessment. Therefore, the estimation of hydrological input, possible changes in river bed

elevation due to sediment transport and the effects on water surface elevations are dissected. Vulnerability analyses and damage estimation tools are methodologically improved by connecting the overtopping probability, the variability of inundation depth and object related damage functions to obtain a damage-probability relationship (Fig.3).
Initially, the hydrology of the catchment is simulated with a semi-distributed precipitation-runoff model. Variability of the hydrograph is obtained by generating numerous scenarios with different initial moisture conditions and by considering different spatial and temporal distributions, durations and amounts of rainfall. The hydrologic model provides runoff scenarios which are subsequently used as an input for the hydraulic and sediment transport model. Additionally, the variability of possible morphological changes due to torrential sediment entry is analysed. For this purpose scenarios with randomly drawn sediment loads from torrential inflows based on probability distribution functions are developed to account for the high variability and unpredictability of torrential sediment input to the system. The calculated morphological changes of the river bed provide a basis to estimate the variability of water surface levels and inundation lines which need to be considered in flood hazard maps and flood risk maps. For each scenario the water table, river bed elevation and the respective inundation lines as well as inundation depths are calculated. Therefore, every single exposed object is linked to a distribution function consisting of estimated damages related to flood inundation depth and inundation probability.

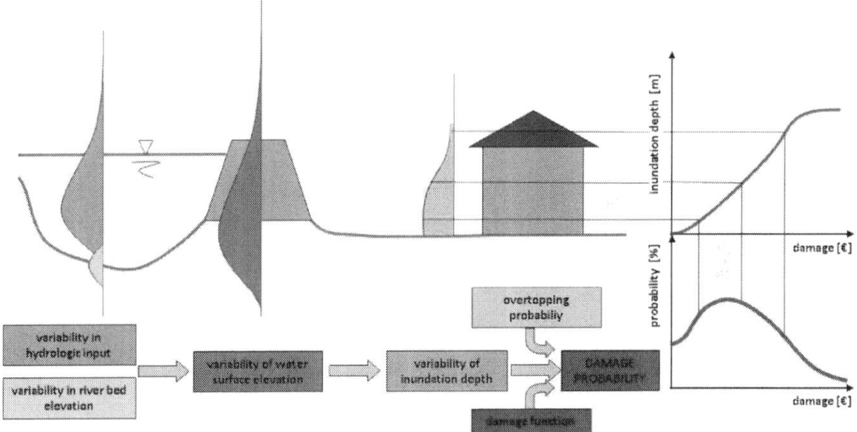

Fig. 3. Scheme of approach to derive the damage probability of vulnerable utilisations

2.1 Hydrologic input generation

The continuous, semi-distributed rainfall-runoff model, COSERO, developed by the Institute of Water Management, Hydrology and Hydraulic Engineering, BOKU (Nachtnebel et al., 1993, Kling, 2002 among others) is applied to the entire Ill catchment. The model accounts for processes of snow accumulation and melt, interception, evapotranspiration, infiltration, soil storage, runoff generation and routing. Separation of runoff into fast surface runoff, inter flow and base flow is calculated by means of a cascade of linear and non-linear

reservoirs. Spatial discretisation relies on the division of the watersheds into sub-basins and subsequently into hydrologic response units (HRUs).
The Ill watershed is divided into 37 sub-basins, based on the location of runoff gauges, anthropogenic diversions and reservoirs, with sub-basin areas ranging from 10 to 200 km^2 (Fig.2). 828 HRUs, with a mean area of 1.6 km^2, are derived by intersection of 200 m-elevation bands with soil type data (Peticzka and Kriz, 2005) and land use data (Fürst and Hafner, 2005).
The model is calibrated and validated based on observed discharge hydrographs of 6 years with continuous daily records and hourly records for 16 flood periods, measured at 14 gauges (Stanzel et al., 2007). Calibrated parameters of gauged sub-basins are transferred to neighbouring ungauged sub-basins. Storage coefficients for base flow and interflow, which correlated well with catchment size for the calibrated sub-basins, are assigned according to this relation. After this, storage coefficients for fast runoff are allocated in order to achieve characteristics of runoff separation into surface flow, interflow and base flow as simulated in neighbouring calibrated sub-basins with similar physical features. Nash-Sutcliffe model efficiencies (Nash and Sutcliffe, 1970) between 0.80 and 0.90 for the calibration period and between 0.75 and 0.85 for the validation period are achieved. Mean relative peak errors of the 16 simulated flood periods range between -15 % and +10 %.
After calibration, the rainfall-runoff model is applied to simulate flood runoff scenarios. Design storms with assumed return periods of 100 years are used as input. The underlying assumption of using design storms with a 100-year recurrence interval is that they may produce flood peaks of the same return period. While this premise can be regarded as appropriate for design purposes, it is clear that a rainstorm with a given return period may cause a flood with a higher or lower return period (Larson and Reich, 1972). This is due to factors affecting the runoff event like the distribution of rainfall in time and space or antecedent soil moisture and river discharge. Therefore, several scenarios, with variations of major influencing factors, are defined. Precipitation scenarios are obtained by varying total precipitation depth, storm duration and temporal and spatial distributions. Each rainfall scenario is combined with three different initial catchment conditions, which are selected from simulated state variables of historical flood periods.
Storm durations of 12 and 24 hours are selected for the assessment. Recorded events leading to floods in the years 2000, 2002 and 2005 showed rainfall duration within this range. These assumptions are also in accordance with the common procedure of testing storm duration up to twice the concentration time which is estimated as being 11 to 13 hours for the Ill catchment (BMLFUW, 2006 b). Precipitation depths of 100-year storms with 12 hours duration are provided by a meteorological convective storm event model (Lorenz & Skoda, 2000). Design storms based on these meteorological modelling results are recommended by Austrian authorities (BMLFUW, 2006 b) and therefore, are a common basis for design flood estimations in Austria. The values given by this model refer to point precipitation. Areal precipitation, to be used as input for rainfall-runoff modelling, is obtained by reducing the point precipitation values with areal reduction factors (ARF). The developers of the convective storm event model recommend two different procedures to determine such factors, both depending on catchment area, precipitation depth and duration of the storm (Lorenz & Skoda, 2000; Skoda et al., 2005). ARF resulting from these two calculations varied considerably and defined the range of ARF values used to reduce mean 12-hour point precipitation depths for the Ill catchment. Precipitation depths of 24-hour storms are based

on statistical extreme value analyses provided by local Austrian authorities and values from the Hydrological Atlas of Switzerland (Geiger et al., 2004).
Total precipitation depth is disaggregated to 15-minute time steps applying three different temporal distributions, with peaks at the beginning, in the middle or at the end of the event. Three different spatial distributions are considered: a uniform distribution, a distribution with higher precipitation in the south and another with higher precipitation in the north of the watershed. The spatial patterns of the two non-uniform distributions correspond with typical distributions of precipitation in the catchment.
The described variations in the parameters - storm duration, areal reduction factors and resulting precipitation depths, temporal distribution and spatial distribution of rainfall - generated 42 precipitation scenarios. The combination with three different initial catchment conditions led to 126 runoff scenarios (Fig.4).

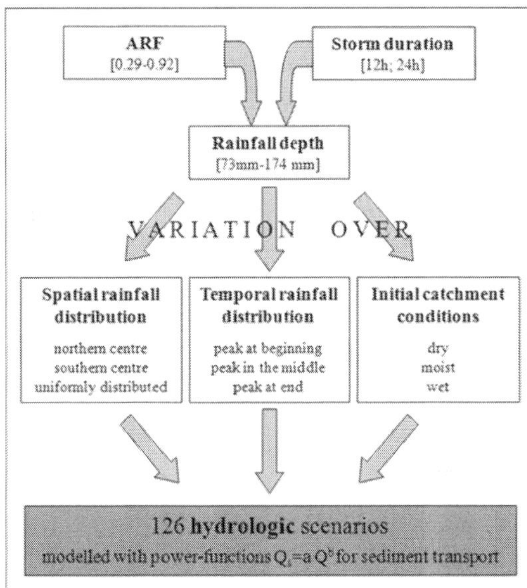

Fig. 4. Derivation of scenarios for hydrologic input variation

2.2 Hydrodynamics and sediment transport

The considered catchment area is characterized by torrential tributaries, hydraulic structures, hydropower plants and complex morphological characteristics. Therefore, it is crucial to apply a model with no restrictions and limitations regarding internal and external boundary conditions. Apart from these demands, a calculation in different fractions of sediment is required.
Satisfying these requests, the software package GSTAR-1D Version 1.1.4, developed by the U.S. Department of the Interior (Huang & Greimann, 2007) is used. GSTAR-1D (Generalized Sediment Transport for Alluvial Rivers - One Dimension) is a one-dimensional hydraulic and sediment transport model for use in natural rivers and man-made canals which applies 16 different sediment transport algorithms. It is a mobile boundary model with the ability to

simulate steady or unsteady flows, internal boundary conditions, looped river networks, cohesive and non-cohesive sediment transport, and lateral inflows.
The model uses cross section data and simulates changes of the river bed due to sediment transport. It estimates sediment concentrations throughout a waterway given the sediment inflows, bed material, hydrology and hydraulics of that waterway. Resulting from the one-dimension solutions for flow simulation the limitations are the neglect of cross flow, transverse movement, transverse variation and lateral diffusion. Therefore, the model cannot simulate such phenomena as river meandering, point-bar formation and pool-riffle formation. Additionally, local deposition and erosion caused by water diversions, bridges and other in-stream structures cannot be simulated (Huang & Greimann, 2007).
The hydrodynamic model is calibrated and validated based on runoff data from seven gauging stations by varying calculated roughness coefficients which refer to 71 sediment samples (Nachtnebel & Neuhold, 2008). Sediment transport is calibrated and validated based on historical cross section measurements (1978-2006) and the respective runoff time series as well as by balancing calculated volumes of transported sediments.
Hydrological input to the model is delivered by the precipitation-runoff model. Boundary conditions as well as initial conditions concerning sediment transport are defined and derived from sediment samples. Model calibration and validation is conducted on a stretch of 4.5 km where reliable historical measurement data is available. Calibration is done for the time period of 1985 to 1991; the validation period is defined with 1991 to 1993. Further efforts to obtain reliable sediment transport volumes are done by comparing calculated sediment volumes with estimated accumulations upstream of a power plant. At this power plant, including a movable weir and a reservoir, days of flushing (to empty the reservoir from sediment accumulations and enabling the highest possible head) are recorded by the operating company. This analysis aims at having evidence of the model's sensitivity for periods of increasing sediment transport. Comparing the calculated accumulations with this information enables a qualitative statement for parameters like conveyance, bed load rate, bed load motion, beginning of bed load motion, etc. This analysis is done for the year 2006, the only year with reliable data sets.
Calibration as well as validation show satisfying results (Fig.5). The coefficients of determination are very high although there are some cross sections which show substantial differences of measured and calculated cross section thalweg points.
Analysing the unsatisfactory cross sections revealed that they are located in river bendings where transverse flow occurs and therefore, transverse sediment transport is highly likely. These effects cannot be simulated by one-dimensional models.
As this study focuses on more conceptual goals the restrictions of model accuracy and reliability are accepted and the model setup is characterised as accurate.
Another focal point of the study is to analyse and quantify potential sediment inputs from torrential tributaries for various flood scenarios (HQ_1, HQ_5, HQ_{30} and HQ_{100}). Considerable uncertainty exhibits from the estimation of the sediment input from torrential inflows. Therefore, an observed flood event from 2005, with an estimated recurrence interval of 100 years, was investigated in more detail.
Local authorities provided estimates of sediment accumulations as well as data of removed sediment volumes in the frame of reconstruction and maintenance works after the flood event. Further, sediment volumes in torrential catchment areas are assessed to classify the validity of model results.

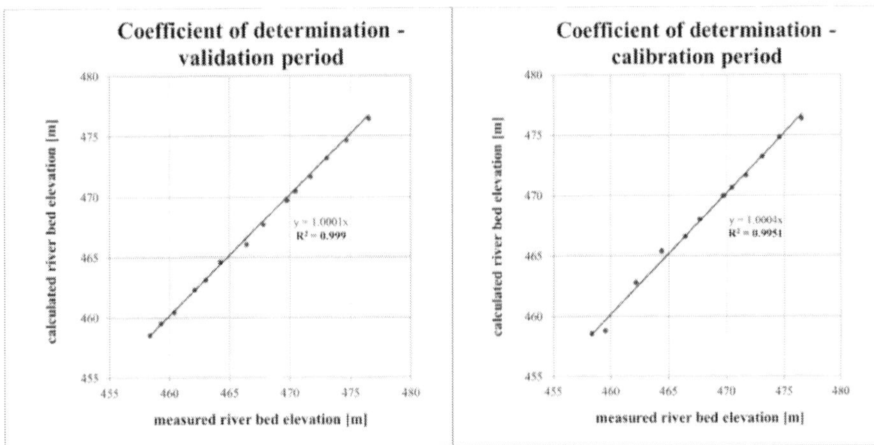

Fig. 5. Comparison of measured and calculated river bed elevations

Sediment transport models are compiled for the main river system and 8 tributaries. Two river bed conditions were defined for each tributary. The first of these assumed a fully-armoured upper layer with a mean layer thickness of 15 cm and the second model scenario calculated a river bed without any armouring. This assumption addresses the range of possible sediment input from its minimum, an armoured upper layer, to its maximum, by calculating the full potential transport rate.

Fig. 6 presents the example of the tributary river Alfenz, where simulated sediment transport loads referring to various water discharges are illustrated. The transport rates [t/h] are calculated for water discharges up to 160 m^3/s (the estimate for a 100-years flood event). Calculated sediment loads for an assumed armoured upper layer are only simulated up to a 30-years flood event (110 m^3/s) as values of shear stress indicated that the armoured layer will be forced open and therefore the transport of the potential transport rate is assumed.

Sediment routing is solved with the Meyer-Peter and Müller formula (1948, Eq. (1)), which is appropriate for alpine gravel-bed rivers like the river Ill and its torrential tributaries:

$$q_b^{2/3}\left(\frac{\gamma}{g}\right)^{1/3}\frac{0.25}{(\gamma_s-\gamma)d}=\frac{(K_s/K_r)^{2/3}\gamma RS}{(\gamma_s-\gamma)d}-0.047 \tag{1}$$

Where γ and γ_s = specific weights of water and sediment, respectively, R = hydraulic radius, S = energy slope, d = mean particle diameter, ρ = specific mass of water, q_b = bed load rate in under water weight per unit time and width, K_s = conveyance, K_r = roughness coefficient and $(K_s/K_r)S$ = the adjusted energy slope that is responsible for bed-load motion.

Based on 8 simulated tributaries (with available measurement data, sediment samples and estimates of potential available sediment volumes in the catchment area) sediment input functions are estimated for 47 unobserved torrents (Neuhold et al., 2007; Nachtnebel & Neuhold, 2008).

Accounting for the high variability and unpredictability of torrential sediment input during flood events, scenarios are developed representing spatio-temporal variability of rainfall, discharge and sediment transport. According to hydrological rainfall patterns (northern

centred, southern centred and uniformly distributed - see Fig. 4) areas of high probability of sediment input are defined related to the river sections [km] 60-40, 40-20 and 20-0.
For these three sections sediment input was randomly defined relying upon sediment input functions (e.g. Fig. 6) and calculated discharge rates. A minimum (armoured upper layer for all tributaries) and a maximum (no armouring for all tributaries) scenario, related to the restricting transport functions (Fig. 7), are simulated. Within these extremes, 10 scenarios were compiled by randomly drawing input capacities of each torrential inflow dependent on the magnitude of the associated flood peak in the torrential sub-catchment.

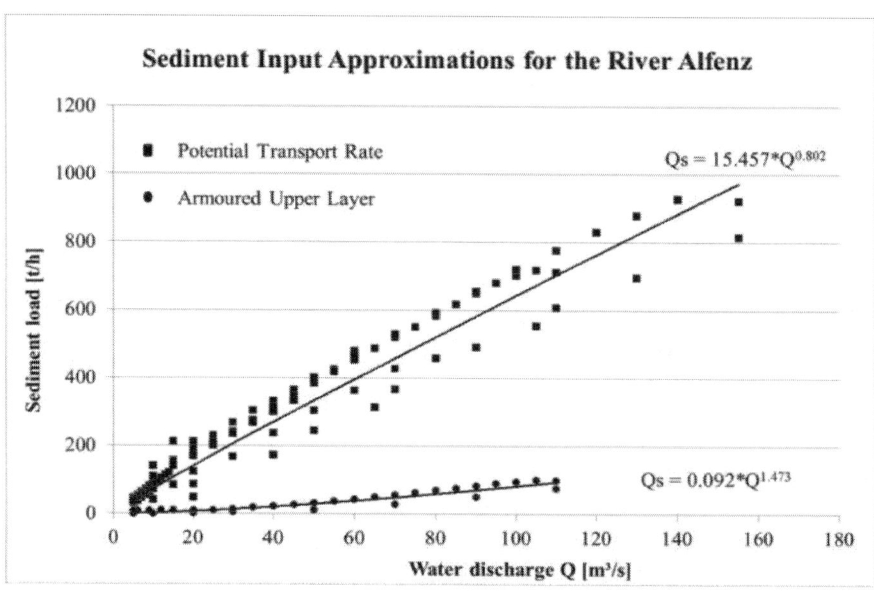

Fig. 6. Upper and lower sediment input boundary condition for the River Alfenz

To obtain realistic input distributions only one of the three sections (60-40, 40-20, 20-0) is allowed to be a dominant source of sediment input. Furthermore, to account for rainfall clusters a boundary condition for the acceptance of a randomly chosen scenario is defined: a minimum percentage of 50 % related to the section's torrential catchment areas had to deliver maximum sediment input (i.e. potential transport rate). The 12 resulting scenarios were simulated with observed runoff data taken from the 2005 flood event with an estimated recurrence interval of 100 years.

2.3 Risk assessment

Accounting for the variability of processes (hydrology, hydrodynamics and sediment transport) on a micro scale, probability distribution functions for object related inundation depths can be obtained. The variability of the water surface elevation (V_{WSE}) is dependent on the variability of the bed elevation (V_{BE}), as well as on the variability of the hydrologic input (V_{HI}).

$$V_{WSE}=f(V_{BE} \mid V_{HI}) \tag{2}$$

Relating the resulting variability of the water surface elevation (Eq. (2)) with the dyke top edge elevation (h), the variability of inundation depth (V_{ID}) can be obtained on a micro scale basis (Eq. (3)).

$$V_{ID}=f(V_{WSE}\,|\,h) \qquad (3)$$

Corresponding to utilisation related damage functions (f_D), typically based on the inundation depth (h_I) and the associated damage (D), a damage probability function (f_{DP}) can be derived by multiplying the damage function (inundation depth dependent) with the variability of the inundation depth (Fig. 3, Eq. (4)).

$$f_{DP}=V_{ID}*f_D(D\,|\,h_I) \qquad (4)$$

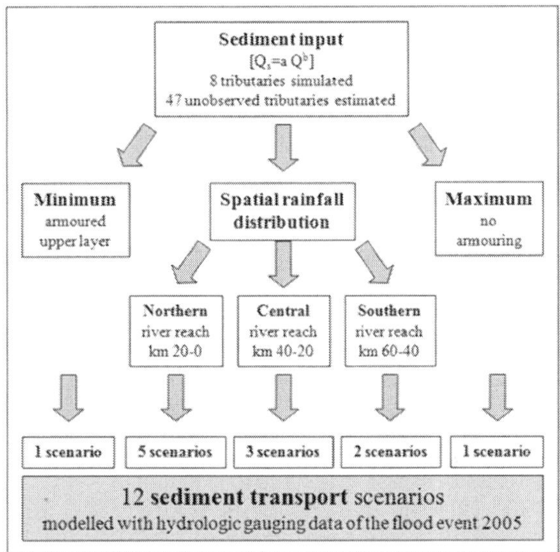

Fig. 7. Derivation of scenarios for sediment input variation

3. Results

Variability and uncertainty is described related to the processes hydrology, hydrodynamics and sediment transport as well as risk assessment based on the scenario analyses. The results of hazard assessment are expressed quantitatively, the results of vulnerability assessment qualitatively.

3.1 Hydrology

Fig. 8 illustrates 126 flood waves resulting of 100-years rainfall events as described in Fig. 4 for the catchment outlet at Gisingen as well as the relative distribution of associated peak discharges. The effects of applied parameter variations, which can be seen as a way of taking into account various uncertainties related to the hydrological assessment of design floods,

are shown in Table 1. Each variation of a single parameter over the full range of applied values - while keeping the others constant - yields a maximum variation in resulting runoff peaks. For a relative measure this value is related to the mean discharge of runoff peaks. The values given in Table 1 are the mean of relative peak variations for all considered scenarios. This mean relative variation shows the sensitivity of the flood simulation to changes in the respective parameter and establishes an evaluation approach for the respective uncertainty.
Regarding the basin outlet at Gisingen, the spatial distribution of rainfall has the smallest impact on flood peaks. Obviously, this impact is much higher at the most-upstream gauges with a smaller catchment area (with either high or low precipitation), with relative runoff peak variations of up to 117 %. The mean variation for all Ill gauges is 41 %. Even though only three different spatial patterns are tested in this study, this shows that the importance of considering uncertainty of spatial rainfall distribution for design flood simulations depends on the spatial focus of the subsequent assessment. Other parameter variations lead to similar runoff peak variations at the basin outlet and at upstream gauges. With 11% relative runoff peak variation, changes in the temporal distribution of rainfall have a markedly lower impact on flood peaks than variations of the initial catchment conditions (27%) The variation of ARF for 12-hour storms has by far the largest effect on simulated flood hydrographs, as it directly alteres the total depth of a precipitation scenario. Storm duration, the second parameter influencing total precipitation depth cannot directly be assessed for the River Ill, because 12-hour and 24-hour storms are determined with different methods and other factors apart from duration influenced the resulting total depth. An evaluation of 2 to 12-hour storms resulting only from the mentioned meteorological convective storm model for Ill tributary sub-catchments shows mean variations in simulated runoff peaks of 20 % (Stanzel et al., 2007). In this analysis also uncertainty related to the estimation of fast runoff model parameters is investigated. Resulting runoff peak variations in tributary rivers are rather small (5 %) - as better observations are available for calibration on the River Ill, the effects of uncertainty in parameter estimation is assumed to be even smaller when regarding the entire basin.

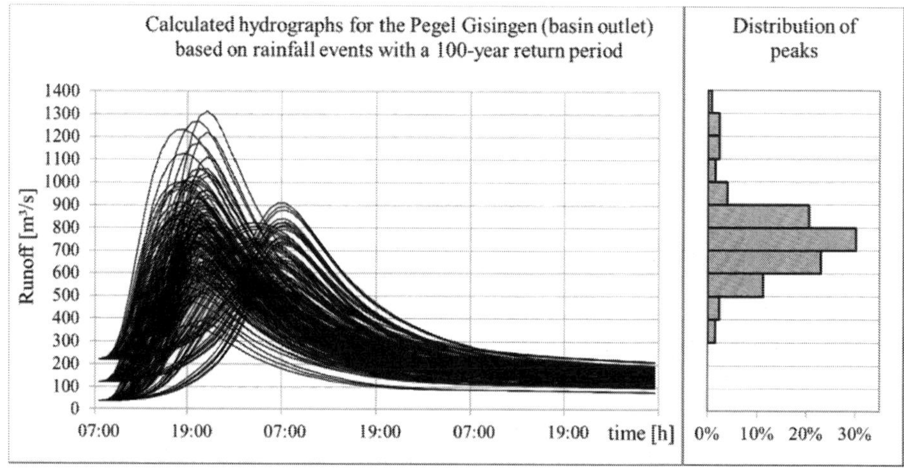

Fig. 8. Calculated hydrographs for 100-year rainfall events and distribution of simulated peak values

In relation to the normative 100-year design value of 820 m³/s at the gauge in Gisingen, the simulated peaks range from 45 % to 160 %. Several peaks are far below as well as over the 90 % confidence interval of statistical extreme value analyses of observed runoff, underlining that 100-year rainfall events produce flood events of different return periods. Yet, the large range of hydrographs shows how much of the possible variability of flood waves is disregarded by a design flood approach.

Varied Parameter	Mean variation of simulated runoff peaks at Gisingen
Spatial rainfall distribution	4 %
Temporal rainfall distribution	11 %
Initial catchment conditions	27 %
Areal reduction factor	88 %

Table 1. Sensitivity of flood peaks due to input variation for Gisingen (basin outlet)

3.2 Hydrodynamics and sediment transport

Hydrodynamic and sediment transport simulation results are illustrated exemplarily for a highly dynamic section (km 30 to 29) chosen from the considered 60 km. The selected river section is characterised by a torrential inflow located at the upper boundary. The sediment input function of this torrential inflow is documented in Fig. 6. The first 300 m of the considered reach are dominated by hydraulic structures (in- and outflow for energy generation, weir and chute) which cause spacious accumulations of sediment due to a reduction of flow velocity and accordingly to lower shear stress. In the case of higher discharge the accumulated sediment moves downstream where a dynamic river bed is encountered.

In Fig.9 the modifications of river bed elevations due to hydrological and sediment input variations are illustrated. The three lines represent the maximum (dark grey), the mean (dashed grey) and the minimum (light grey) calculated bed elevation changes resulting from varying the discharge by means of 126 scenarios (Fig. 4). The inflow of the tributary immediately upstream km 30 leads to locally calculated accumulations of roughly 0.80 m. The black vertical lines indicate the station of considered cross sections and display the range of calculated bed elevation changes due to randomly selected sediment input of torrential inflows. The magnitude is based on the simulation of 12 input scenarios (Fig. 7).

Fig. 10 outlines the differences (max/min) of water surface elevation and embankment elevation. Continuous lines correspond to the orographic right-hand hinterland with numerous utilisations such as private housing. Thick lines define the limits due to hydrological input variation and the thin ones, the limits due to sediment input scenarios. Corresponding to the orographic left-hand side, with no vulnerable utilisations, results are represented by grey dashed lines (thick for hydrology and thin for sediment input). The value 0.00 represents a water surface elevation equal to the dyke top edge. Overtopping occurs when displayed lines show positive values.

Due to hydrologic input variation (126 scenarios - 25 % of them exceed the design water level, see Fig. 8), a high probability of overtopping is indicated. Considering sediment input variation (12 scenarios) based on discharge data of a 100-year flood (2005) only the lower part of the section is subjected to inundation. From chainage 29,100 m to 29,000 m even the minimum values of calculated water surface elevations lead to inundation of the flood plain. Therefore, damages have to be expected for floods lower the design value of the protection scheme (recurrence interval of 100 years, including freeboard).

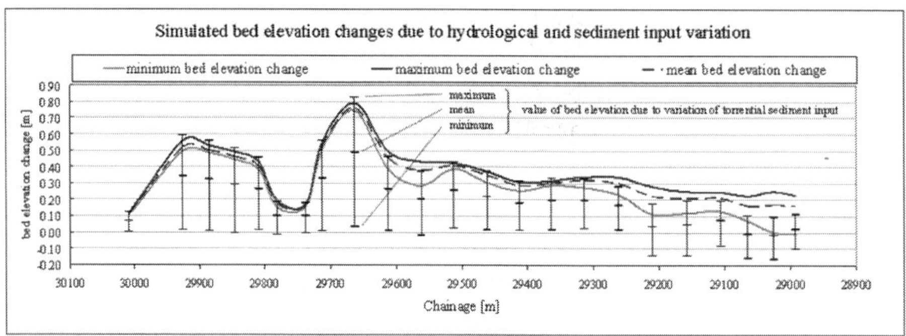

Fig. 9. Changes of river bed elevations due to hydrological and sediment input variation

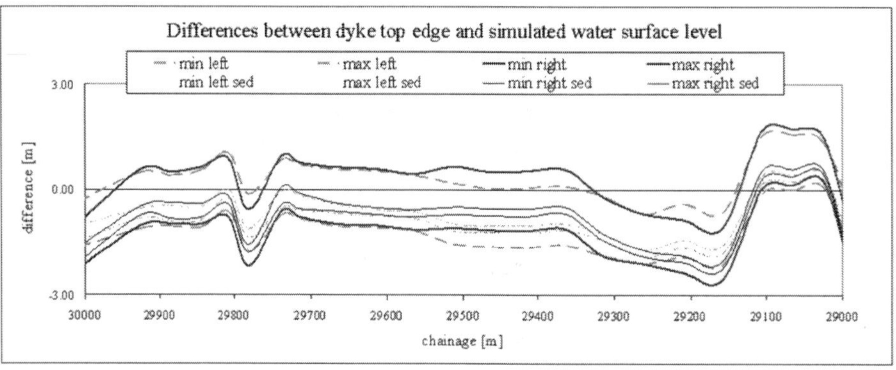

Fig. 10. Differences of water surface elevations and dyke top edge

3.3 Risk assessment

The associated uncertainty of results obtained by design-flood-based procedures (BMFLUW, 2006 a) is emphasized by the overtopping probability caused by 138 considered scenarios (126 hydrologic scenarios, 12 sediment transport scenarios; Fig. 11). Alongside the River Ill settlements and utilisations are mainly protected by dykes and natural barriers with an estimated flood safety up to a recurrence interval of 100 years. Fig. 11 outlines the probability of overtopping along the 60 km due to variation of discharge input (126 scenarios).

The calculated overtopping probability of 12.27 % indicates that 7.4 km are not protected against floods caused by 100-year rainfall events which have not been previously identified as such. In the frame of this study affected utilizations are not elaborated in detail. The analysis of the section displayed in Fig. 9 and Fig. 10 (km 30-29) proves that there are also settlements in the inundated areas.

Referring to the results of the hydrological input variation, it has to be underlined, that considered discharges resulting from 100-year rainfall events lead to as much as 160 % of the applied design value discharge (normative 100-year flood event) for the gauge furthest downstream.
Analysing scenarios by means of sediment input variation obtained by an observed 100-year flood event in the year 2005, the overtopping probability equals 1.59 % for the entire reach. Nevertheless, at 40 cross sections dykes or barriers are overtopped and therefore most likely to break.

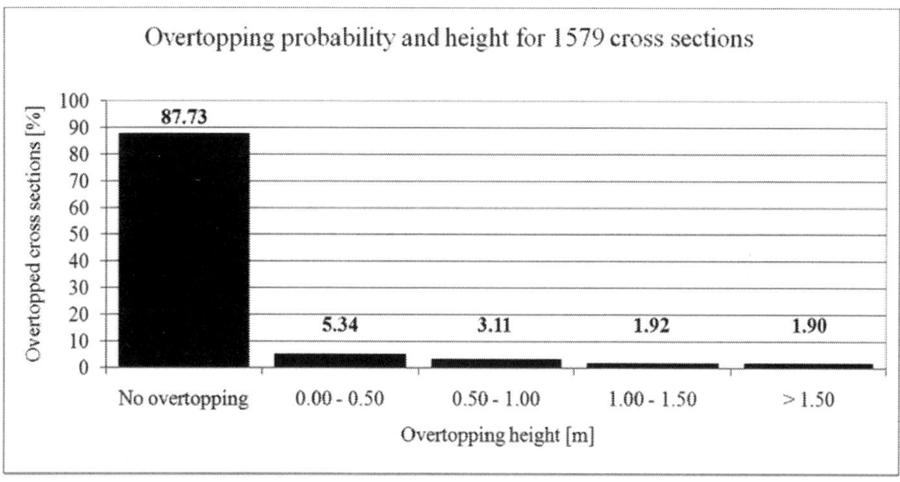

Fig. 11. Overtopping probability and height

4. Conclusion

Integrating river bed dynamics to flood risk assessment enables a refined, more realistic, estimation of water tables and inundation lines. Obviously, uncertainty increases by including additional processes such as sudden changes of the river bed. However, the opportunity to identify related variability is provided. The presented method provides an identification of process related variability and uncertainty, considering hydrologic input generation, hydrodynamic and sediment transport.
To obtain reliable results it is essential to analyse scenarios from "no damage" up to the "worst case". State of the art approaches usually consider a small number of design flood events (in Austria: HQ_{30}, HQ_{100}, HQ_{300}) to assess the overall flood risk which are usually in between "no damage" and "worst case". Thinking about protected areas a HQ_{300} is by no means appropriate to assess worst case scenarios - in these cases flood risk will be under-estimated remarkably. Additionally, normatively defined scenarios usually rely upon uniform recurrence intervals of flood peaks at any location in the catchment area independent from the overall size, i.e. they disregard spatial and temporal rainfall characteristics. Therefore, the presented survey adapts state of the art approaches by

substituting the scenario approach (a few normatively defined design floods) through a multi scenario approach by means of variation of inputs (hydrographs and sediment load).

Further, vulnerability analyses and damage estimation tools were improved methodologically by interrelating the overtopping probability, the variability of inundation depth and a damage function to obtain a damage-probability relationship (Fig. 3). Hence, uncertainty and sensitivity are implicitly comprised in the probability distribution function of the expected damage.

Discharge input scenarios are obtained by rainfall-runoff simulations with different 100-year rainfall events. Sediment input scenarios are simulated based on a flood event with an estimated recurrence interval of 100 years by randomly drawing loads of torrential inflows. A sensitivity analysis indicates that the discharge input variation leads to flood peaks as high as 160 % of the normative 100-year design flood. A higher probability of inundations of vulnerable utilizations like settlements, infrastructure, etc. results from discharge input variation (12.3 %) than from sediment input variations (1.6 %). Regarding the magnitude of bed elevation changes, however, the influence of sediment input variation is found to be much higher than the influence of discharge input variations. Consequently, the derivation of sediment input functions appears to be the most important task wherever the incorporation of sediment transport calculations or estimations are applicable. In this context scarce data availability seems to be the restricting factor (Nachtnebel & Neuhold, 2008). Therefore, an upgrade of continuous sediment gauges as well as the volumetric survey of accumulations, especially after flood events, is desirable. By means of an extended data base the derivation of sediment input functions as well as calibration and validation of sediment transport models would be more feasible and should be adaptable to further river types and scales.

Results indicate that damage has to be assumed where safety was expected. This clearly implies that flood risk assessment approaches should be revised with respect to a multi-scenario and multi-process approach to obtain a more realistic basis for decision making.

5. References

BMFLUW (Bundesministerium für Forst- und Landwirtschaft, Umwelt und Wasserwirtschaft), 2006 a: Richtlinien zur Gefahrenzonenausweisung für die Bundeswasserbauverwaltung: Fassung 2006. Wien.

BMLFUW 2006 b: Niederschlag-Abfluss-Modellierung. Arbeitsbehelf zur Parameter-ermittlung (in German). Vienna.

Fürst, J. & Hafner, N. (2005): Land cover. In: Federal Ministry of Agriculture, Forestry, Environment and Water Management (Ed.). Hydrological Atlas of Austria, 2nd Edition. Vienna.

Geiger, H., Röthlisberger, G., Stehli, A. & Zeller, J. (2004): Extreme point rainfall of varying duration and return period. In: Swiss National Hydrological and Geological Survey (Ed.) 2004. Hydrological Atlas of Switzerland. Bern.

Huang, J.V. & Greimann, B. (2007): US Department of Interior, Bureau of Reclamation, Technical Service Center, Sedimentation and River Hydraulics Group. User´s Manual for GSTAR-1D. 1-817.

Kling, H. (2002): Development of tools for a semi-distributed runoff model. Master thesis, IWHW BOKU, Vienna, Austria.

Larson, C.L. & Reich, B.M. (1972): Relationship of observed rainfall and runoff recurrence intervals. In: Schulz, E.F., Koelzer, V.A. and Mahmood, K., (Eds.). Floods and Droughts, Water Resources Publications, Fort Collins, pp. 34–43.

Lorenz, P. & Skoda, G. (2000): Bemessungsniederschläge kurzer Dauerstufen (D ≤ 12 Stunden) mit inadäquaten Daten (in German). Report of the hydrographical service in Austria, Nr. 80, Vienna.

Meyer-Peter, E. & Müller, R. (1948): "Formula for bed-load transport" Proc. of the Int. Assoc. for Hydraulic Research, 2nd Meeting, Stockholm.

Nachtnebel, H.P., Baumung, S. & Lettl, W. (1993): Abflussprognosemodell für das Einzugsgebiet der Enns und Steyr (in German). Report, IWHW, BOKU Austria.

Nachtnebel, H.P. & Debene, A. (2004):"Schwebstoffbilanzierung im Bereich von Stauräumen an der österr. Donau". FloodRisk: Endbericht im Auftrag des Lebensministeriums. In: Habersack H., Bürgel J. and Petraschek A.: Analyse der Hochwasserereignisse vom August 2002- Flood Risk, 2004, BMLFUW, Wien, Österreich.

Nachtnebel, H.P. & Neuhold, C. (2008): Schutzwasserbauliche Bestandserhebung Ill, Arbeitspaket 5: Hydraulik/Geschiebe/Schwebstoffe. IWHW, Wien.

Nash, J.E. & Sutcliffe, J.V. (1970): River Flow Forecasting through Conceptual Models. Journal of Hydrology, 10 / 3.

Neuhold, C.; Stanzel, P. & Nachtnebel, H.P. (2007): Modelling morphological changes during flood events utilised as impact on flood risk assessment. In: European Geosciences Union, Geophysical Research Abstracts, Volume 9.

Neuhold, C.; Stanzel, P. & Nachtnebel H.P. (2009): Incorporating river morphological changes to flood risk assessment: uncertainties, methodology and application. Nat. Hazards and Earth Syst. Sci., 9, 789-799

Neuhold, C.; Stanzel, P. & Nachtnebel H.P. (2010 a): Integrating river bed dynamics to flood risk assessment. In: Nachtnebel H.P.; Cruz, A.M.; Amendola, A.; Mechler, R. & Tatano, H. (Eds). Abstract-Volume: 1st Annual Conference of the International Society for Integrated Disaster Risk Management - IDRiM 2010

Neuhold, C.; Stanzel, P. & Nachtnebel H.P. (2010 b): Analysing uncertainties associated with flood hazard assessment. In: EGU, Geophysical Research Abstracts, Volume 12.

Neuhold, C. & Nachtnebel, H.P. (2011): Assessing flood risk associated with waste disposals: methodology, application and uncertainties. Nat Hazards 56:359-370

Peticzka, P. & Kriz, K. (2005): General soil map. In: Federal Ministry of Agriculture, Forestry, Environment and Water Management (Ed.). Hydrological Atlas of Austria, 2nd Edition. Vienna.

Skoda, G., Weilguni, V. & Haiden, T. (2005): Heavy Convective Storms - Precipitation during 15, 60 und 180 minutes. In: BMFLUW (Ed.) 2005. Hydrological Atlas of Austria, 2nd Edition. Vienna.

Stanzel, P., Neuhold, C. & Nachtnebel, H.P. (2007): Estimation of design floods for ungauged basins in an alpine watershed. In: EGU, Geophysical Research Abstracts, Volume 9.